Phil. Trans. R. Soc. Lond. B. **280**, 103–374 (1977) [103]
Printed in Great Britain

THE CHANGING ENVIRONMENTAL CONDITIONS IN GREAT BRITAIN AND IRELAND DURING THE DEVENSIAN (LAST) COLD STAGE

A DISCUSSION ORGANIZED JOINTLY FOR
THE ROYAL SOCIETY AND THE ROYAL IRISH ACADEMY
BY G. F. MITCHELL, F.R.S., P.R.I.A. AND R. G. WEST, F.R.S.

(*Discussion held* 1 *and* 2 *April* 1976 – *typescripts received* 21 *October* 1976)

(Including 1 plate and 4 pullouts; 1 microfiche inside back cover)

CONTENTS

Phil. Trans. R. Soc. Lond. B. **280**, 105 (1977) [105]
Printed in Great Britain

Introductory remarks

By J. S. Sawyer, F.R.S.
Meteorological Office, London Road, Bracknell, Berkshire

In the last year or two there has been a remarkable increase in the interest, both popular and scientific, in the subject of climatic change. This stems from a recognition that even a highly technological society is vulnerable to the effects of climatic fluctuations and indeed may become more so, as margins of surplus food production are reduced, and nations become more interdependent for their food supply. In this respect our concern is with quite small changes – a degree (Celsius) or less in temperature and 10 % or so in rainfall. Probably we may discount some of the more alarmist suggestions of an imminent and rapid change towards near glacial conditions as these are based on very sketchy evidence.

However, whatever the time-scale of climatic fluctuations with which we are concerned, we may hope to learn a great deal which is relevant to the factors which will control our future climate from the study of its more extreme vagaries in the past. Information relevant to the weather in such extreme periods is coming forward in increasing detail and volume from a wide range of disciplines. The variety of the evidence, its lack of precision as a strict measure of climate, and the number of different sources all make it difficult for an individual to build up a clear picture of past climates. However such a picture is needed, if explanations and interpretation are to be possible. Ideally one would need a synchronous picture of the climate of the whole world at selected epochs in the past. Various international programmes are directed to forming such pictures.

Today's meeting is another step towards this objective. It concentrates on a specific period, the last cold stage. Even if the extent of our discussion is not global, the welcome participation of our Irish colleagues will ensure that we have the geographical variation of the climate in mind over a region where there must have been strong contrasts of climate during extensions of the ice sheets.

In the end it will fall to the meteorologists to explain how past climates came about, and perhaps to say what the future holds. This may be why our societies have chosen to have a meteorologist as chairman for this session. If so, their choice is a little premature. Despite progress which is being made towards the calculation of the climate from the basic laws of physics, I doubt if we are yet near the stage when we can identify the causes of glacial periods with any certainty, or trace the physical and dynamical processes which determined the distribution of the ice. However, I hope that at the end of our two days of discussion we shall all have a clearer picture of the facts that need to be explained.

Phil. Trans. R. Soc. Lond. B. **280**, 107–118 (1977) [107]
Printed in Great Britain

CHRONOLOGY, CLIMATE, AND MARINE RECORD

The Devensian Stage: its development, limits and substages

By F. W. Shotton, F.R.S., Hon. M.R.I.A.
Department of Geology, University of Birmingham, Birmingham B15 2TT, U.K.

The name Devensian for the last glacial stage of the British Pleistocene arose from the attempt of the Quaternary subcommittee of the Stratigraphic Committee of the Geological Society of London to define the divisions of the British Quaternary and to draw up correlation tables. The name of the stage derives from the location of the stratotype. Although the stage is broadly equivalent to the Weichselian of NW Europe and the Wisconsinan of North America, exact equivalence of the boundary limits and of substages in the British sequence to overseas divisions should not be assumed.

As a means of correlation, radiocarbon dating is of great use in the later half of the stage, and a few 'enriched' dates help in a few millennia before 50000 B.P. No other methods of absolute dating have yet been applied, and the possibility of doing these on the land-based Devensian deposits is remote.

Significant recognizable events in a glacial stage are provided by interstadials. Claimed interstadials in the Weichselian of NW Europe are briefly examined, and the evidence for interstadials in the British Isles is summarily presented and compared as far as possible with the continental succession. On the question of when, in the Devensian, the British area was physically 'glacierized', it is claimed that so far there is no evidence for this in the Early Devensian. The timing and duration of the well-established Late Devensian glaciation is discussed, and a word of caution is sounded against regarding this as synchronous throughout the area.

Introduction to the Devensian

Any discussion of the environmental conditions of Great Britain and Ireland during the Devensian demands first an explanation of the stage name which is relevant only to these islands. It labels the period which includes the last glacierization of the country, and the question immediately arises why one of the terms Weichselian or Wisconsinan, which cover roughly the same period of time in Europe and North America respectively, was not used. When a committee of the Geological Society of London undertook the task of producing a correlation of the British Quaternary (Mitchell, Penny, Shotton & West 1973) it attempted:

(*a*) to draw up a fully comprehensive succession of major climatic stages in Great Britain and Ireland,

(*b*) to relate sequences in different parts of the country to this succession, and

(*c*) to provide a framework for the extension of this correlation outside the British Isles.

It followed from this brief that the limits of the major stages and the subdivisions within them should be determined on British stratigraphy, and take British names. The stratotype for the last glacial stage was selected at Four Ashes in Staffordshire, within the area once peopled by the British tribe of the Devenses, and hence arose the now generally accepted stage name. Although Devensian may roughly coincide with the Weichselian as it is known in western Europe, there is no implication that either their upper or lower limits coincide, nor that the smaller climatic divisions within each stage necessarily correspond with each other.

The division of the Quaternary into stages and substages is based wholly on the concept of climatic cycles. In any region it may be possible to work out a succession in time of estimated variations in mean annual temperature, accompanied by other superimposed and less easily quantified fluctuations such as continentality, oceanicity, aridity or pluviality. The colder periods, if accompanied by adequate precipitation, often led to the growth of the continental ice sheets, and then a glaciation or 'Glacial Stadial' may be inferred. If an amelioration of climate leads to shrinkage of the glaciers, this is an 'Interstadial'; but if the decay of glaciers is drastic, this is an 'Interglacial'. Decisions on nomenclature of this kind must be highly subjective, particularly since the nature and extent of climatic change usually have to be deduced from their effect upon biota, rather than from the waxing and waning of past glaciers. If it can be demonstrated to a Briton that glaciers covered most of Scotland, Ireland, Wales and England down to the Midlands as recently as about 16000 years ago, he needs no convincing about the Devensian glaciation. Since he is also well satisfied with the congenial climate in which he lives, he will believe that he is now in an interglacial (Flandrian). It might be less easy to convince an inhabitant of Zaire that his forefathers of 600 generations back lived in a glacial stage, or to persuade a penguin on the Ross Ice Shelf that he is now enjoying an interglacial. When scientists in different parts of the world reach back and attempt to fix the upper and lower limits of the last glacial stage (or, indeed, any earlier one), they find themselves with a choice of the criteria which indicate climatic change, and have to make decisions about their relative importance, and the points in time when they achieved maximum significance.

A region as small as the British Isles has today enough variation in latitude, altitude and proximity to the ocean to produce profound differences in the floral and faunal assemblages of different regions. In the past when climates were changing rapidly these patterns were further distorted by the reaction of plant or animal species to the change. Some would adapt to the new conditions, but many would have to change their areas of colonization in order to continue in an environment where they could thrive. The populations might move northward or southward in search of unchanged temperatures, eastward or westward if precipitation and cloud cover affected their viability, up or down hill to seek unchanged temperatures and precipitation. Interpretations along these lines may still miss the more subtle factors in the tangled pattern of existence. So although the biologist can make sound deductions about ecology, if he chooses some empirical factor to mark a time division – for example, the first arrival of a broad-leaved tree such as oak or alder, or the colonization by a particular beetle, or where the rain of tree pollen exceeds that of herbs – inevitably that postulated boundary will be diachronous. Synchroneity of boundaries biologically deduced cannot really be expected; at the best it can be hoped that the departures from this ideal is small, as it could be if we can assess and make allowance for the effect of geographical position upon the contemporary biota.

The upper and lower limits of the Devensian

If the Devensian limits are to accord with the principles of the Stratigraphic Code, then the base must be clearly defined, but the top merely occurs under a clearly defined base for the overlying Flandrian. Everyone accepts that Pollen Zone III (Younger Dryas of NW Europe), indicating a rigorously cold climate, should be retained within the Devensian. It has been the custom to start the Flandrian with Pollen Zone IV, which begins at a radiocarbon date of about 10250 B.P. (Godwin, Walker & Willis 1957; Godwin & Willis 1959; Hibbert, Switsur

& West 1971). In fact the III/IV boundary is often not easy to fix, and there is a striking vegetation change at the end of Zone IV, when hazel (*Corylus*) spreads with dramatic suddenness and *Pinus* begins to dominate over *Betula*. This occurs at about 9600 B.P., a date which still precedes the advent in force of broad-leaved deciduous trees. Work on beetles at two Midland sites shows that a thermophilous fauna began to move in around 10300 B.P. (Ashworth 1972), and was firmly established at 9500 B.P. (Osborne 1974), in both cases ahead of the woodland vegetation that might have been expected with them. There are thus reasonable grounds to start the Flandrian in lowland England at the III/IV junction which here is around 10250 B.P. Nevertheless, the Quaternary subcommittee felt that there was no reason to expect these zonal boundaries, defined either by plants or insects, to remain fixed in time across the whole length of the British Isles and throughout the great variation in land height. Since there was a good radiometric clock available, to define the beginning of the Flandrian in terms of radiocarbon years avoided any problem of diachronism. A figure of convenience, 10000 years, close to the III/IV boundary of the English lowlands, was chosen because it conformed to the starting date of the Holocene recommended by the Holocene Commission of INQUA. Hence we may say that the Devensian ends at a date of 10000 *radiocarbon years* using the Libby half-life for ^{14}C of 5568 years as the standard for measurement.

The base of the Devensian cannot be so conveniently defined, for here we are beyond the range of radiocarbon. There are alternative methods of absolute dating, applicable with different degrees of reliability to various periods of the Quaternary succession, but none has proved useful or possible at the beginning of the British Devensian. Indirect use of magnetic reversals is impossible, because the whole time range of the Devensian is small compared with the scale of the magnetic changes. Complete lack of volcanic activity (in the whole of the British Quaternary, not only in the Devensian) prevents use of the potassium/argon methods even if these could be used for an antiquity of about 100000 years. For the same reason, fission-track dating need not be considered. The uranium series has rarely been consistent – except in the case of corals – and gives no promise for the Devensian. It is perhaps too early to judge the technique of dating based on the epimerization of amino acids (see Turekian & Bada 1972 for a summary of the method), but as an assumption of constant temperature is basic to the method, its application to the Devensian will be difficult. So we have to look to biological stratigraphy in an attempt to fix the transition from the preceding inter-glacial.

In fact this cannot be done in a satisfactory manner. The last interglacial, which precedes the Devensian, is the Ipswichian. Its type section is at Bobbitshole near Ipswich, but there are about 12 other studied sites east of a line from the Solent to the Wash (see Sparks & West 1970), and a few west of this line. Four floral zones have been set up with changing vegetation marking progress from cool temperate, through climax forest and back to more open woodland associated with declining temperature. If any of these sites continued in an unbroken succession to conditions which could be described as early glacial, the definition of the base of the Devensian would be clear. It would be at the base of this post-Ipswichian zone, however it might be characterized. It is, however, an extraordinary fact that none of the sites includes all Ipswichian zones, the sequence of four zones having been built by up overlap, and only one site (Wretton) is overlain by Early Devensian deposits; and here there is a great gap in which the latest part of zone Ip III, the whole of Ip IV and some of the Devensian is not represented by organic deposits. Deposits at Crayford ascribed to Ip IV might include the Early Devensian, but the age of these is not universally agreed.

The best stratotype for the Devensian is undoubtedly the Four Ashes section north of Wolverhampton, studied by A. V. and Anne Morgan (Morgan, A. V. 1973; Morgan, Anne 1973). This site is unique in having deposits of Late, Middle, and Early Devensian above an Ipswichian horizon. The only section at all comparable in importance is at Tattershall in Lincolnshire (Girling 1974), which has a fuller development of the Ipswichian, but, within the overlying gravels which may span the whole of the Devensian, fewer diagnostic organic horizons and no Late Devensian till.

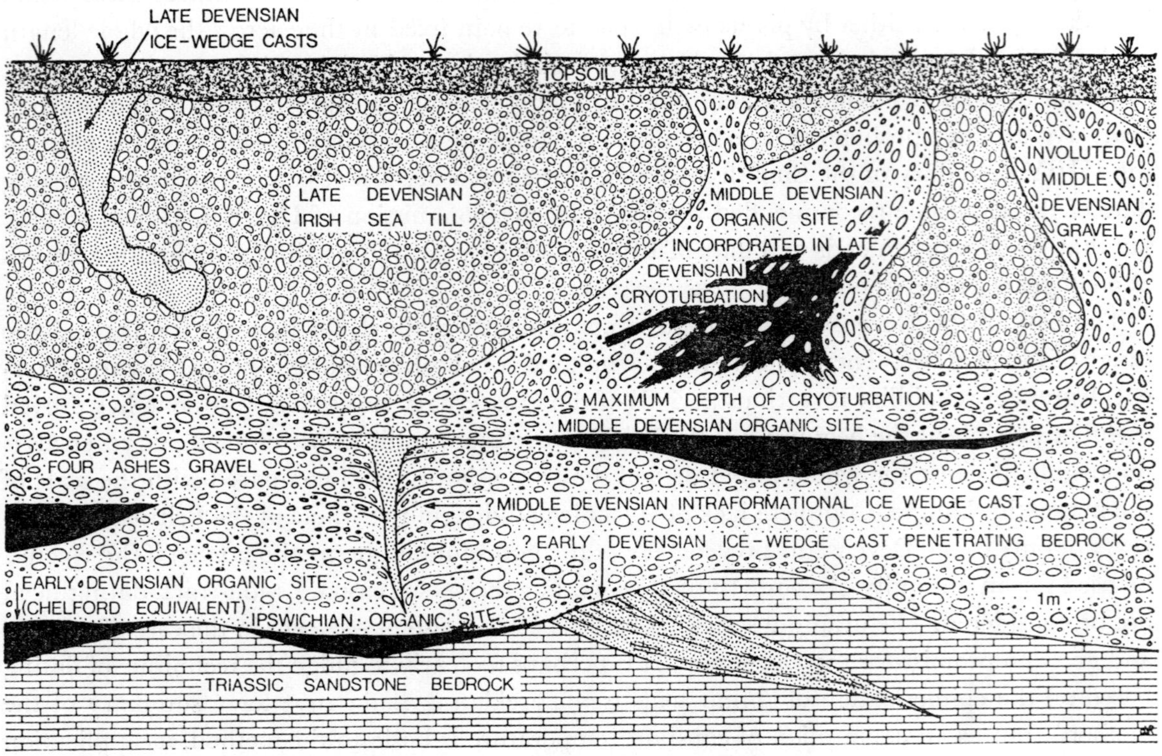

Figure 1

The dominant member of the Four Ashes sequence is a gravel varying in thickness between 0.5 and 4.5 m, but containing numerous small lenses of organic silt and peat, which have been subjected to faunal and floral analysis and radiocarbon dating. On the basis of dates alone, the Middle Devensian is well represented (6 determinations between 30000 and 42530 years). Evidence for the Early Devensian depends very much upon the recognition of the Chelford Interstadial (see below), but this is well substantiated. Above the Four Ashes gravel rests the Late Devensian till of the Irish Sea glaciation, and beneath the gravel at one point occurred an undoubted Ipswichian peat. A. V. Morgan summarized the essential features of the succession without attempting to show the 45 separate organic lenses which had been recognized and this diagram, with his permission, is reproduced as figure 1. The implications of this sequence are discussed later.

Interstadials claimed in the Weichselian of NW Europe

Based entirely upon palaeobotany, and almost wholly on palynology, a number of horizons in countries between north France and Denmark have been considered to show relatively short-term ameliorations of climate, and so to be ranked as interstadials. They are shown in table 1, together with the dates which have most recently been ascribed to them.

Table 1

claimed interstadial	dates B.P.
Allerød	numerous, between 11000 and 11800
Bølling	between about 12000 and 13000
Denekamp	*ca.* 30000
Hengelo	*ca.* 39000
Moershoofd	43000–50000
Odderade	*ca.* 58000
Brørup	61000–63500
Amersfoort	65000–68000

The interstadials are not similar in their ecological and climatological implications which are discussed below.

The Allerød Interstadial, as interpreted in Denmark, demands the close proximity of pine and birch woodlands, but the Bølling episode can be interpreted as one of shrub tundra. It is claimed that the pollen record of what is usually called the Older Dryas period, intervening between the Allerød and Bølling, shows a short-lived return to a colder climate which justifies the making of two interstadials. There is some dispute about this concept of two separate interstadials, and also about the date of the maximum amelioration of climate, opinion being less divided on the east side of the North Sea where little but botanical work has been done, than in Great Britain and Ireland where insect studies have supplemented palynology. However, other contributors to this discussion will debate the relation of the British Late Devensian Interstadial (or interstadials) to the Late Weichselian oscillations of NW Europe. What is not disputed is that Bølling, Older Dryas and Allerød together constitute a major interstadial between the rigours of the preceding glacierization and the return to arctic conditions in Zone III.

In the earlier third of the Weichselian (Early Pleniglacial) three interstadials are claimed, and the lower two (Amersfoort and Brørup) are well substantiated (Andersen 1961; Zagwijn 1961). At Amersfoort in the Netherlands and Brørup Bog in Denmark, there are two organic horizons which lie above the Eemian, and which each show a strong increase of tree pollen which must indicate contemporary boreal forest. The lower one is dominated by *Betula* and *Pinus*, the upper by *Betula*, *Pinus*, *Picea* and *Alnus*. The sequences in the two localities parallel each other, and the Danish name has been accepted for the later interstadial and the Dutch name for the earlier.

Both these interstadials are beyond the normal reach of radiocarbon dating but the Groningen laboratory, using the enrichment technique, gave them positive dates. Recently Grootes has made new determinations on newly collected material in the hope that the older ones might be confirmed, or at least corrected for the minimal contamination that may have been present.

These dates are listed below, the earlier series being shown in italics and any given with the 'greater than' symbol being unenriched. Grootes has not yet fully published his dates but a short abstract (Grootes 1976) gives a diagram from which they may be read with reasonable precision. They are quoted below without standard deviations or laboratory numbers.

Brørup I/S.	At Brørup	*58740 ± 1000* GRO 1729
		59430 ± 1000 GRO 1470
	At Amersfoort	> *53000* GRO 1280 and 1285
		61000 at the end, 63000 at the beginning.
Amersfoort I/S.	At Amersfoort	> *53000* GRP 1221
		63500 ± 1100 GRO 1398
		66500 at the peak, 68000 at the beginning.

These 'enriched' old dates must still be treated with caution, and regarded only as minima. It is perhaps not generally realized how little of its original radioactivity is retained in a piece of organic matter 68000 years old. It is 0.024 %. Broecker (1965, p. 749) raised doubts about the accuracy of the original Amersfoort date of 63500, on the grounds that the 'date' of the alkali-extracted humate gave a positive figure of 42300, which indicated that there had been contamination originally. The new Gröningen determinations were done to examine this criticism, and it is claimed that the slightly enhanced figures now presented are probably close to the truth.

The Odderade Interstadial claimed by Averdieck (1967) in Schleswig Holstein depends to some extent upon acceptance of the Brørup Interstadial below it. The date of 58000 appeared at first sight to offer little chance of distinguishing it from the Brørup by absolute dating alone, but now that the latter's age has been advanced to around 64000 years, separate identities appear more probable.

Within the Middle Weichselian (the Interpleniglacial of some authors) three interstadials are claimed, but they are different in character from those of the Early Weichselian. In no case does the pollen indicate anything but a shrub tundra environment, and the main justification for regarding them as interstadial (which implies a measurable amelioration of climate) is that the organic horizons are separated from each other by mineral sediments which are cryoturbated. No great claims are made for the extent of the climatic amelioration in these interstadials, van der Hammen, Maarleveld, Vogel & Zagwijn (1967) postulating an improvement in mean July temperature from 5 to 10 °C. Zagwijn (1974) regards the Moershoofd Interstadial as requiring more pollen analyses before its interstadial nature can be accepted without question.

Radiocarbon dates for the Moershoofd period range between 43000 and 50000 years. The two later interstadials are more closely defined, though in the case of the Hengelo, there is still an apparent spread in dates. Seven dates taken in the three localities of Hengelo, Ruigekluft and Denekamp range between 36000 ± 600 and 40500 ± 1500, but the youngest date may be affected by contamination. Nevertheless, the range of this relatively minor interstadial would be from about 37500 to 40000, with its peak around 39000. The Denekamp Interstadial is dated around 30000 B.P.

One of the characteristics of NW Europe is the paucity of datable organic deposits between the Denekamp period and the Oldest Dryas around 13000 B.P. This period, which embraces the maximum advance of the North European Ice Sheet, is interpreted ecologically as polar desert, and clearly invites comparison with the same length of time in Britain and Ireland.

Devensian interstadials

(*a*) *Evidence for the Amersfoort Interstadial*

The only British deposits which might be in the Amersfoort Interstadial occur in Zones C and D of Wretton in Norfolk, and they have been labelled as the Wretton Interstadial (West *et al.* 1974). The pollen diagrams suggest the lower half of an interstadial, in that over the short range of the sediment, there is an upward increase in tree pollen and a decline in grasses. There seems little doubt from the stratigraphy that this horizon is post-Ipswichian, and also low in the Devensian, but the argument that it coincides in time with Amersfoort is not conclusive. The pollen diagram for site WX is interpreted as beginning in Ipswichian Zone II, and then passing without obvious break into the possibly Amersfoort horizon. There must, however, be a considerable sedimentary gap if Ip III and Ip IV are absent, and in this case one cannot deduce how close the interstadial is to the bottom of the Devensian. Support for an Amersfoort age is given by the tree pollen (see West *et al.* 1974, p. 403), which is predominantly *Betula*, *Pinus* and *Alnus* with very little *Picea*, and, if the next overlying tree pollen-rich zone is to be correlated with Brørup, the case for equivalence between the Wretton and Amersfoort interstadials is strengthened. However, there is a division of opinion about the correlation of this higher zone (Coope, in appendix to West *et al.* 1974). No radiocarbon date has been attempted on the Wretton interstadial. It could only be done after diffusion-enrichment, and in view of the doubts about the accuracy of the Amersfoort measurement, it is doubtful whether any result would be conclusive.

(*b*) *Brørup Interstadial and the Chelford Interstadial*

The well-known bed of peat with trunks and other macro-remains of birch, pine and spruce in the yellow sands of Chelford, Cheshire, was the first deposit to be correlated with the Brørup Interstadial (Simpson & West 1958). It demonstrably antedates the Upper Boulder Clay of Cheshire, now known to be Late Devensian, but it cannot be proved to be post-Ipswichian by superposition on Ipswichian organic deposits. Simpson & West described the flora, with *Betula* cf. *pubescens* and cf. *verrucosa*, *Pinus sylvestris* and *Picea abies* the dominant trees, and considered correlation with Brørup to be the most likely possibility. At the time, the Gröningen laboratory had not finished the calculation of an 'enriched' ^{14}C date, but shortly after this was given as 57000. It has since been recalculated at 60800 ± 1500 (GrN 1480). This has been taken as additional proof that the Chelford Interstadial of England is the same as the Brørup of Holland and Denmark. Coope (1959) described the abundant insect fauna, and so established an independent assemblage zoning for the Chelford Interstadial. As might be expected, the insect record agreed with the palaeobotany in evaluating the climate, and was consistent with the vegetational environment.

It is because the two types of evidence conflict at Wretton that there is doubt about the horizon attributed there to the Chelford Interstadial. The pollen is interpreted to mean an amelioration of climate resulting in birch-conifer woodland, whereas Coope (in appendix to West *et al.* 1974) interprets the beetles as indicating a deterioration in climate leading to treeless tundra.

There seems no doubt that the Chelford Interstadial is recognizable in the type Devensian site at Four Ashes (Morgan, Anne 1973), even though the evidence is somewhat circumstantial.

The many small sporadically occurring, organic lenses included:

(*a*) A group with an arctic insect fauna, all in the upper part of the gravels, and three of them dated at 30 500 ± 440 (Birm-195), 30 655 ± 730 (Birm-25) and 36 340 ± 730 (Birm-24).

(*b*) A group with mixed fauna but including a number of warm climate stenotherms, and a great reduction in the cold stenotherms. Three samples here gave dates of $38\,500^{+1300}_{-1100}$ (Birm-170), $40\,000^{+1400}_{-1200}$ (Birm-196) and $42\,530^{+1345}_{-1115}$ (Birm-56). This group was correlated with the Upton Warren Interstadial (*sensu stricto*).

(*c*) A small number of lenses in the lowest metre of the gravel, in two of which the beetle fauna was conspicuously arctic, closely resembling group (*a*), but older than group (*b*), since one sample gave a date of > 43 500 (Birm-74).

(*d*) Two organic lenses at the base of the gravel, and presumably therefore older than (*c*). They gave a beetle fauna which closely resembles that of Chelford, including species dependent upon the presence of trees of birch and pine and/or spruce, and the two species *Scolytus ratzeburgi* and *Blastophagus piniperda* which were hitherto only known from Chelford. Professor West (personal communication) states that the palynology confirms a Chelford horizon.

(*e*) A third lens also at the base of the gravels, devoid of insects but with abundant thermophilous plants. Clearly an interglacial, it has been confirmed as Ipswichian by Professor West (personal communication). A 'precautionary' radiocarbon date was done (Birm-171), which was, as expected, > 45 000.

The Chelford Interstadial (and that of Brørup, assuming synchroneity) is thus clearly represented at Four Ashes. A flora from Beetley in West Norfolk, worked on by Dr Linda Phillips, is stratigraphically not much later than Ipswichian. Pollen of *Betula*, *Pinus* and *Picea* and an abundance of *Picea* needles indicate the existence of woodland of Chelford type. The site is therefore ascribed to this interstadial.

The only other possible claimant to Brørup status is in Borehole V 050 which penetrates the infilling of the 'Fosse Dangeard' in the English Channel, and which can use its position well within British territorial waters to justify its reference here. Morzadec-Kerfourn (1975) considered that the pollen of the organic deposits indicated a period towards the end of the Brørup Interstadial, but did not exclude the possibility of a later date with the pollen largely reworked. If the age is truly Brørup, then the occurrence of salt marsh deposits (for that is how they are interpreted) as low as −89 m N.G.F. provides an interesting figure for an Early Devensian sea level (if there has been no subsequent tectonic disturbance).

(*c*) *The Upton Warren Interstadial*

The Middle Devensian was defined between the radiocarbon age limits of 26 000 and 50 000 years (Mitchell *et al.* 1973). It lies between the Late Devensian which includes the manifestations of British glacierization, and the Early Devensian with broadly corresponds to the 'Early Würm' of Europe. Because an episode of interstadial nature had already been established at an age of about 42 000 years at Upton Warren in Worcestershire (Coope, Shotton & Strachan 1961), the Middle Devensian has been alternatively referred to as the Upton Warren Interstadial Complex, and would be deemed to include additional climatic ameliorations such as equivalents of the Hengelo and Denekamp. Further work has substantiated the existence of a short but marked interstadial in the Devensian between about 43 000 and 40 000 B.P., in marked contrast to the later part of the Middle Devensian which was consistently very cold.

So the label of Upton Warren Interstadial is used in a limited sense here for this shorter period.

The Upton Warren site lies on a river terrace correlated with the Main Terrace of the River Severn. There are three organic bands lying within gravels which show clear evidence of contemporary cryoturbation. A sample from the middle organic bed sent to Gröningen was dated at 41 500 ± 1200 (GRO 595) after acid pretreatment only. A second portion of the same sample was dated after alkali-treatment at 41 900 ± 800 (GRO 1063), and the alkali extract was dated at > 40 000. This indicated the absence of contamination, and an age near to 42 000 years may be considered to be reliable.

Nothing in the fauna or flora suggested a climate which approached interglacial. The abundant mammals included *Bison, Coelodonta antiquitatis, Mammuthus primigenius, Rangifer tarandus* and *Dicrostonyx torquatus*. Pollen indicated a treeless environment, but there were several plant taxa which belied the description of this as tundra. Mollusca favoured the interpretation of a cold climate, but the insect list included species now restricted to north Scandinavia and north Russia along with others now south of latitude 60°. Fauna and flora therefore both included taxa seemingly incompatible, a fact which could be explained if adjustment to a rapidly changing climate was taking place.

The Upton Warren Interstadial has been recognized now at five other places. It has already been indicated how Anne Morgan established this at Four Ashes, by recognizing the appearance of warm stenotherm beetles between dates of $38\,500^{+1300}_{-1050}$ and $42\,530^{+1345}_{-1115}$. Frances Bell (1970) compared the flora of a silt bed at Earith with other Devensian floras, and considered it closest to Upton Warren and Sidgwick Avenue, Cambridge (which is undated). She also classified it as interstadial on the grounds that some of the plants demanded summer mean temperatures up to 16 °C. The ascription to the Upton Warren Interstadial is supported by the radiocarbon date of $42\,140^{+1890}_{-1530}$ (Birm-88).

The beetle fauna described by Coope & Angus (1975) from the Upper Floodplain Terrace of the Thames at Isleworth is considered by them to be more thermophilous than any which have been mentioned above, necessitating summer temperatures around 18 °C, and winter temperatures only slightly below those of the present day. They would regard it as marking the peak of the Upton Warren Interstadial, fitting to a radiocarbon date of $43\,140^{+1520}_{-1280}$ (Birm-319). Characteristically, the environment appears to have been treeless.

Finally, the work of Maureen Girling (1971) at Tattershall in Lincolnshire, only as yet summarily reported on, brings out the great rapidity of the climatic change which led up to this interstadial. Two silt beds occur in the post-Ipswichian gravels, very close to each other in direct upward succession. The lower has a conspicuously cold beetle fauna and two radiocarbon dates, $42\,100^{+1400}_{-1100}$ (Birm-398) and $44\,300^{+1600}_{-1300}$ (Birm-408). The large standard deviations overlap, but a figure between 43 000 and 44 000 is likely. The fauna in the upper silt has changed dramatically with the introduction of typical Upton Warren thermophiles. Radiocarbon dates are $43\,000^{+1300}_{-1100}$ (Birm-341) and 42 200 ± 1000 (Birm-409), which are not statistically separable from the two in the lower bed, but they illustrate how rapid the climatic change up to the interstadial must have been. This, and the short duration of the mild period, evidently conspired to inhibit the establishment of trees.

(d) *The latter half of the Middle Devensian*

It is clear that the Upton Warren Interstadial does not fit with the date given to Hengelo, but is rather earlier, though there is some overlap. There is a growing body of evidence that the climate deteriorated very rapidly after the Upton Warren Interstadial to become arctic, and eventually to pass to polar desert before the advent of the Late Devensian glaciers. There are now many dates between about 39000 and 26000 measured in deposits which contain fauna and/or flora leading to climatic deductions. No. 2 Terrace of the Warwickshire Avon, and its correlative in the tributary Carrant Brook, have at least eight dates on organic horizons ranging from 38000 to 26000 years. In the Trent drainage we have three for Four Ashes (36340, 30655 and 30500), and one in the Tame Valley (32160). Standlake Common in Oxfordshire gave 29500, Queensford in Oxfordshire 39300, Sutton Courtenay in Oxfordshire 33190 (with a very large standard deviation), Syston in Leicestershire 37420, Thrapston in Northants 25780, Great Billing in Northants 28230, Kirkby-on-Bain in Lincolnshire 34800, Oxbow in Yorkshire 38500 (on a mammoth tusk), Leadenhall Street, London, on a bone of *C. antiquitatis*, 29450, and as far afield as Derryvree, Co. Fermanagh, 30500 and Castlepook Cave, Co. Cork, 33500. In all these cases wherever flora has been studied, tundra is indicated, and where fauna has been worked on, an arctic environment can be deduced. It would be misleading to affirm that the small Denekamp oscillation does not exist in the British record. Before this could be said, one would have need to study a continuous sequence extending between, say, 32000 and 28000 years, and such a sequence is as yet unknown. All that we are entitled to say is that none of the mid-Weichselian interstadials of the continent has yet been recognized in the British Isles, and that our one clear interstadial appears to lie between the claimed dates for the Moershoofd and Hengelo.

Glaciation in the Devensian

With the increase in the number of, and knowledge about, the organic beds of the Devensian and radiocarbon dates extending back towards 50000 years, there is now no evidence for an Early Devensian glaciation in Great Britain or Ireland. There are situations, as at Derryvree (Colhoun, Dickson, McCabe & Shotton 1972), where two tills are separated by Middle Devensian sediments, but so far nowhere that a lower till is underlain by deposits of Chelford, Amersfoort or Ipswichian date. Till of Late Devensian age is widespread in Ireland, Wales, Scotland and England north of the Midlands, and it is of interest to examine what we know about the duration of the main glaciation, as distinct from late readvances.

It is possible, even probable, that the spread of ice over the country did not occur in a single surge, but was effected by lobes which advanced and retreated over different areas at different times. Such appears to have been the case in the Late Wisconsinan of North America. If, however, we hope to date these glacial episodes, it can only be done by obtaining limiting dates on organic deposits above and below tills, or associated with fluvioglacial sediments emanating from the glaciers. We are still far from having enough conclusive information on this score. One of the most surprising dates was effected on the Dimlington moss layer, which occurs in silt-filled pools on the weathered surface of the Basement Till, and is overlain by the layered Drab and Purple Tills of the Yorkshire coast. American dating produced a figure of 18500 ± 400 (I-3372) (Penny, Coope & Catt 1969). A second determination on independently collected material strikingly confirmed this figure, 18240 ± 250 (Birm-108). This has led to a widespread quoting of a duration for the main Late Devensian glaciation as lying between 18000 and about 15000 years ago. This may not be the whole story. The problem of the relation between the 'Irish Sea Glacier' of the Midlands and the Severn Main Terrace still remains an unsolved puzzle. The Main Terrace is, because of its erratics, clearly the outwash below the Ironbridge Gorge of the Irish Sea Glacier. It has no radiocarbon dates. Topographically it grades with Avon No. 2 Terrace (which is not glacial outwash) and in this there are numerous dates, none newer than 32000, though there are figures as late as 26000 from the same terrace along the Carrant Brook. Admittedly many of these dates refer to organic material near the base of the terrace deposits. Even more confusing is the Upton Warren date of 42000, in the middle of terrace gravels on the River Salwarpe, tributary of the Severn and apparently grading into the Severn Main Terrace. It must also be recalled that of the seven dates from the gravels underlying the till at Four Ashes, the youngest was just over 30000. Much of this evidence is negative, and we urgently need to find datable material within gravels of the Severn Main Terrace which carry the characteristic Lake District and Scottish erratics of the Irish Sea Glacier. Until we do this, we are unlikely to know whether the ice advance down the Irish Sea and across the Cheshire Plain was earlier or synchronous with that along the Yorkshire coast. It may be significant that a layer of the moss *Drepanocladus revolvens* at Glen Ballyre in the Isle of Man has four determined dates, three of them separate measurements on one sample (Birm-270, a, b, c) and one on a sample collected at a different time (Birm-213). The four dates are bracketed between 18400 ± 500 and 18900 ± 330 and something close to 18650 must be correct. The deposit is near the base of a succession which post-dates the Late Devensian glaciation, yet the date is slightly older than the one which antedates the East Yorkshire tills. The possibility of 'hard water error' has been suggested (Shotton & Williams 1973), since the moss is capable of photosynthesis under water as well as subaerially. Even so, the correction would have to be considerable to abolish the suggestion of non-synchronous advances and retreats of lobes of the Late Devensian glacier.

References (Shotton)

Andersen, S. T. 1961 Vegetation and its environment in Denmark in the Early Weichselian Glacial. *Geol. Surv. Denmark*, II Series, 75.

Ashworth, A. C. 1972 A late-glacial insect fauna from Red Moss, Lancashire, England. *Ent. scand* **3**, 211–224.

Averdieck, S. T. 1967 Die vegetationsentwicklung des Eem-Interglazials und des Frühwürm-Interstadials von Odderade, Schleswig-Holstein, in C.O. Gripp, *Frühe Menschkeit und Umwelt*, **2**, 101–125, Köln-Graz.

Bell, Frances G. 1970 Late Pleistocene floras from Earith, Huntingdonshire. *Phil. Trans. R. Soc. Lond.* B **258**, 347–378.

Broecker, W. S. 1965 Isotope geochemistry and the Pleistocene climatic record. Lamont Geol. Obs. contribution 792 in *The Quaternary of the United States* (eds H. E. Wright & D. G. Frey), pp. 737–53. Princeton: Princeton University Press.

Colhoun, E. A., Dickson, J. H., McCabe, A. M. & Shotton, F. W. 1972 A Middle Midlandian freshwater series at Derryvree, Maguiresbridge, County Fermanagh, Northern Ireland, *Proc. R. Soc. Lond.* B **180**, 273–292.

Coope, G. R. 1959 A late Pleistocene insect fauna from Chelford, Cheshire. *Proc. R. Soc. Lond.* B **157**, 70–86.

Coope, G. R., Shotton, F. W., & Strachan, I. 1961 A Late Pleistocene fauna and flora from Upton Warren, Worcestershire. *Phil. Trans. R. Soc. Lond.* B **244**, 279–421.

Coope, G. R. & Angus, R. B. 1975 An ecological study of a temperate interlude in the middle of the last glaciation, based on fossil coleoptera from Isleworth, Middlesex. *J. Anim. Ecol.* **44**, 365–391.

Girling, Maureen, A. 1974 Evidence from Lincolnshire of the age and intensity of the mid-Devensian temperate episode. *Nature, Lond.* **250**, 270.

Godwin, H., Walker, D. & Willis, E. H. 1957 Radiocarbon dating and postglacial vegetational history: Scaleby Moss. *Proc. R. Soc. Lond.* B **147**, 352.

Godwin, H. & Willis, E. H. 1959 Radiocarbon dating of the Late-glacial Period in Britain. *Proc. R. Soc. Lond.* B **150**, 199–215.

Grootes, P. M. 1976 Chronology of the early part of the Last Glacial. Abstract, *Fourth European Colloquium of Geochronology, Cosmochronology and Isotope Geology*, p. 44. Amsterdam.

Hammen, T. van der, Maarleveld, G. C., Vogel, J. C. & Zagwijn, W. H. 1967 Stratigraphy, climatic succession and radiocarbon dating of the last glacial in the Netherlands. *Geol. Mijnb.* **45**, 79.

Hibbert, F. A., Switsur, V. R. & West, R. G. 1971 Radiocarbon dating of Flandrian pollen zones at Red Moss, Lancashire. *Proc. R. Soc. Lond.* B **177**, 161.

Mitchell, G. F., Penny, L. F., Shotton, F. W. & West, R. G. 1973 A correlation of Quaternary deposits in the British Isles. *Geol. Soc. Lond., Spec. Rep. No. 4*, 99 pp.

Morgan, Anne, 1973 Late Pleistocene environmental changes indicated by fossil insect faunas of the English Midlands. *Boreas*, **2** (4), 173–212.

Morgan, A. V. 1973 The Pleistocene geology of the area north and west of Wolverhampton, Staffordshire, England. *Phil. Trans. R. Soc. Lond.* B **265**, 233–297.

Morzadec-Kerfourn, M. T. 1975 Palynology of the Quaternary sediments in borehole V050. Appendix in Destombes, J. P., Shephard-Thorn, E. R. & Redding, J. H. A buried valley system in the Strait of Dover, *Phil. Trans. R. Soc. Lond.* A **279**, 243–256.

Osborne, P. J. 1974 An insect assemblage of the Early Flandrian age from Lea Marston, Warwickshire, and its bearing on the contemporary climate and ecology. *Quaternary Res.* **4**, 471–486.

Penny, L. F., Coope, G. R. & Catt, J. A. 1969 Age and insect fauna of the Dimlington Silts, East Yorkshire. *Nature, Lond.* **224**, 65–67.

Shotton, F. W. & Williams, R. E. G. 1973 Birmingham University Radiocarbon Dates. VI. *Radiocarbon* **15** (1), 1–12.

Simpson, I. M. & West, R. G. 1958 On the stratigraphy and palaeobotany of a late-Pleistocene organic deposit at Chelford, Cheshire. *New. Phytol.* **57**, 239–250.

Sparks, B. W. & West, R. G. 1970 Late Pleistocene deposits at Wretton, Norfolk. I. Ipswichian interglacial deposits. *Phil. Trans. R. Soc. Lond.* B **258**, 1–30.

Turekian, K. K. & Bada, J. L. 1972 The dating of fossil bones. In *Calibration of hominid evolution* (eds W. W. Bishop & J. A. Miller), pp. 171–185. Scottish Academical Press and University of Toronto Press.

West, R. G., Dickson, Camilla A., Catt, J. A., Weir, A. H. & Sparks, B. W. 1974 Late Pleistocene deposits at Wretton, Norfolk. II. Devensian deposits. *Phil. Trans. R. Soc. Lond.* B **267**, 337–420.

Zagwijn, W. H. 1961 Vegetation, climate and radiocarbon datings in the late-Pleistocene of the Netherlands. Part 1. Eemian and Early Weichselian. *Mem. Geol. Found. Netherlands* N.S. **14**, 15–45.

Zagwijn, W. H. 1974 Vegetation, climate and radiocarbon datings in the late Pleistocene of the Netherlands. Part II. Middle Weichselian. *Med. Rijks. Geol. Dienst.* N.S. **25** (3), 101–110.

Phil. Trans. R. Soc. Lond. B. **280**, 119–142 (1977) [119]
Printed in Great Britain

Glacial/Interglacial response rate of subpolar North Atlantic waters to climatic change: the record in oceanic sediments

By W. F. Ruddiman†, C. D. Sancetta‡ and A. McIntyre§

† *Ocean Floor Analysis Division, U.S. Naval Ocean Research and Development Activity, Chesapeake Beach, Maryland* 20732, *U.S.A.*

‡ *Department of Oceanography, Oregon State University, Corvallis, Oregon* 98301, *U.S.A.*

§ *Department of Earth and Environmental Sciences, Queens College of the City, University of New York, Flushing, New York* 11367, *U.S.A.*
and *CLIMAP Project, Lamont-Doherty Geological Observatory of Columbia University, Palisades, New York* 10964, *U.S.A.*

Estimated rates of response of northeast Atlantic surface waters to large-scale palaeoclimatic changes have been reconstructed in two forms: (1) as changes through time of estimated temperature at selected points in space, and (2) as movements through space of the polar front during particular intervals in time. Three periods of rapid change between the glacial and interglacial climatic extremes were examined: a short cold episode during isotopic stage 7 (the next-to-last interglaciation); the deglacial warming into the last interglaciation at the isotopic stage 6/5 boundary; and the most recent deglacial warming into the present interglaciation at the isotopic stage 2/1 boundary.

Changes in sea-surface temperature of 7–11 °C (estimated from transfer function analysis) are characteristically registered in these cores in a few thousand years. The corresponding temporal rates of cooling or warming recorded usually average 1–5 °C/1000 years (a) for the complete climatic shift. During local passage of the polar front, these rates are even higher. Regional advance and retreat rates of the polar front along a NW/SE axis from Cape Farewell, Greenland, to Spain characteristically range from 200 to 1600 m/a during these intervals. These estimates represent the rates of change presently recorded in the sediments. The actual (faster) rates of palaeo-oceanographic change in the overlying North Atlantic surface waters will only be determined once the smoothing effects of vertical mixing can be removed.

Introduction

The world ocean is important both as (1) a permanent geologic monitor of late Quaternary climatic change and (2) an integral part of the land–ice–water–air system in which the changes occur. One critical factor in the kinematics of that system is the response rate of the surface ocean to climatic change. Modern annual oceanic responses are but small oscillations around the interglacial configuration and do not bear directly on larger scale variations and longer terms rates of response. In this paper we will primarily focus on the rate of change of the surface ocean between the extreme glacial and interglacial configurations in both time and space.

The record left in the sediments by subpolar North Atlantic surface waters shows a strong response to past oscillations of mean global climate. This response has been interpreted as temporal changes in the palaeobiota and in estimated sea-surface palaeotemperature at selected points in space (Bramlette & Bradley 1941; McIntyre, Ruddiman & Jantzen 1972; Sancetta, Imbrie & Kipp 1973). It also reflects spatial changes in watermass and oceanic frontal positions during selected time intervals (McIntyre *et al.* 1972; Ruddiman & McIntyre 1973, 1976). These studies show that palaeotemperature changes and watermass migrations in the subpolar North Atlantic are at least as large as in any other area of the world ocean.

These strong changes, combined with the high quality of the sedimentary record, make the subpolar North Atlantic an ideal location to define the maximum oceanic response rates recorded in deep-sea sediments. With this as our objective, we have taken from the record of the last 200000 years three intervals registering prominent and rapid palaeoclimatic shifts between the extreme glacial and interglacial oceanic configurations. By examining faunal evidence of these changes in several cores, we have attempted to overcome local biases at individual core sites. Future studies will be necessary to remove the blurring effects imposed by the mixing activities of burrowing organisms in order to derive the actual rates of oceanic response.

Selection of climatic levels for study

In the subpolar North Atlantic, changes in percentage of the single foraminiferal species with polar watermass preferences – *G. pachyderma*, left-coiling – are an excellent first-order indicator of oceanic sensitivity to climatic change. Changes in abundance of this species as a percentage of the total foraminiferal fauna are shown in figure 1 for a portion of one core spanning the last 600000 years (Ruddiman & McIntyre 1976). High percentages indicate low temperatures and salinities (glacial climates), and low percentages signify higher temperatures and salinities (interglacial climates).

The palaeoclimatic changes that we will examine are located by arrows in figure 1. They fall within faunal stages numbered to equate with the isotopic stages of Emiliani (1955, 1966) and Shackleton & Opdyke (1973). The correspondence of isotopic and faunal stages throughout the last 600000 years has been demonstrated by Ruddiman & McIntyre (1976). The intervals that we have selected are: (1) an intensely cold climatic episode within interglacial stage 7; (2) the glacial-to-interglacial change across the stage 6/5 boundary (isotopic termination II of Broecker & van Donk (1970), followed by a short (stage 5d) cooling about 17000 years later; and (3) the glacial-to-interglacial shift across the stage 2/1 boundary (termination I), including the very brief Younger Dryas cooling within the lowest portion of isotopic stage 1. Because the Younger Dryas cold episode lasted less than 1000 years, the initial sampling resolution of about 10 cm (2500 years) in figure 1 missed it entirely.

All three intervals have broad climatic significance. The stage 7 cold episode during the next-to-last interglaciation represents one potential analogue for future climate. It is the fastest complete interglacial-to-glacial-to-interglacial oceanic event that we have verified in this suite of northeast Atlantic cores. The stage 6/5 boundary was selected because the fauna consistently showed the fastest complete deglacial termination of oceanic glacial conditions of any of the major Brunhes climatic oscillations (Ruddiman & McIntyre 1976). The stage 5d cooling that followed represents a second analogue for future climate. The stage 2/1 boundary warming was chosen because it is the most recent deglacial termination, is accessible to ^{14}C dating, and is thus directly comparable to continental data during the same period. In addition, the subsequent Younger Dryas cold episode at 10200 a B.P. is an important indication that a very brief but intense climatic change can still be discriminated in cores with high deposition rates.

The cores in which we have examined one or more of these changes are located in figure 2 and table 1. These cores have relatively high deposition rates (2.4–15 cm/1000 a), virtually undissolved carbonate fractions, and excellent long-term stratigraphy.

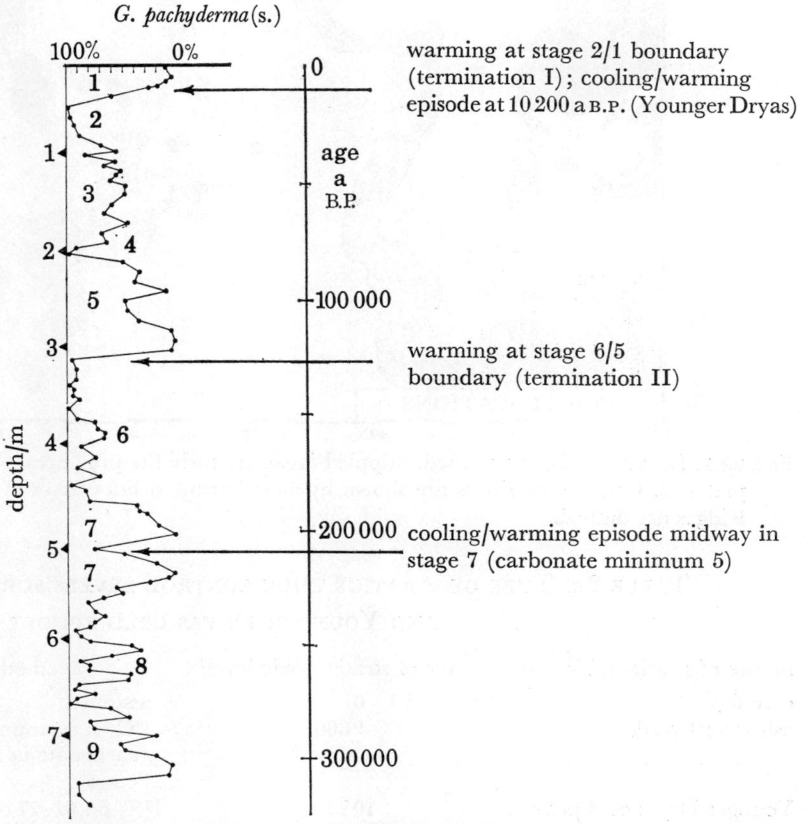

FIGURE 1. Placement of three studied intervals in long-term record of climatic change. Polar fauna as a percent of total planktonic foraminifera larger than 149 μm after Ruddiman & McIntyre (1976). Large numbers next to curve are designations of faunal/isotopic stages. Small numbers to left are metres of depth in core K708-7.

TABLE 1. LOCATIONS AND DEPTHS OF NORTHEAST ATLANTIC CORES

core	north latitude	west longitude	depth/m
K708-1	50° 00′	23° 45′	4053
K708-4	49° 59′	25° 01′	3346
K708-6	51° 34′	29° 34′	2469
K708-7	53° 56′	24° 05′	3502
K708-8	52° 45′	22° 33′	4009
RC9-225	54° 59′	15° 24′	2334
RC9-228	52° 33′	18° 45′	3981
V23-81	54° 15′	16° 50′	2393
V23-82	52° 35′	21° 56′	3974
V23-83	49° 52′	24° 15′	3871

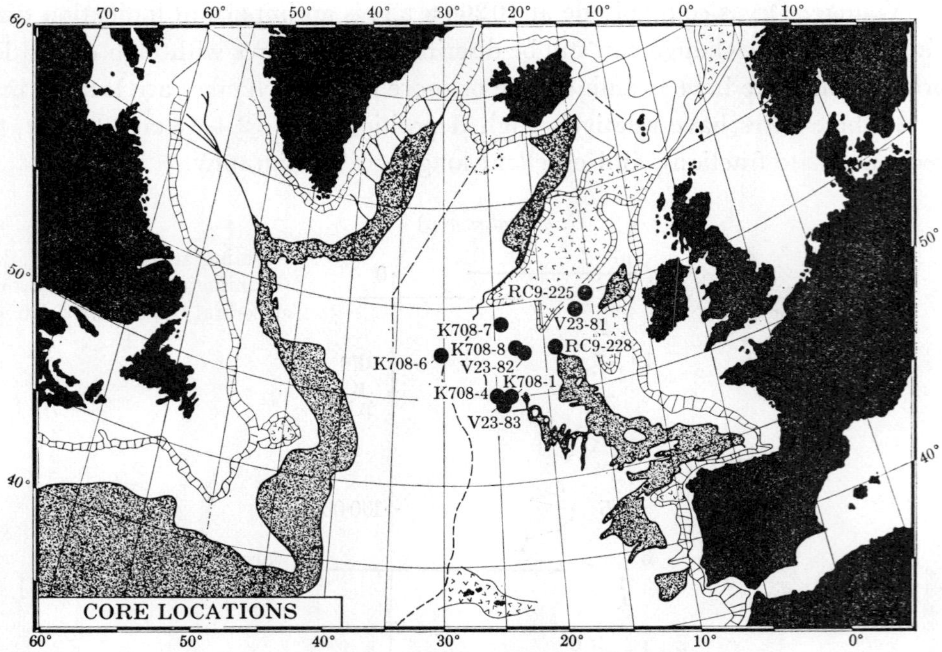

FIGURE 2. Locations of cores studied. Stippled areas are turbidite provinces. Regions with V pattern are shallow plateaux. Continental slopes are shown by short hachured lines. Axes of Mid-Atlantic and Iceland–Faeroe Ridges are dashed.

TABLE 2*a*. TYPE OF STRATIGRAPHIC CONTROL LEVELS FOR STAGE 2/1 WARMING AND YOUNGER DRYAS COLD EPISODE

nature of stratigraphic level	age of stratigraphic level/a	method	references
core top	0	assumed	—
ash zone 1 peak	9300	^{14}C on surrounding carbonate in several cores	1, 2
Younger Dryas cold peak	10220	^{14}C on 61–77 cm in core V23-82	1, 3
faunal termination 1	~13500	interpolation between ^{14}C dates above and below faunal change in several cores	1, 3, 4

1. Ruddiman & McIntrye (1973).
2. Ruddiman & Glover (1975).
3. Sancetta *et al.* (1973).
4. McIntyre *et al.* (1972).

TABLE 2*b*. DEPTHS OF STAGE 2/1 CONTROL LEVELS IN CORES STUDIED

	depth in core/cm				
control level (age)	K708-1	V23-82	RC9-228	V23-81	RC9-225
core top (0)	0	0	0	0	0
peak of ash zone 1 (9300)	62	63	—	—	70
Younger Dryas cold peak (10220)	(70†)	70 (^{14}C date)	(92†)	(157†)	(75†)
faunal termination 1 (~13500)	(85†)	88	120	213	93
other ^{14}C dates	—	—	—	40 (4010 a)	—

(†) These levels not used as time control to plot figures 7–8.

Stratigraphy and temporal plotting assumptions

For the climatic change at the stage 2/1 boundary, we used actual ^{14}C dates for primary control,† as well as faunal or mineralogic levels correlative with ^{14}C-dated horizons in other cores (table 2). We then made simple linear interpolations between all listed control levels to transform each sample depth into an estimate of actual age.

The major uncertainty in the plotted levels lies in the age of the faunal warming near the stage 2/1 boundary. Ruddiman & McIntyre (1973) calculated an extrapolated age of 13500 a B.P. from overlying sedimentation rates in several cores, but did not date the transition directly using ^{14}C. We recently obtained a ^{14}C age of 14825 ± 365 a B.P. for the section of core V23-81 from 200–185 cm, which would place the faunal boundary (212 cm) at 16 300 a B.P. We regard this date as unrealistically old, probably because of contamination by older glacial-age carbonate, and have used the 13500 a B.P. date.

Table 3*a*. Type of stratigraphic control levels for calculating and testing long-term sedimentation rates in northeast Atlantic cores

nature of stratigraphic level	age of stratigraphic level	method of age determination	references
stage 6/5 boundary (faunal equivalent of isotopic termination II)	126000 a B.P. ($\pm$ 6000 a)	interpolation from $^{230}Th/^{234}U$ and $^{231}Pa/^{235}U$ control	1, 2
stage 2/1 boundary (faunal equivalent of isotopic termination I)		see table 2*a*	
ash zone 1 peak		see table 2*a*	
Barbados 1 high sea level; isotopic stage 5*a* (faunal equivalent)	82000 a B.P. ($\pm$ 4000 a)	$^{230}Th/^{234}U$ $^{231}Pa/^{235}U$	1

1. Broecker *et al.* (1968).
2. Broecker & Ku (1969).

Table 3*b*. Depths of long-term control levels (cm) and rates of sedimentation (cm/1000 a)

	cores							
	K708-1	K708-4	K708-6	K708-7	K708-8	RC9-225	V23-82	V23-83
stage 6/5 boundary (127000 a B.P.)	858	358	445	307	550	920	805	675
approx. stage 2/1 boundary (9300 B.P.)	62	34	28	28	45	70	63	70[a]
mean rate of sedimentation (stage 6/5–stage 2/1)	6.77	2.75	3.54	2.37	4.21	7.22	6.30	5.33
Barbados 1 high sea level faunal equivalent (82000 a B.P.)	460	235	290	215	365	600	470	400

[a] Faunal termination 1 (13500 a B.P.) used instead of ash zone 1.

† Stuiver & Borns (1975) have shown that no age correction is necessary for ^{14}C ages from deep-sea sediments due to equal and exactly offsetting errors caused by (1) biological fractionation and (2) mixed layer composition.

TABLE 4. COMPARISON OF INTERGLACIAL AND GLACIAL SEDIMENTATION RATES IN cm/1000 a FOR NORTHEAST ATLANTIC CORES

core	rate during last glaciation	rate during last interglaciation	ratio interglacial/glacial
K708-1	5.47	9.29	1.70
K708-4	2.75	2.76	1.00
K708-6	3.60	3.44	0.95
K708-7	2.57	2.04	0.79
K708-8	4.40	4.11	0.93
RC9-225	7.29	7.11	0.98
V23-82	5.60	7.44	1.33
V23-83	4.78	6.11	1.20
	$\bar{x}$ = 4.56	$\bar{x}$ = 5.29	$\bar{x}$ = 1.16

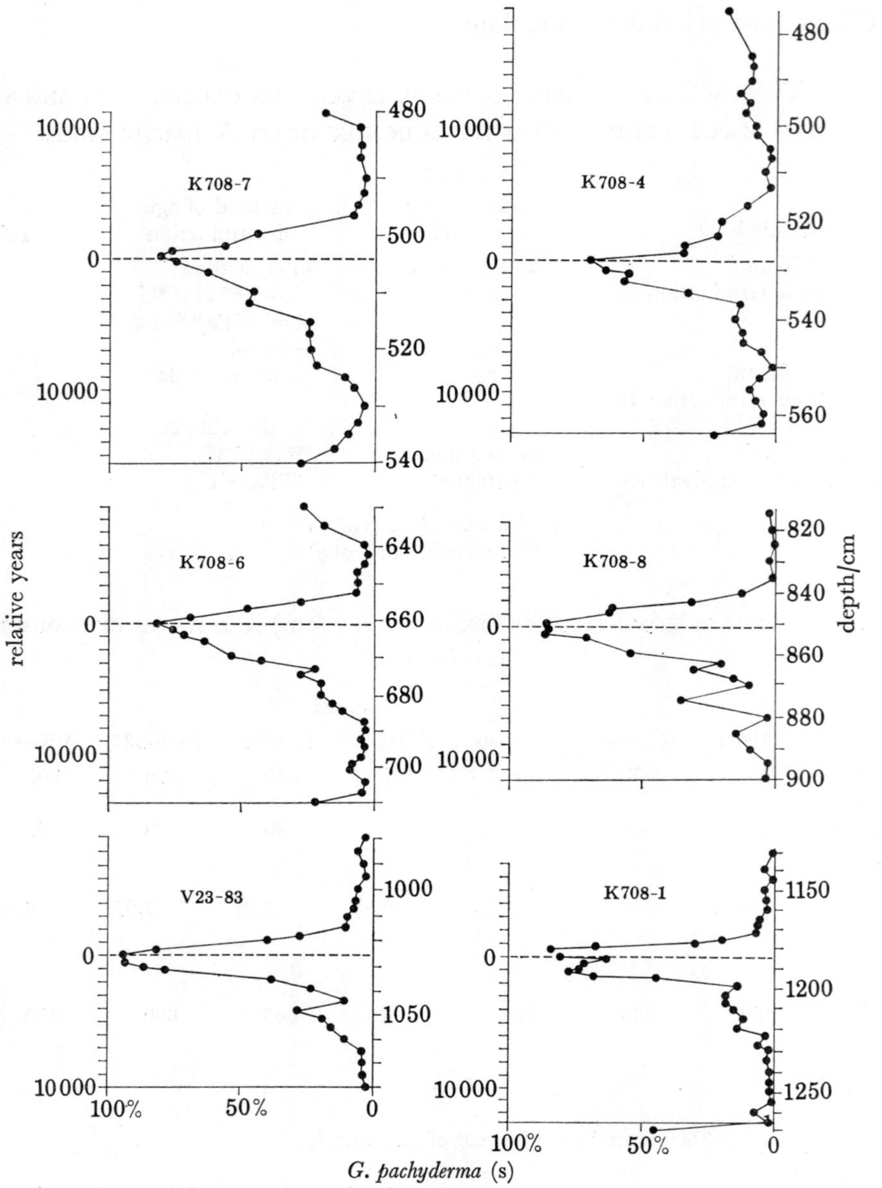

FIGURE 3. Down-core percentages of *G. pachyderma* (s.) (●) as a percent of total planktonic foraminifera larger than 149 μm during cold episode in middle of stage 7. Time plotted in relative years increasing in both directions away from midpoint of cold episode set at 0 a. Time plotting assumptions in table 3 and text. Core depth in centimetres shown on right.

For the two earlier episodes (in stage 7 and at the stage 6/5 boundary), we made no attempt to specify absolute ages, although the cold episode in stage 7 occurred at roughly 195000 a B.P. (Ruddiman & McIntyre 1976) and the stage 6/5 boundary at about 127000 a B.P. (Broecker & Ku 1969; Ku, Bischoff & Boersma 1972). Instead, we chose to express the duration of these changes in a relative time framework. To do this, we determined in each core the long-term rate of sedimentation during the late Quaternary and assumed that this rate prevailed during the two intervals examined.

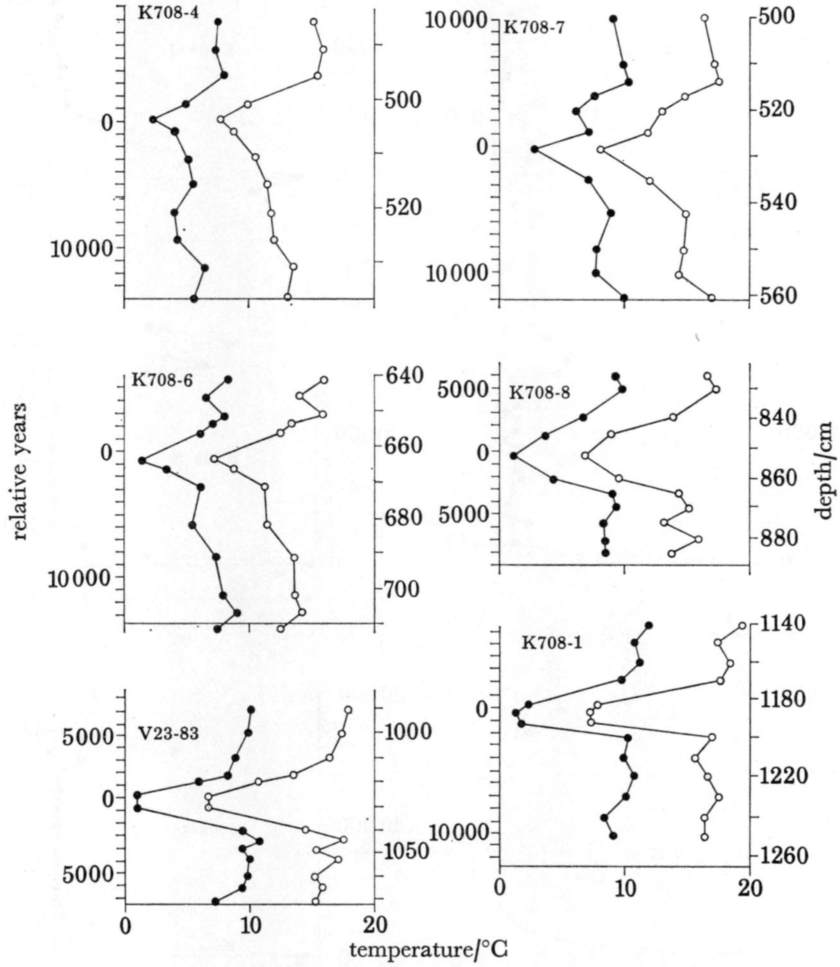

FIGURE 4. Estimates of summer (○) and winter (●) palaeotemperature (°C) during stage 7 cold episode. Time and depth plotted as in figure 3. Estimates based on transfer function equation F13B-4CE.

We selected the climatic cycle between stage boundaries 6/5 and 2/1 as the best long interval to provide a mean sedimentation rate (table 3*a*). The isotopic stage 6/5 boundary has been dated in other oceanic cores at 126000 ± 6000 a B.P. (Broecker & van Donk 1970) and correlated with North Atlantic faunal, floral and lithologic changes (McIntyre *et al.* 1972; Sancetta *et al.* 1973; Crowley 1975). The faunal stage 6/5 boundary is at most about 1000 years older than its isotopic counterpart in this region of the subpolar North Atlantic.† For the

† In these cores, the faunal stage 2/1 boundary precedes the isotopic stage 2/1 boundary dated at low latitudes by about 2000 years. Since the stage 6/5 deglaciation appears to have been faster, we estimate no more than a 1000-year lead time.

younger control level we used in most cases ash zone 1 at 9300 a B.P. (Ruddiman & Glover 1972). In one core (V23-83) where the ash zone was not examined, we used the faunal stage 2/1 boundary estimated at ~13500 a B.P. by Ruddiman & McIntyre (1973). Control levels and average sedimentation rates for this long interval (stage 6/5 to 2/1) are listed in table 3*b*.

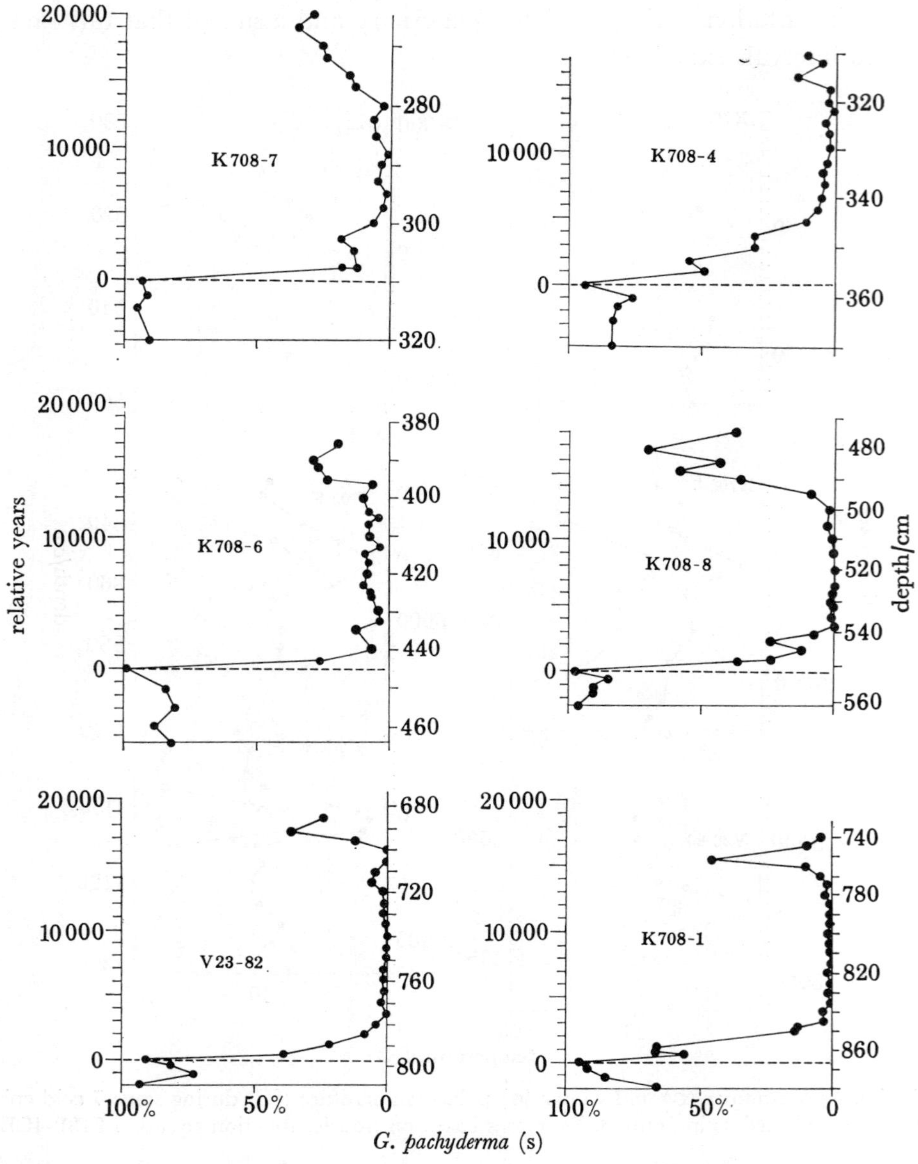

FIGURE 5. Down-core percentages of *G. pachyderma* (s.) (●) as a percent of total planktonic foraminifera larger than 149 μm during climatic warming at stage 6/5 boundary. Time plotted in relative years increasing upwards from uppermost full-glacial sample set at 0 a. Time plotting assumptions in table 3 and text. Core depth in centimetres shown on right.

To test the assumption of constant sedimentation rates, we used an additional control level located just below the stage 5/4 boundary and roughly midway within the interval used to calculate the mean sedimentation rates (table 3*b*, figure 1). We previously correlated this peak with the Barbados 1 coral reefs dated by U/He methods and indicative of a high stand of sea level at 82000 a B.P. (McIntyre *et al.* 1972; Mesolella, Matthews, Broecker & Thurber

1969). This correlation has been supported by other marine stratigraphic work (Sancetta *et al.* 1973; Crowley 1975).

The 82000 year level is equivalent to substage 5*a* of Shackleton (1969) and lies just below the stage 5/4 boundary. Beneath the stage 5a level lies an earlier interglacial interval (127000–82000 a B.P.) of highly calcareous coccolith-foraminiferal ooze mainly derived from overlying surface waters. Above it is a glacial interval (75000–10000 a B.P.) dominated by moderately

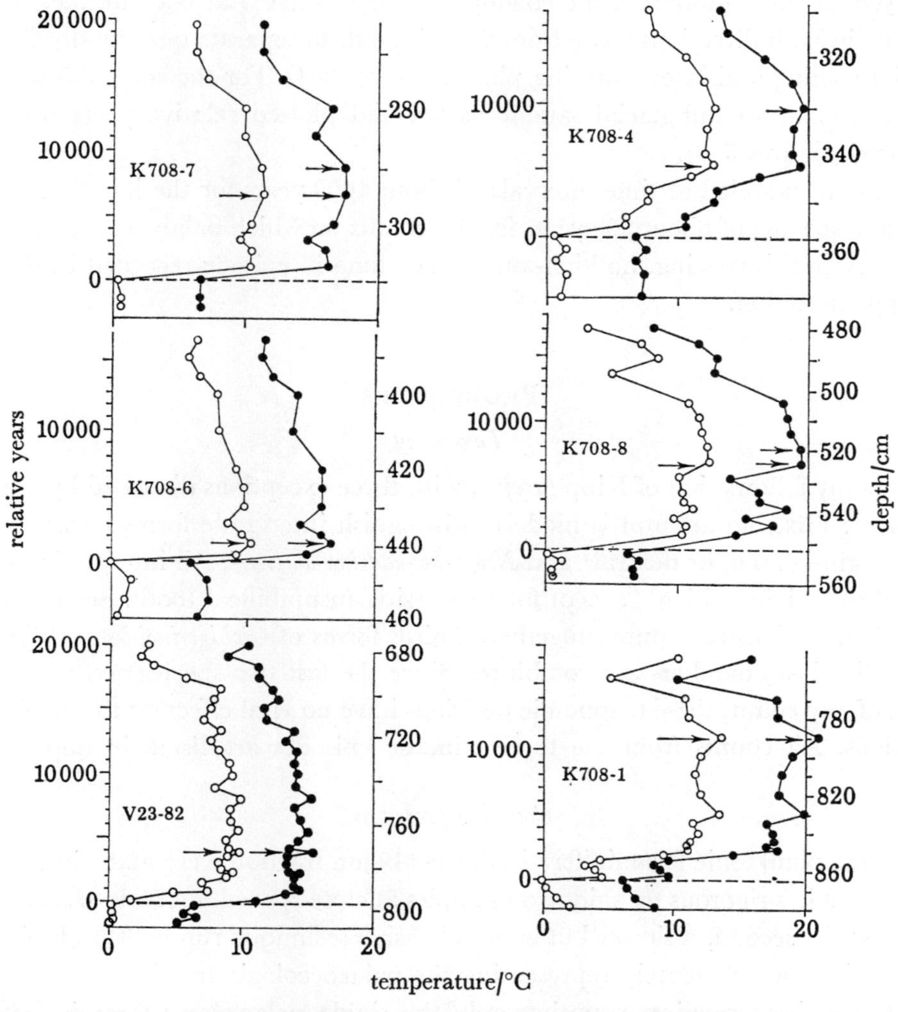

Figure 6. Estimates of summer (●) and winter (○) palaeotemperature (°C) during stage 6/5 warming. Time and depth plotted as in figure 5. Estimates based on transfer function equation F13B-4CE.

calcareous sandy lutites ice-rafted from continental sources. These two contrasting lithologic units provide a meaningful test of constancy in depositional rates during varying climate. As shown in table 4, the sedimentation rates are usually constant to within about 30 % and often to within 10 % (4 out of 8 cores). The interglacial mean for all cores is 16 % larger than the glacial average, indicating that sedimentation rates stay reasonably constant during changing climates on at least a long-term scale.

A finding of nearly constant sedimentation rates seems at odds with the gross changes in lithologic composition and sediment source during these two climatic intervals. Nevertheless,

Ruddiman & McIntyre (1976) have shown that the constant rates of sedimentation indicate a coincidental but almost exactly balanced input of the two major components. Increases in the absolute input of one component balance and offset decreases in the absolute input of the other.

Because the sedimentation rates in these cores are reasonably constant over long intervals of time (10^5 a), we have assumed that they are also constant over the shorter intervals (10^3–10^4 a) encompassing the two early climatic episodes. We then plotted these episodes in units of relative time. We set the midpoint of the cooling episode in stage 7 at 0 a (not B.P.) and plotted relative years in both directions away from the midpoint to separate out visually the durations of the initial cooling and later warming phases (figures 3–4). For the stage 6/5 boundary, we selected the uppermost full-glacial sample as 0 a and plotted relative years upwards away from this level (figures 5–6).

The cores were sampled at time intervals of about 1000 years for the initial counts of polar fauna and at a spacing of about 200 years for the counts on which palaeotemperature estimates were made. With this spacing, no high-amplitude climatic episode recorded in the sediments should escape detection.

PROCEDURES

Taxonomy

The taxonomy follows that of Kipp (1976), with three exceptions discussed by Ruddiman & Glover (1975). First, no attempt is made to distinguish intergrade forms between *Globigerina pachyderma* (right-coiling, or dextral) and *Neogloboquadrina dutertrei*. All intergrades are assigned to *G. pachyderma* (right-coiling), except for those with an umbilical tooth. Second, *Globorotalia menardii* and *G. tumida* are counted together. Third, forms of *Globigerinoides sacculifer* with and without sac-like last chambers are combined. Since the last two species account for no more than 0.5 % of any count, these taxonomic decisions have no vital effect on the palaeoecological interpretations. All counts from the three climatic episodes are listed in appendixes 1–3.†

Counting procedure

Counts of the planktonic foraminifera in the > 149 μm fraction were made in two ways: the first, a faster and less rigorous technique on samples at close-spaced intervals to search for short-term changes; the second, a slower but comprehensive technique run on samples indicated by the first technique as adequately representing the palaeoecologic trends.

The first technique requires counting only the single polar-water foraminifer *Globigerina pachyderma* (left-coiling, or sinistral) as a percentage of all foraminifera present in the > 149 μm fraction. Using portions of a sample aliquot strewn across a tray marked with perpendicular lines, counts were made along a diagonal series of boxes leading from the tray margin to its centre. The number counted ranged from 150 to 610 and averaged 275.

The technique used to count full aliquots of > 149 μm planktonic foraminifera for palaeotemperature analysis is that of Imbrie & Kipp (1971) and Sancetta *et al.* (1973). The number of individuals counted in all samples averaged 414 and ranged from 191 to 629.

† Appendixes 1–6 appear on the accompanying microfiche.

Transfer function analysis

The basic methods of transfer function analysis were first derived by Imbrie & Kipp (1971) and later modified slightly by Imbrie, van Donk & Kipp (1973). To convert the foraminiferal counts listed in appendixes 1–3 to estimates of palaeotemperature, we used transfer function equation F13B-4CE (Ruddiman & Glover 1975).† Equation F13B-4CE is based on the same 191 core tops as the F13 set of Kipp (1976) but incorporates the three taxonomic changes. Relevant statistics in Ruddiman & Glover (1975) show lower standard errors and higher correlation coefficients than equation F13 on estimates of seasonal palaeotemperature, but higher standard errors and lower correlation coefficients on palaeosalinity estimates. None of the differences were large. The standard errors of estimate for equation F13B-4CE are ± 1.157 for winter palaeotemperature (T_w) and $+1.357$ for summer palaeotemperature (T_s). This represents roughly 10 % of the observed range of temperature variation (appendixes 4–6). Communalities (defined in Imbrie *et al.* 1971) for most samples were very high (appendixes 4–6), indicating that excellent analogues for most palaeoassemblages can be found in the model generated from the set of 191 core tops and used to calibrate the surface-sediment fauna against the modern oceanographic environment.

Berger & Gardner (1975) have discussed problems in separating past variations in temperature from those in salinity. If the strong correlation in the modern ocean between temperature and salinity were broken or shifted in some way, any palaeoestimating technique using the modern ocean for calibration would produce errors in palaeoestimates. In such a case, the ecological variable that is causally dominant would be best estimated, with larger errors for secondary variables. In this paper, we chose to make estimates only of palaeotemperature, the variable normally considered dominant (Imbrie & Kipp 1971).

The temperature estimates in this paper have an additional problem; there is an artificial lower limit due to the monospecific composition of the fauna in polar waters. Because there is no variation possible beyond the essentially 100 % of *G. pachyderma* (s.) found just behind the modern Polar Front, no temperature estimates lower than -0.1 °C in winter and 5.6 °C in summer can be made for older levels. All estimates at or near these limits should be understood to represent 'less than or equal to' values. Additional work along the lines initiated by Kennett (1968, 1970) and by Vilks (1975) on morphologic variations within the species *G. pachyderma* (s.) might extend the range of temperature estimates to somewhat colder values.

Observed rates of change: temporal

Down-core changes in estimated palaeotemperature during the specified intervals of time (figures 4, 6 and 8) are the basis for establishing the recorded rates of change in oceanic temperature at points in space. Procedurally, we reconstructed these rates of change in two ways. First, we measured the full temperature change and time duration between the coldest (full-glacial) and warmest (full-interglacial) sample in a chosen interval. This yields the mean rate of change across the entire climatic transition (table 5).

Unfortunately, interglacial temperature estimates tend to vary within a rather narrow range around a mean value for long intervals of time. The warmest sample usually does not stand out above the other temperature estimates by an amount exceeding the standard error of

† Equation F13 was used for estimates in cores V23–81 and V23–82.

TABLE 5. AVERAGE AMPLITUDES, DURATIONS AND RATES OF OCEANIC COOLINGS AND WARMINGS RECORDED FOR THREE CLIMATIC EPISODES

	estimated summer temp./°C†	estimated winter temp./°C†	average duration of change between peaks/a†	estimated rate of cooling °C/1000 a†‡	estimated rate of warming °C/1000 a†‡
stage 7 episode					
initial warm peak	16.0	9.4	9200 (4400)	0.9 (1.4)	—
cold episode peak	7.3	1.7	5800 (2900)	—	1.6 (2.3)
later warm peak	17.3	9.8			
stage 6/5 boundary					
stage 6 cold peak	6.2	0.2	7000 (1700)	—	1.7 (5.2)
stage 5 warm peak (5e)	18.4	11.7			
stage 2/1 boundary: entire transition					
stage 2 cold peak	6.6	0.9	10100 (2800)	—	1.0 (2.6)
stage 1 warm peak	16.4	10.8			
stage 2/1 boundary: separate parts					
initial stage 2 cold peak	6.6	0.9	2700 (1500)	—	2.9 (3.9)
last stage 2 warm peak	14.3	9.0	700 (400)	11.0 (13.2)	—
Younger Dryas cold peak	7.4	1.8	7700 (500)	—	1.1 (12.1)
stage 1 warm peak	15.2	10.1			

† Values for the two oldest climatic episodes based on six-core averages calculated from data in figures 4 and 6. Values for stage 2/1 boundary based on five-core average, except for Younger Dryas cold peak, which is based only on cores V23-81 and RC9-225.

‡ Values on parentheses apply to the 75 % of temperature change occurring toward or away from the coldest sample. Based on the average temperature change of the winter and summer temperature estimates.

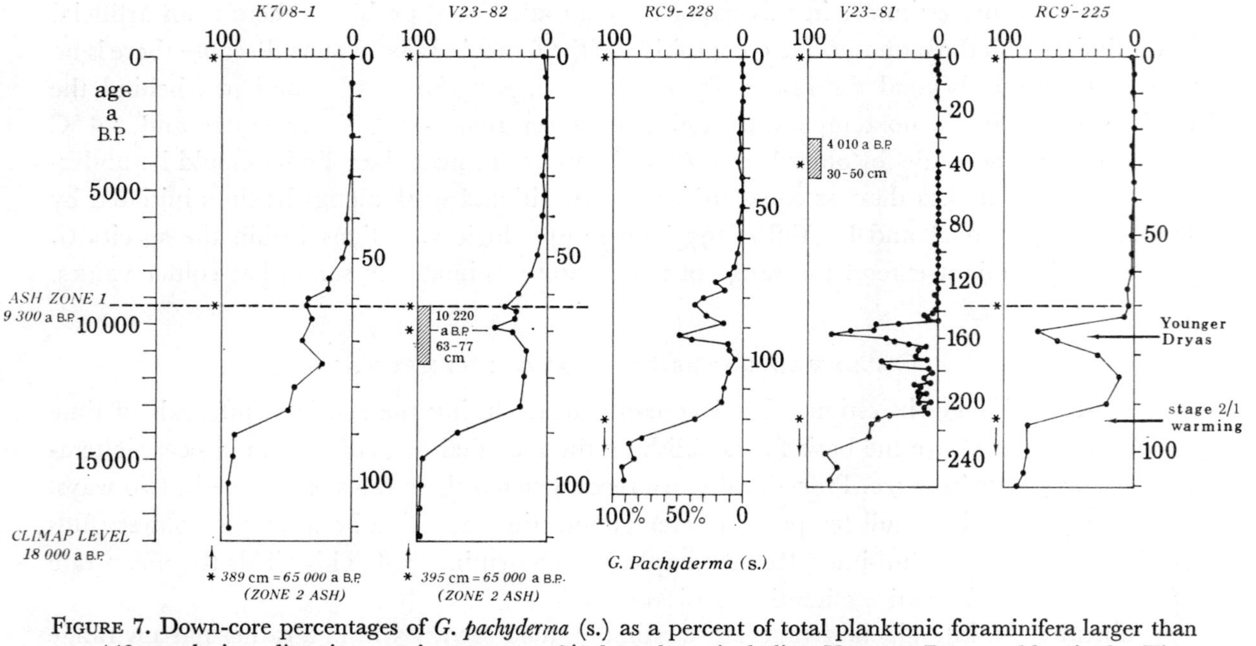

FIGURE 7. Down-core percentages of *G. pachyderma* (s.) as a percent of total planktonic foraminifera larger than 149 μm during climatic warming at stage 2/1 boundary, including Younger Dryas cold episode. Time plotted in absolute ^{14}C a B.P. by interpolation between levels listed in table 2.

estimate (±1.2 °C). In these cases, the choice of a 'warmest' sample thus seems somewhat arbitrary, and its interpolated age from core to core can vary considerably (for example, note the discrepancies among the arrows in figure 8 locating the warmest Holocene temperature estimates). In addition, the largest part of the temperature changes in most cores is recorded in a relatively short period of time, usually occurring near the cold peak.

Consequently, we have also measured the amount of time required for 75 % of each warming or cooling to take place (table 5, values in parentheses). We then calculated the substantially faster rates of change across these shorter intervals.

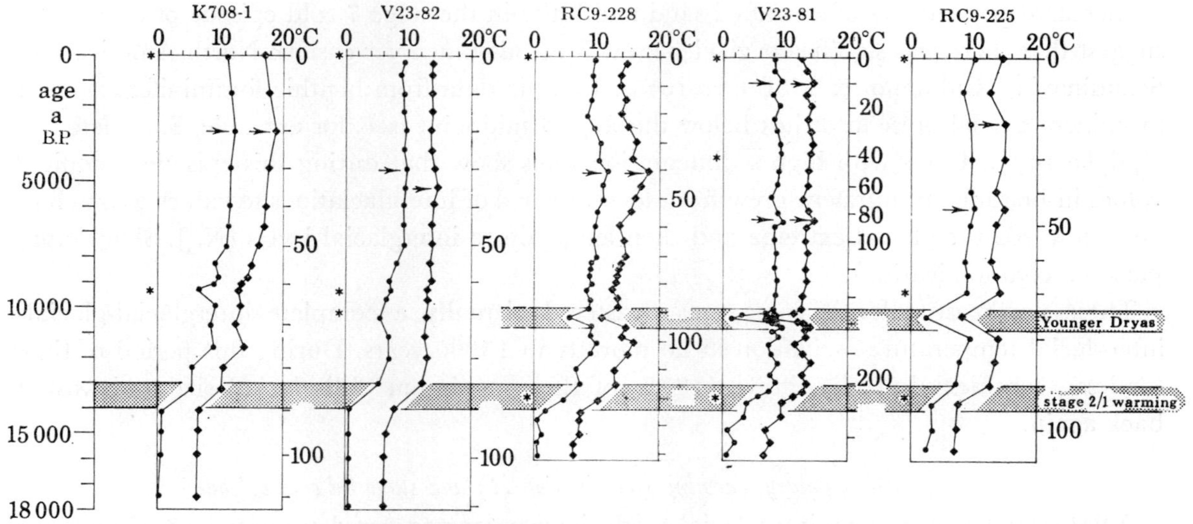

FIGURE 8. Estimates of summer (●) and winter (◆) palaeotemperature (°C) during stage 2/1 warming and Younger Dryas cold episode. Time and depth plotted as in figure 7. Estimates based on transfer function equation F13B-4CE. Arrows show warmest stage 1 estimates in all five cores.

Because post-depositional mixing has presumably lowered the amplitudes and lengthened the durations of the original oceanic variations, we stress that all these rates are merely our best compilation of the smoothed record left in the sediments. They are probably not the actual rates of oceanic response.

Stage 7 episode

The stage 7 cooling episode is slightly asymmetrical, with a slower initial cooling and a faster subsequent warming (figures 3–4). The estimated summer temperature (averaged in six cores) started at 16.0 °C early in stage 7, fell to 7.3 °C in the middle of the cold episode, and warmed to 17.3 °C late in stage 7 (figure 4). At the same time, estimated winter temperature cooled from 9.4 to 1.7 °C and then warmed to 9.8 °C. These changes reflect an initial gradual decrease, followed by a more rapid increase in polar fauna (figure 3).

The initial temperature shift from the warmest sample in the bottom half of stage 7 to the peak cold sample is recorded in an average of 9200 years, with the later change from the cold peak to the warmest overlying sample in 5800 years (figure 4, table 5). We also calculate that 75 % of the initial cooling is registered in an average of 4400 years, and an equal portion of the later warming in an average of 2900 years (table 5).

For the entire initial cooling in the stage 7 episode, the recorded rates of temperature change averaged just under 1 °C/1000 a (table 5). Along the steeper slopes near the cold peak, the rates rise to slightly higher values. Similarly, for the later peak-to-peak warming in the upper half of the stage 7 episode, the recorded rates of change average 1.6 °C/1000 a (table 5) but reach peak values above 2 °C/1000 a.

The duration of the stage 7 cold episode appears briefest in cores K708-1 and V23-83, longest in core K708-7, and intermediate in the three other cores. We infer that these apparent differences in part actually reflect relative glacial/interglacial changes in sedimentation rate. The long-term changes listed in table 4 correctly predict that glacial episodes will be somewhat under-represented (shorter in appearance) in cores K708-1 and V23-83, over-represented in K708-7, and reflected fairly accurately for the other three cores.

Increased percentages of ice-rafted sand and lutite in the stage 7 cold episode of these cores suggest that there was significant glacier growth at this time over eastern North America and Scandinavia (Ruddiman & McIntyre 1976). Isotopic data from benthic foraminifera show a prominent glacial pulse at or just below the stage 7 midpoint (see, for example, Shackleton & Opdyke 1973). Cores with high sedimentation rates show that during this episode isotopic δ values in benthic foraminifera grew from levels typical of interglaciations to values as much as 80 % towards the glacial extreme and then fell again to interglacial levels (N. J. Shackleton, personal communication).

These northeast Atlantic cores thus registered virtually a complete interglacial-glacial-interglacial temperature oscillation in no more than 15000 years. During this period of time (or less), continental glaciers grew to 80 % of their maximum (full-glacial) size and wasted back again.

Stage 6/5 boundary warming (termination II) and stage 5d cold episode

A faster and even stronger interglacial-to-glacial warming occurred at the stage 6/5 boundary (figures 5–6). From estimated winter temperatures of 0.2 °C and summer temperatures of 6.2 °C in the coldest levels late in stage 6, the warming brought estimated sea-surface temperatures to 11.7 °C and 18.4 °C, respectively. This peak-to-peak change is recorded in an average of 7000 years; the first 75 % of the change is registered in just 1700 years (table 5). The corresponding rates of warming average 1.7 °C/1000 a for the entire peak-to-peak shift but reach 5.2 °C/1000 a in the more rapid change in the initial portions of the warming (figures 5–6).

The stage 6/5 boundary warming carried global climate from full-glacial conditions to an unusually extreme interglacial state. Isotopic data from Shackleton & Opdyke (1973) indicate that the uppermost portion of stage 6 may not be quite as intensely glacial as earlier stage 6 levels or as the stage 2 isotopic maximum. Still, a fully glacial oceanic climate late in stage 6 must be inferred from the totally dominant polar faunas at these levels (figure 7). Isotopic data indicate that during stage 5e the mass of continental ice was smaller than during any other Brunhes interglaciation (Shackleton 1969). This is supported by a stage 5e fauna and flora warmer than those of any interglaciation in the last 600000 years (Ruddiman & McIntyre 1976).

In summary, we find that a full glacial-to-interglacial oceanic warming in the northeast Atlantic, and apparently correlative with a total wasting of continental ice, was registered in no more than 6500 years, with most of the change recorded in less than 2000 years.

The stage 5d cold pulse that ended the extreme interglacial warmth of stage 5e reached its peak intensity an average of 17000 years after the beginning of termination II. Using slightly different stratigraphic assumptions from those in this paper, Sancetta *et al.* (1973) calculated an age of 110000 a B.P. for the peak cooling in core V23-82. The age of stage 5d in the 6 cores in this study ranges from 109000 to 111000 a B.P., with a standard deviation of just ±1100 a around the average of 110000 a B.P.

Stage 2/1 boundary warming (termination I and Younger Dryas)

Considered in its entirety, the deglacial warming into the Holocene was less abrupt than that at stage 6/5. The warmest temperatures were not reached until late in stage 1, about 9200 years after the beginning of deglacial warming at 13500 a B.P. (figures 7–8). The lengthened interval required to reach the peak interglaciation may be related to a reversal of deglacial warming centred at about 10200 a B.P. This short cold episode was detected and dated in core V23-82 and correlated to the Younger Dryas (Ruddiman & McIntyre 1973; Sancetta *et al.* 1973). It occurred at a time when the deglacial warming of the northeast Atlantic was not complete, and it briefly returned that ocean to an almost glacial state.

Viewed as a single climatic warming, the entire stage 2/1 shift registered a rise in mean summer temperatures from 6.6 to 16.4 °C and in winter temperatures from 0.9 to 10.8 °C in 9200 years (table 5). This represents a mean rate of 1.0 °C/1000 a. Roughly 75 % of that change occurred in the 2800-year span between 13700 and 11900, at a mean rate of 2.6 °C/1000 a.

Viewed in terms of its separate parts, the long interval of deglaciation at the stage 2/1 boundary actually consists of three separate and rapidly recorded climatic changes: (1) an initial warming at about 13500 a B.P.; (2) a cooling beginning after 11500 a B.P. and culminating at about 10200 a B.P.; and (3) a later warming occurring for the most part before 9300 a B.P. but not reaching peak interglacial warmth until about 5000 a B.P.†

This general chronology is largely supported by climatic evidence in immediately adjacent continental regions. Coope (1975) used insect assemblages recorded in peat layers on Great Britain to estimate climatic changes at land sites only 10–15° of longitude (500–750 km) east of oceanic cores V23-81 and RC9-225. He inferred a strong and abrupt warming at about 13000 a B.P. (bracketed between a glacial insect assemblage dated at 14460 a B.P. at the Glanllynnau site and a warm coleopteran fauna dated at 12940 a B.P. from the site at Robert Hill). Mangerud & Gulliksen (1975) quote work in preparation on coastal molluscs asserting that warm North Atlantic waters reached western Norway before 12600 a B.P. These estimates substantiate a deglacial warming at about 13500 a B.P. that affected the easternmost North Atlantic and westernmost Europe and Scandinavia. If this is correct, it challenges the conventional palaeoclimatic interpretations of European (or at least British) pollen data. Specifically, it can be argued that the British arboreal pollen response to an actual atmospheric climatic warming that occurred about 13500 a B.P. lagged as much as 1500 years or more behind the faster response of the insects (Coope 1975), the aquatic pollen (Iversen 1954), the coastal molluscs (Mangerud & Gulliksen 1975) and the open-ocean planktonic foraminifera (Ruddiman & McIntyre 1973).

Coope (1975) inferred a more gradual cooling into the Younter Dryas peak, culminating shortly before 10000 a B.P., followed by a rapid warming out of the Younger Dryas. This equates more closely with the conventional European interpretation of the Younger Dryas cooling, as does the oceanic record.

In the oceanic record, estimated winter ocean temperatures (calculated from the average

† In other contributions to the symposium, the initial warming is referred to as the 'Windermere Interstadial' (Britain) and the Woodgrange Interstadial (Ireland). The later warming into stage 1 is the 'Flandrian' (Britain) and the Littletonian (Ireland). Stage 2 glacial conditions are 'Late Devensian'.

of all five cores) rose during the initial (13500 a B.P.) warming from 0.9 to 9.0 °C, as summer sea-surface temperature increased from 6.6 to 14.3 °C.† This part of the deglacial warming was recorded in an average of 2700 years, for a mean rate of warming of 2.9 °C/1000 a (figure 8, table 5). The first 75 % of the warming, however, is registered in 1500 years, for a rate during that period of almost 4 °C/1000 a.

Analysis of the short Younger Dryas cooling is more complex: the episode has different peak intensities in the five cores sampled in detail (figures 7–8). This mostly reflects actual differences in the overlying surface waters during the Younger Dryas. The climatic changes discussed so far have minimized regional differences because they all document abrupt shifts from one climatic extreme to another. As a result, all cores have shown virtually the same climatic response. Because the Younger Dryas cold episode was more moderate in intensity, it affected the three northern cores far more severely than the two southern cores. This will be explained from a geographic viewpoint in the next section.

Mixing may have altered in a differential manner the record of the Younger Dryas cooling in these five cores. If so, the cores with highest sedimentation rates should be least mixed and register the smallest loss of resolution of short events. Some such effect on the cores is suggested in figures 7 and 8, which show strongest Younger Dryas peaks in the three cores to the north with highest sedimentation rates; however, core RC9-225 to the north has a rate of deposition comparable to the two southern cores and yet registers a much more intense Younger Dryas peak. We conclude that the north-to-south temperature gradient is for the most part real, and that the Younger Dryas cooling was far more intense in cores north of 53° N.

An additional complication is climate-related variations in sedimentation rate. Even though we demonstrated constancy of sedimentation rate for most of the cores in this study, the core with the highest depositional rate (V23-81) clearly breaks the pattern. The Würm sedimentation rate from 75000 to 13500 a B.P. is 8.8 cm/1000 a. From 13500 a B.P. to the Younger Dryas at 10200 a B.P., the rate rises to 21 cm/1000 a; from the Younger Dryas to the ^{14}C date at 4010 a B.P., it remains high at 18.9 cm/1000 a; and from the ^{14}C date to the present it falls to 10 cm/1000 a. Because the sedimentation rates vary by a factor of two or more due to climate, we think that the duration of the Younger Dryas cooling on a time plot could be somewhat compressed relative to the remainder of the Holocene record in this core. This is analogous to the differing apparent durations of the stage 7 cooling discussed earlier.

Weighing all these complexities, we have used only the two far northern cores (V23-81, RC9-225) to compute mean rates of change. During the warm-to-cold shift into the Younger Dryas, estimated winter temperatures cooled from 8.2 to 1.8 °C and summer temperatures from 14.3 to 7.4 °C (figure 8; table 5). This change is recorded in 700 years, giving a mean rate of cooling of 11 °C/1000 a. Recorded rates of cooling near the cold peak reached values of over 13 °C/1000 a. During the subsequent warming away from the Younger Dryas, estimated winter temperatures rose from 1.8 to 10.1 °C and summer temperatures from 7.4 to 15.2 °C (figure 8; table 5). The total change peak-to-peak is registered in 7700 years, giving a mean rate of warming of about 1.1 °C/1000 a. Most of the change ($\sim$6 °C) is recorded within 500 years, giving rates of change of over 12 °C/1000 a.

† Roche, McIntyre & Imbrie (1975) reported palaeotemperature estimates on coccolith assemblages in the upper sections of cores RC9-228 and V23-81. Recent work by one of us (A.M.) shows that these estimates were several degrees Celsius too high, in part because of the lack of a cold end-member species. Work is underway to try to incorporate such an end-member into a new transfer function equation.

In summary, peak-to-peak rates of warming and cooling for North Atlantic surface waters ranging from 1 to 5 °C are registered in these cores for three intervals of relatively rapid climatic change. During shorter intervals, the recorded rates ranged as high as 13 °C/1000 a.

Inferred rates of polar front migration

To this point, we have documented down-core changes in estimated sea-surface temperature at selected points in space. From a geographic perspective, these changes also reflect large-scale horizontal migrations of water masses and frontal systems. In the North Atlantic, these migrations generally occur along the NW/SE axis documented by Ruddiman & McIntyre (1973) for the last deglacial warming (figure 9).

The most prominent feature in this complex march of isotherms back and forth across the North Atlantic is the polar front (convergence). Its position varies from the full-glacial position along 41° N to 44° N documented by CLIMAP (McIntyre, Moore *et al.* 1976) to the modern location as a coastal current restricted to the continental shelf of Greenland (figure 9). The intermediate positions shown in figure 9 have been defined by reconstructions at the levels of two ash layers deposited at 9300 a B.P. and roughly 65000 a B.P. (Ruddiman & Glover 1972, 1975, in preparation).

These geological reconstructions suggest that although the intensity (thermal gradient) of the polar front has varied with time, it has retained the strongest gradient in the subpolar North Atlantic. Consequently, passage of this front across a given core site will constitute the dominant thermal event within any general climatic cooling or warming. As a result, these areal migrations will leave in the temporal record of any one core short intervals of rapid temperature change imposed upon general periods of more gradual cooling or warming.

We suggest that the most rapid rates of temperature change calculated for the three climatic intervals in this paper occurred during periods of passage of the polar front. Such rates may be locally very rapid while the response in adjacent waters is slower. To develop an integrated geographical perspective, we have thus examined the rates of horizontal migration of the polar front in the subpolar North Atlantic.

As in the previous procedures for studying temporal rates of change, we have broken the spatial rates of change into two categories: (1) the average long-term rates of change from one climatic extreme to the other, and (2) the faster short-term rates prevailing during more rapid portions of the climatic transitions.

Three parameters must be established to measure recorded rates of polar front migration during the three climatic episodes we are examining: (1) a reference axis along which to measure the migrations; (2) polar frontal positions along the reference axis during the periods of interest, and (3) the timing (absolute or relative) of changes between these positions.

Choice of a reference axis

The polar front moves in part as a line hinged in the western North Atlantic and sweeping out an increasingly broad arc towards the northeastern North Atlantic (figure 9). In the western North Atlantic southeast of Newfoundland, movement of the front has been contained between the continental margin and the relatively stable western boundary current of the subtropical gyre. As a result, rates of migration in the western Atlantic have been relatively slow. To define maximum rates of migration, we chose as a reference axis a great circle line from Cape Farewell, Greenland, to the northwestern tip of the Iberian Peninsula, a distance of about 2800 km (figure 9). Frontal movements along this axis in the northeast Atlantic were particularly large and the rates of migration very rapid.

Geographic positions of the polar front

Geographical positions at the climatic extremes are relatively easy to establish for the three climatic changes we are studying†. Polar faunal percentages of 90 % or higher are found during the cold episode in stage 7 and in the glacial levels at the beginning of the stage 6/5 and 2/1 warmings. These polar-dominated faunas indicate polar front locations well southeast of the cores in this study, and probably close to the glacial position at 18000 a B.P. defined by CLIMAP (McIntyre, Moore *et al.* 1976). These inferred extreme glacial positions near the coast of the Iberian Peninsula are shown in figure 9.

The location of the Younger Dryas polar front is bracketed north of cores K708-4 and V23-82, slightly south of RC9-225 and V-23-81, and very near the location of core RC9-228. Other data show that it also lies slightly north of cores K708-7 and K708-8. This evidence seems to demand the serpentine twist to the Younger Dryas polar front shown in figure 10.‡ This is a signficant departure from the usual pattern of advance and retreat shown in figure 9.

The extreme northwestern retreat positions of the polar front during warm interglaciations are also reasonably easy to infer. We have already documented the polar front positions at 11000 and 6000 a B.P. (Ruddiman & McIntyre 1973). Work in progress with the CLIMAP project has shown that the polar front in isotopic stage 5e must have been restricted to a band on the East Greenland Continental Shelf even thinner than that during stage 1 (figure 9). The stage 7 interglaciation was not so intense as that in stage 5e; we infer that the polar front positions both below and above the cold episode were close to the modern position (figure 9).

We do not have direct evidence of where the polar front was located during most of the shorter intervals of more rapid temperature change registered in these northeast Atlantic cores. We do, however, know one of these positions positively – the deglacial retreat position at 9300 a B.P., shortly after the Younger Dryas readvance. This position is based on the zone 1 ash reconstruction of Ruddiman & Glover (1975). The 9300-year datum in the northeast Atlantic cores occurs at roughly the level of 75 % warming from the glacial stage 2 maximum to the interglacial stage 1 maximum. The polar front retreat at this point was roughly 90 % complete (figure 9).

Based on the polar front position at this important datum, we have assumed that during any other advance or retreat in which these northeast Atlantic cores show temperatures roughly 75 % toward the interglacial extreme, the polar front location was also (at least) 75 % toward the interglacial extreme near Greenland. To calculate the short-term migration distances, we thus multiplied the total migration distances (peak-to-peak between the climatic extremes) by 75 %.

† Faunal data from 15 additional cores were used to supplement the inferences on polar front locations made here.

‡ This position predicts interesting consequences for the intensity of Younger Dryas cooling on the British Isles, pointing to a much larger downwind cooling in the maritime climate over Scotland than in the southern British Isles. It also indicates that warm North Atlantic Drift flow into the Norwegian Sea ceased during the Younger Dryas.

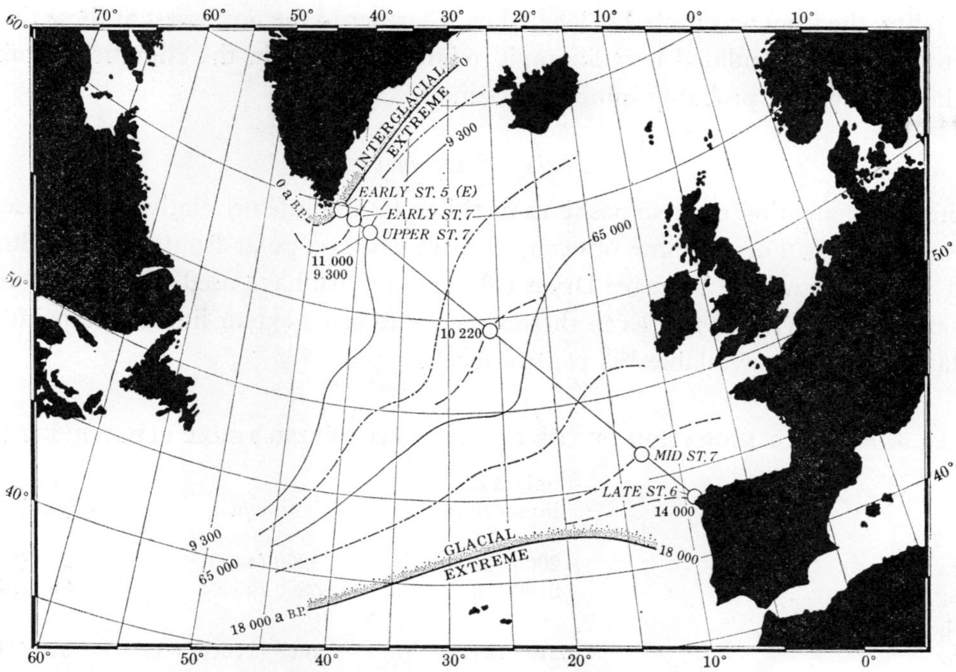

FIGURE 9. Past migrations of the North Atlantic polar front. Straight line is reference axis along which migrations are measured. Curving lines give general advance and retreat pattern of the polar front, including specific positions during the climatic intervals under study. Circles show intercepts of front with reference axis. Positions are keyed as follows. For the cold episode in stage 7 (*ca.* 195000 a B.P.), the positions 10000 a before, at, and 5000 a after the cold peak are shown as early st. 7, mid st. 7 and upper st. 7. For the stage 6/5 boundary warming (*ca.* 127000 a B.P.), the last cold level just prior to the change is shown as late st. 6 and the warmest level at the end of the change as early st. 5 (E). Positions during the stage 2/1 deglacial warming and Younger Dryas cold episode are specified in absolute ^{14}C a B.P.

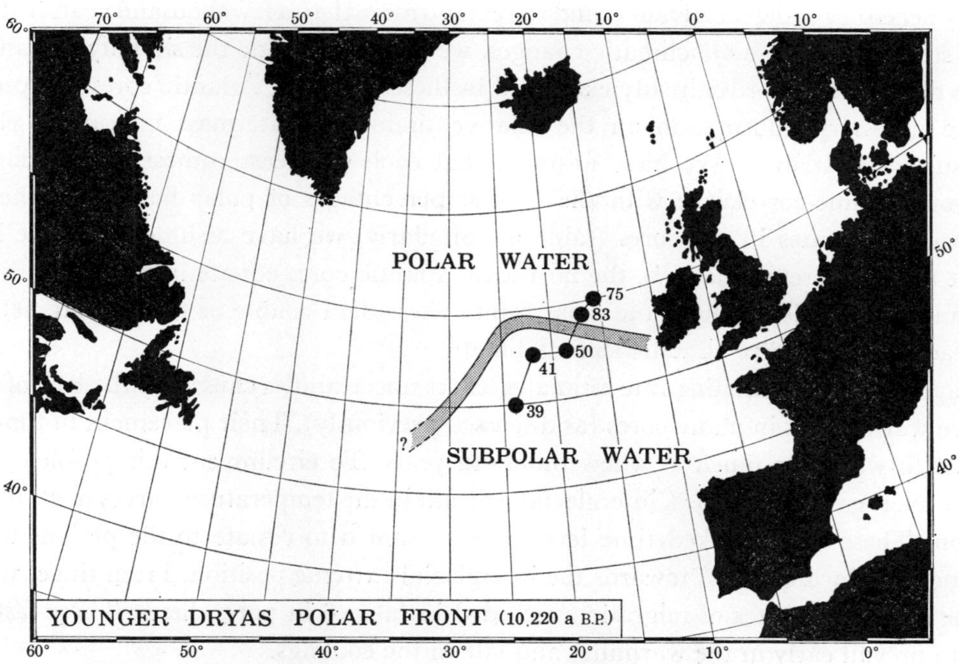

FIGURE 10. Location of the Younger Dryas Polar Front (stippled pattern). Percentages of polar fauna (*G. pachyderma*, s.) as total of planktonic foraminifera larger than 149 μm at Younger Dryas cold peak are shown for five cores also plotted in figure 7.

In actuality, the front was probably located even farther to the northwest at these times; thus the short-term rates calculated here for rapid migrations early in the climatic warmings and late in the coolings are probably minimum estimates.

Timing of migrations

The time of occupation of these positions in either an absolute or relative time sense is more difficult to establish. For the time of occupation of extreme polar front positions during the stage 2/1 warming and the Younger Dryas cold episode, we have used the ages published by Ruddiman & McIntyre (1973). Even these ages are to some extent linear interpolations and extrapolations between available ^{14}C control levels.

TABLE 6. RATES OF MIGRATION OF THE POLAR FRONT DURING THREE CLIMATIC EPISODES

	migration distance/km	time/a	rate of migration/(m/a)
mid-stage 7			
cooling	2000 (1500)	9200 (4400)	220 (340)
warming	2100 (1575)	5800 (2900)	360 (540)
stage 6/5 boundary			
warming	2700 (2025)	7000 (1700)	390 (1190)
stage 2/1 boundary: entire transition			
warming	2300 (1725)	10100 (2800)	230 (620)
stage 2/1 boundary: separate portions			
initial deglacial warming	2000 (1500)	2700 (1500)	740 (1000)
Younger Dryas cooling	800 (600)	700 (400)	1140 (1500)
later stage 1 warming	1100 (825)	7700 (500)	140 (1650)

No absolute dating technique can measure events 100000 to 200000 years old with the accuracy necessary to detect advances and retreats no more than a few thousand years in duration. For the stage 7 and stage 6/5 climatic changes, we have thus made the simplifying assumption that the down-core records already examined in these northeast Atlantic cores can provide us with the necessary information on the relative timing of watermass migrations elsewhere in the subpolar Atlantic. We have assumed that each southeasternmost penetration of the polar front is time-correlative with the highest percentages of polar fauna and the coldest temperature estimates in the cores (table 6). Similarly, we have assumed that the levels of warmest temperature estimates in the northeast Atlantic cores equate in time with the northwesternmost retreat locations of the polar front. These data enable us to calculate net rates of change across each entire climatic shift (table 6).

This approach to calculating rate estimates leaves uncertainties caused by the lack of strongly definitive warm peaks in many cores (as discussed previously). Their placement in time can be uncertain to within as much as a few thousand years. To circumvent this problem, we also chose the average levels of 75 % interglacial warmth in the temperature curves of each climatic transition. These easily picked time levels were assumed to equate to the periods when the polar front was located 75 % towards the interglacial extreme position. From these, we calculated the short-term rates of migration recorded in table 6 in parentheses. These faster rates tended to prevail early in the warmings and late in the coolings.

Rates of migration

First, we have calculated peak-to-peak rates for each climatic transition. During the cooling episode in stage 7, the polar front advanced a total distance of 2000 km southeastward in 9200 years and then retreated 2100 km northwestward in 5800 years.† The corresponding mean rates of migration for the peak-to-peak changes are 220 m/a (= km/1000 a) for the entire initial cooling and 360 m/a for the full later warming (table 6). During the more rapid stage 6/5 warming, the polar front retreated 2700 km in 7000 years, for a mean retreat rate of 390 m/a. There was a total retreat of 2300 km in 10000 years during the entire stage 2/1 warming; this gives a mean retreat rate of 230 m/a. In short, recorded advance and retreat rates of the polar front from one climatic extreme to the other averaged from 200 to 400 m/a and were sustained over several millennia.

The very brief Younger Dryas cooling produced a total southeastward polar front advance of 800 km in 700 years, followed by a northwestward retreat of 1100 km in 7700 years. The advance rate averaged 1140 m/a; the retreat occurred at a mean rate of 140 m/a.

The recorded rates of migration during selected portions of each of these climatic changes were considerably faster. During the last 4400 years of the cooling episode in stage 7, the front advanced 1500 km at a mean rate of 340 m/a. During the initial warming out of this cold peak, there was a retreat of 1575 km in 2900 years, a rate of migration of 540 m/a. The extremely rapid warming early in the stage 6/5 transition produced a polar front retreat of 2025 km in 1700 years, an average rate of 1190 m/a. The entire stage 2/1 warming began 14000 a B.P. and was 75 % completed by 11900 a B.P. During this period, the polar front retreated 1725 km, for a mean rate of 620 m/a for the fastest part of the warming.

As for the separate advances and retreats, the deglacial retreat early in the stage 2/1 warming carried the polar front northwestward some 1500 km in 1500 years; this gives a mean retreat rate of 1000 m/a. The very brief Younger Dryas cooling produced a southeastward polar front advance of 600 km in 400 years, followed by a northwestward retreat of 825 km in 500 years. The advance rate averaged 1500 m/a; the early retreat rate averaged 1650 m/a.

In summary, polar front advances and retreats in the more rapid portions of these large climatic changes are recorded at average rates of 300–1600 m/a over periods of a millennium or more.

Significance of vertical mixing in blurring the actual rates of change

In the North Atlantic, vertical mixing is the greatest single impediment to the exact reconstruction of oceanic palaeoclimates. Infauna burrow into the uppermost 5–30 cm of sediment, which in the Atlantic usually contain a record of climatic change spanning anywhere from 3000 to 15000 years. The vertical exchange of sediment that results from this burrowing has the effect of blurring and smoothing the climatic record. The amplitudes of climatic oscillations are reduced, with the greatest losses of original amplitudes occurring for the shortest (highest frequency) oscillations. Prominent climatic transitions that were originally very abrupt may be smoothed into considerably more gradual forms. All of this reduces the oceanic rates of change recorded in the sediments.

The problem of quantitatively modelling vertical mixing in order to remove its effects is complex, and work on this is presently underway. The purpose of this paper is to document as

† Mean times are taken from calculations in last section and represent an average value for all cores examined.

carefully as possible the rates of change preserved in the sediments. When precise quantification of mixing is available, these oceanic data will then be 'unmixed' in a quantitative way. Until then, the actual (more rapid) rates of change must largely remain a matter of conjecture, as must detailed comparisons to those portions of the continental record lying within the range of ^{14}C dating. In particular, we see the need for work to improve the oceanic and continental records in the northeast Atlantic and on Great Britain. This is one of the few areas of the world where a really detailed oceanic record of high quality is available near a continent whose climate is dominantly maritime.

Summary and conclusions

1. We examined several northeast Atlantic cores for evidence of rates of change during three prominent palaeoclimatic shifts during the last 200000 years.

2. During a cooling episode midway through interglacial isotopic stage 7, the northeast Atlantic cores registered virtually a complete interglacial-to-glacial-to-interglacial oscillation in no more than 15000 years.

3. At the isotopic stage 6/5 deglaciation (termination II), a full glacial-to-interglacial oceanic warming was recorded in no more than 6500 years; 75 % of the shift was registered in less than 2000 years.

4. The slower overall deglaciation at the isotopic stage 2/1 boundary was related to a short reversal of deglacial warming centred at 10200 a B.P.; during this Younger Dryas cooling, the North Atlantic very briefly returned to a nearly full-glacial state.

5. The cores register average rates of change in estimated surface water palaeotemperature of 1–5 °C/1000 a for the entire peak-to-peak transition; the rates range from 2 to 13 °C/1000 a during shorter intervals of faster change early in the warmings and late in the coolings.

6. The recorded rates of polar front movement along an axis running from Cape Farewell, Greenland, to the Northern Iberian Peninsula average 200–400 m/a for the several thousand years of change between the climatic extremes; the rates range as high as 400–1650 m/a during shorter intervals of more rapid change.

7. Further work to remove the mixing effects of burrowing organisms will be necessary to derive the actual (faster) rates of oceanic surface water change.

We wish to thank: Barbara Molfino for processing the transfer function estimates; Carol Fruik for illustrations; Gloria Garner and Marny Prout for assistance in the laboratory. Portions of this research were supported by NSF Grants GX-28671 and IOO 75-19627, as well as ONR Grant N-00-0-14-67-A-0108-0004. NSF Grants GA 10635 and 19690 to the Lamont-Doherty Geological Observatory Core Laboratory of Columbia University also aided our efforts.

References (Ruddiman *et al.*)

Berger, W. H. & Gardner, J. V. 1975 On the determination of Pleistocene temperatures from planktonic foraminifera. *J. Foram. Res.* **5**, 102–113.

Bramlette, M. N. & Bradley, W. H. 1941 Geology and biology of North Atlantic deep-sea cores between Newfoundland and Ireland, Part 1: lithology and geological interpretations. *U.S. Geol. Survey Prof. Pap.* 196-A, 1–55.

Broecker, W. S. & Ku, T. L. 1969 Caribbean cores P6304-8 and P6304-9: new analysis of absolute chronology. *Science, N.Y.* **166**, 404–406.

Broecker, W. S. & van Donk, J. 1970 Insolation changes, ice volumes, and the O^{18} record in deep-sea cores. *Rev. Geophys. Space Phys.* **8**, 169–198.

Broecker, W. S., Thurber, D. L., Goddard, J., Ku, T., Matthews, R. K. & Mesolella, K. J. 1968 Milankovitch hypothesis supported by precise dating of coral reefs and deep-sea sediments. *Science, N.Y.* **159**, 297–300.

Coope, G. R. 1975 Climatic fluctuations in northwest Europe since the last interglacial, indicated by fossil assemblages of Coleoptera. In *Ice ages: ancient and modern* (eds A. E. Wright & F. Moseley), pp. 153–168. Liverpool: Seel House Press.

Crowley, T. 1975 Fluctuations of the eastern North Atlantic Gyre during the last 150000 years. Ph.D. Thesis, Brown University, 282 pp.

Emiliani, C. 1955 Pleistocene temperatures. *J. Geol.* **63**, 538–578.

Emiliani, C. 1966 Paleotemperature analysis of the Caribbean cores P6304-8 and P6304-9 and a generalized temperature curve for the last 425000 years. *J. Geol.* **74**, 109–126.

Imbrie, J. & Kipp, N. G. 1971 A new micropaleontological method for quantitative paleoclimatology: application to a late Pleistocene Caribbean core. In *The Late Cenozoic glacial ages* (ed. K. Turekian), pp. 71–182. New Haven: Yale University Press.

Imbrie, J., van Donk, J. & Kipp, N. G. 1973 Paleoclimatic investigation of a late Pleistocene Caribbean deep-sea core: comparison of isotopic and faunal methods. *Quat. Res.* **3**, 10–38.

Iversen, J. 1954 The late-glacial flora of Denmark and its relation to climate and soil. *Danmarks Geol. Unders* II *Raekke* **80**, 87–119.

Kennett, J. P. 1968 Latitudinal variation in *Globigerina pachyderma* (Ehrenberg) in surface sediments of the southwest Pacific Ocean. *Micropaleontology* **14**, 305–318.

Kennett, J. P. 1970 Comparison of *Globigerina pachyderma* (Ehrenberg) in Arctic and Antarctic areas. In *Contributions from the Cushman Foundation for foraminiferal research* **21**, part 2, 47–49.

Kipp, N. G. 1976 New transfer function for establishing past sea-surface conditions from sea-bed distribution of planktonic foraminiferal assemblages in the North Atlantic. *Geol. Soc. Am. Mem.* **145**, 3–42.

Ku, T. L., Bischoff, J. L. & Boersma, A. 1972 Age studies of Mid-Atlantic Ridge sediments near 42° N and 20° N. *Deep-Sea Res.* **19**, 233–247.

McIntyre, A., Ruddiman, W. F. & Jantzen, R. 1972 Southward penetrations of the North Atlantic Polar Front: faunal and floral evidence of large-scale surface water mass movements over the last 225000 years. *Deep-Sea Res.* **19**, 61–77.

McIntyre, A., Moore, T. C. & the CLIMAP Project Team 1976 The surface of the ice-age earth. *Science, N.Y.* **191**, 1131–1137.

Mangerud, J. & Gulliksen, S. 1975 Apparent radiocarbon ages of recent marine shells from Norway, Spitsbergen, and Arctic Canada. *Quat. Res.* **5**, 263–274.

Mesolella, K. J., Matthews, R. K., Broecker, W. S. & Thurber, D. L. 1969 The astronomical theory of climate change, Barbados data. *J. Geol.* **77**, 250–274.

Roche, M. B., McIntyre, A. & Imbrie, J. 1975 Quantitative paleo-oceanography of the late Pleistocene–Holocene North Atlantic: coccolith evidence. In *Late Neogene epoch boundaries* (eds T. Saito & L. H. Burckle), pp. 199–225. New York: Micropaleontology Press.

Ruddiman, W. F. & Glover, L. K. 1972 Vertical mixing of ice-rafted volcanic ash in North Atlantic sediments. *Geol. Soc. Am. Bull.* **83**, 2817–2836.

Ruddiman, W. F. & Glover, L. K. 1975 Subpolar North Atlantic circulation at 9300 yr BP: faunal evidence. *Quat. Res.* **5**, 361–389.

Ruddiman, W. F. & McIntyre, A. 1973 Time-transgressive deglacial retreat of Polar water from the North Atlantic. *Quat. Res.* **3**, 117–130.

Ruddiman, W. F. & McIntyre, A. 1976 Northeast Atlantic paleoclimatic changes over the last 600,000 years. *Geol. Soc. Am. Mem.* **145**, 111–146.

Sancetta, C., Imbrie, J. & Kipp, N. G. 1973 Climatic record of the past 130000 years in North Atlantic deep-sea core V23-82: correlation with the terrestrial record. *Quat. Res.* **3**, 110–116.

Shackleton, N. J. 1969 The last interglacial in the marine and terrestrial records. *Proc. R. Soc. Lond.* B **174**, 135–154.

Shackleton, N. J. & Opdyke, N. D. 1973 Oxygen isotope and paleomagnetic stratigraphy of equatorial Pacific core V28-238: oxygen isotope temperatures and ice volumes on a 10^5 year and 10^3 year scale. *Quat. Res.* **3**, 39–55.

Stuiver, M. & Borns, H. W. Jr. 1975 Late Quaternary marine invasion in Maine: its chronology and associated crustal movement. *Geol. Soc. Am. Bull.* **86**, 99–104.

Vilks, G. 1975 Comparison of *Globorotalia pachyderma* (Ehrenberg) in the water column and sediments of the Canadian Arctic. *J. Foram. Res.* **5**, 313–325.

Discussion

G. Manley (3 *Whitwell Way, Coton, Cambridge*). This establishment by Professor McIntyre and his colleagues of the characteristics of the North Atlantic in glacial times is indeed gratifying as it confirms the general scheme that I presented in the *Geographical Journal* (vol. 117, p. 62) in 1951 as a likely explanation on 'best-fit' principles, of what was at that time known about the climate. The further extension of the cold water as far south as the latitude of northern Spain that he has demonstrated is helpful with regard to mountain glaciation and the vegetation of western France. We now need to elucidate the factors – meteorological or oceanographical, or both – that led to the renewed forward surge of warm mid-Atlantic surface water past Ireland into the Norwegian Sea. Such surges provide an economical way of explaining many minor climate episodes. The squeezing of the subpolar and transitional water masses that he has shown brings to mind the present east–west banding of the surface water masses in the North Pacific about 45°–50° N.

A. Dreimanis (*Geology Department, The University of Western Ontario, London, Canada*). McIntyre reported the beginning of the late-glacial warming of northern Atlantic Ocean about 13500 a ago, while the terrestrial record from Britain, as indicated by the beetle studies (Shotton, Coope) suggests a delay of the beginning of this warming episode to about 13000 a B.P. Farther inland, in the Baltic region, the late glacial warming began again at least several centuries prior to 13000 a B.P., as indicated by some interstadial sites. The most detailed palaeobotanical investigations have been published in the early sixties from the Raunis Interstadial, northern Latvia, and subsequently three ^{14}C dates have been obtained on it: 13390 ± 500 (Mo-196), 13250 ± 160 (TA-177), and 13320 ± 250 (Ri-39) a B.P. Of a similar age, more than 13000 ^{14}C a B.P., is also the Meiendorf Interval in northwestern Germany, and the Vintapper Interstadial of southern Sweden.

In the Great Lakes Region of North America, the Cary-Port Huron or Mackinaw Interstadial culminated 13300 a B.P., and several of its correlatives, or at least rapid glacial retreats began prior to 13000 a B.P., in other areas of North America.

If a more or less synchronous climatic warming began prior to 13000 a B.P. in the Atlantic ocean and widely scattered areas over the continental Europe and North America, a question arises of why the beginning of the Late-Glacial Interstadial was delayed in Britain for several centuries.

H. H. Lamb (*Climatic Research Unit, University of East Anglia, Norwich NR4 7TJ*). What chances are there of obtaining further data points for the ocean surface temperatures in ice-age times in the Atlantic nearer the European seaboard? This area must be of great interest in connection with the climates prevailing in the British Isles and all western and northern Europe.

Phil. Trans. R. Soc. Lond. B. **280**, 143–168 (1977) [143]
Printed in Great Britain

Ice cores and climatic change

By G. de Q. Robin
Scott Polar Research Institute, Cambridge

[Plate 1]

The paper deals primarily with the use of stable isotopic ratios to determine the former climate of ice sheets. Studies of temperature profiles throughout ice sheets have shown that for at least several thousand years, changes of isotopic δ ratios have been proportional to changes of surface temperatures; this relationship is discussed in terms of the physical processes involved. It is considered reasonable to use a similar relation for earlier periods in Antarctica, but in Greenland the relation may have varied with time. When determining past climates from the isotopic record, allowances have to be made for changes in the flow and thickness of ice sheets during major glacial periods. These factors are considered in relation to major ice cores from Vostok and Byrd stations in Antarctica and from Camp Century in Greenland. Vostok is the simplest case glaciologically, Camp Century the most complex. On purely glaciological grounds it appears that the ice age gave way to present-day climates some 10000 ± 1000 a B.P., the coldest period being 20000 ± 3000 a B.P., when the climate in Antarctica was 6–8 °C colder than at present. Glaciological data suggest a duration of 50000 to 100000 years for the last ice age. Before this period, climates in Greenland and Antarctica appear to have been around 2–3 °C warmer than at present.

Introduction

The large polar ice sheets consist of atmospheric precipitation that has fallen over a period of some hundreds of thousands of years. In the present paper we discuss information contained in ice cores that have been recovered from Camp Century in Greenland, and from Byrd Station and Vostok in Antarctica. The data so far available covers a period up to about 100000 years ago.

The historical record present in an ice core takes a number of different forms. For example, the ice contains pockets of air, trapped in bubbles and preserved in the ice sheet, which can be recovered for analysis; the composition of the trapped air records the composition of the Earth's atmosphere at the time the bubble was caught in newly forming ice. Impurities, both soluble and insoluble, are present in small quantities of a few parts per million but are clearly distinguishable owing to the extreme purity of the body of the ice. But the most important approach, in the present context, is to study variations of the isotopes of oxygen and hydrogen with depth in the ice itself.

Perhaps the first and least ambiguous contribution from the study of ice cores is that one can measure variations of different parameters from one ice core as a function of depth. We see for instance (Gow & Williamson 1971) that the frequency of dust bands in the ice at Byrd Station increased dramatically at the same time as isotopic δ values showed a general climatic warming at the end of the last ice age. Thompson, Hamilton & Bull (1975) showed similarly that the same warming was accompanied by a decrease in the number of dust particles less than 0.65 μm in size in the ice core. Ragone, Finelli, Leung & Wolf (1972) have studied the variation of soluble impurities with depth in the same Byrd core. There is a vast amount of meticulous work to be done in this field.

For the purpose of the present study, however, we are concerned with collecting facts from various terrestrial environments in order to put together a picture of climatic changes during the Devensian period. In using records from ice cores for this purpose, two major questions need to be answered. First, how closely do isotopic δ values preserve a record of past temperature changes at the surface of the ice sheet? Secondly, a complex question: how do we allow for effects due to carriage of the ice from its point of deposition on the ice sheet to the site of the ice core, bearing in mind that the deepest ice may have been deposited hundreds of kilometres from the coring site and tens of thousands of years ago, and that during this time the size and shape of the ice sheet may have undergone considerable changes. Unless these factors can be estimated with reasonable accuracy, it is difficult to distinguish between the effects of climatic change and the effects of motion within the ice sheet when we interpret changes in isotopic δ values.

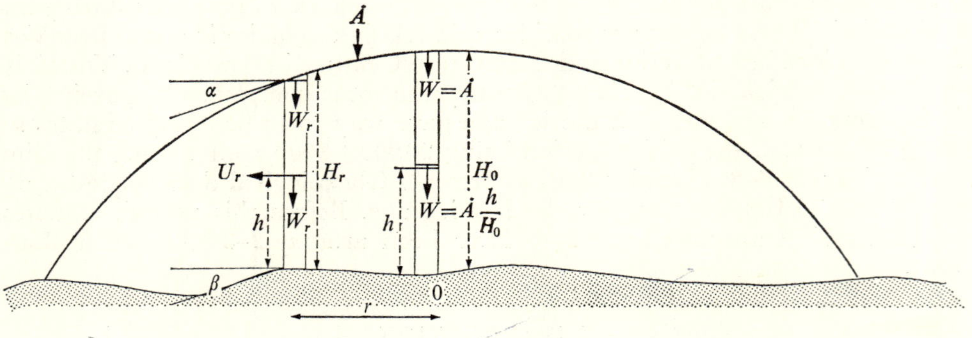

at distance r

$$\begin{cases} U_r = \dfrac{1}{H_r}\displaystyle\int_0^r \dot{A}\,\mathrm{d}r \\[2ex] W_r = (\dot{A} + (\alpha - \beta)U_r)\,h/H_r \end{cases}$$

FIGURE 1. Parameters for a two-dimensional cross-section of an ice sheet in a steady state.

MODELLING OF ICE SHEETS

In answering the above points we need to draw on our knowledge of the flow of ice sheets and its effect on temperature distribution. We find that the theory developed for a steady-state ice sheet (that is, an ice sheet of constant size and climatic conditions) gives a good first order prediction of observed temperatures in deep boreholes. A brief comparison of theory and observation will help to show the degree of confidence we may put in the models of ice-sheet flow which are an essential aid to our interpretation of isotopic profiles.

Some parameters for a two dimensional cross section of an ice sheet in a steady state are shown in figure 1. We treat a vertical column in the ice, and assume it remains vertical and undergoes uniform vertical strain throughout the column when it is deformed. At the centre of the ice sheet ($r = 0$), the steady-state assumption implies that the vertical velocity (W_0) at any height h above the bed is given by

$$W_0 = \dot{A}h/H_0 \tag{1}$$

where $\dot{A}$ is the rate of accumulation of ice on the surface and H_0 the ice thickness.

At a distance r from the centre, the horizontal velocity of our column (U_r), under steady-state conditions will be

$$U_r = \frac{1}{H_r}\int_0^r \dot{A}\,\mathrm{d}r \qquad (2)$$

and the vertical velocity (W_r) at height h above the bed will be

$$W_r = (\dot{A} + (\alpha - \beta)U_r)h/H_r \qquad (3)$$

where α and β are the slopes of the upper and lower surfaces of the ice sheet.

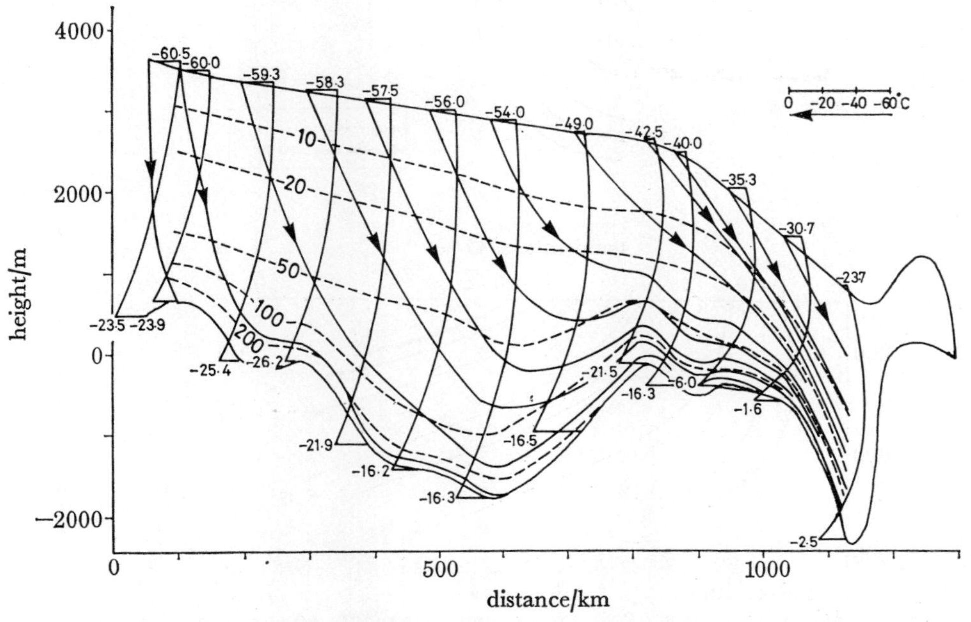

FIGURE 2. Particle paths (⟶), isochrons (10^3 a) (- - - -) and temperature–depth profiles (——) along flowline from Vostok to Wilkes Station calculated on a steady-state model by Budd *et al.* (1971 *a*).

We can match this simple model to heat-flow equations. The rate of change of temperature with time $\partial\theta/\partial t$ at any fixed point in a polar ice mass is determined by a vector equation involving the temperature of ice moving past the point with velocity V through an advection term $\rho c V\,\nabla\theta$, where ρ is the ice density and c its specific heat capacity; by the heat flux divergence at that point $\nabla\,(k\nabla\theta)$, where k is the thermal conductivity of the ice and by the rate of heat generation per unit volume Q within the ice mass. This can be written

$$\rho c\partial\theta/\partial t = \rho c V\nabla\theta + Q - \nabla(k\nabla\theta). \qquad (4)$$

A detailed solution to (4), even under steady-state conditions is very complex, and it is necessary to introduce simplifications. The equation can be reduced to one dimension by confining our attention to a single vertical column travelling with the ice, and feeding in appropriate strain rates from (3), as was done in Robin (1955), in order to estimate the temperature distribution in the Greenland ice sheet. The advent of faster computing techniques has been used by Budd, Jenssen & Radok (1971 *a*) to compute temperature distributions along many flowlines in the Antarctic ice sheet. These can be seen in figure 2, which also gives information of the depth of former surface layers after a given time interval.

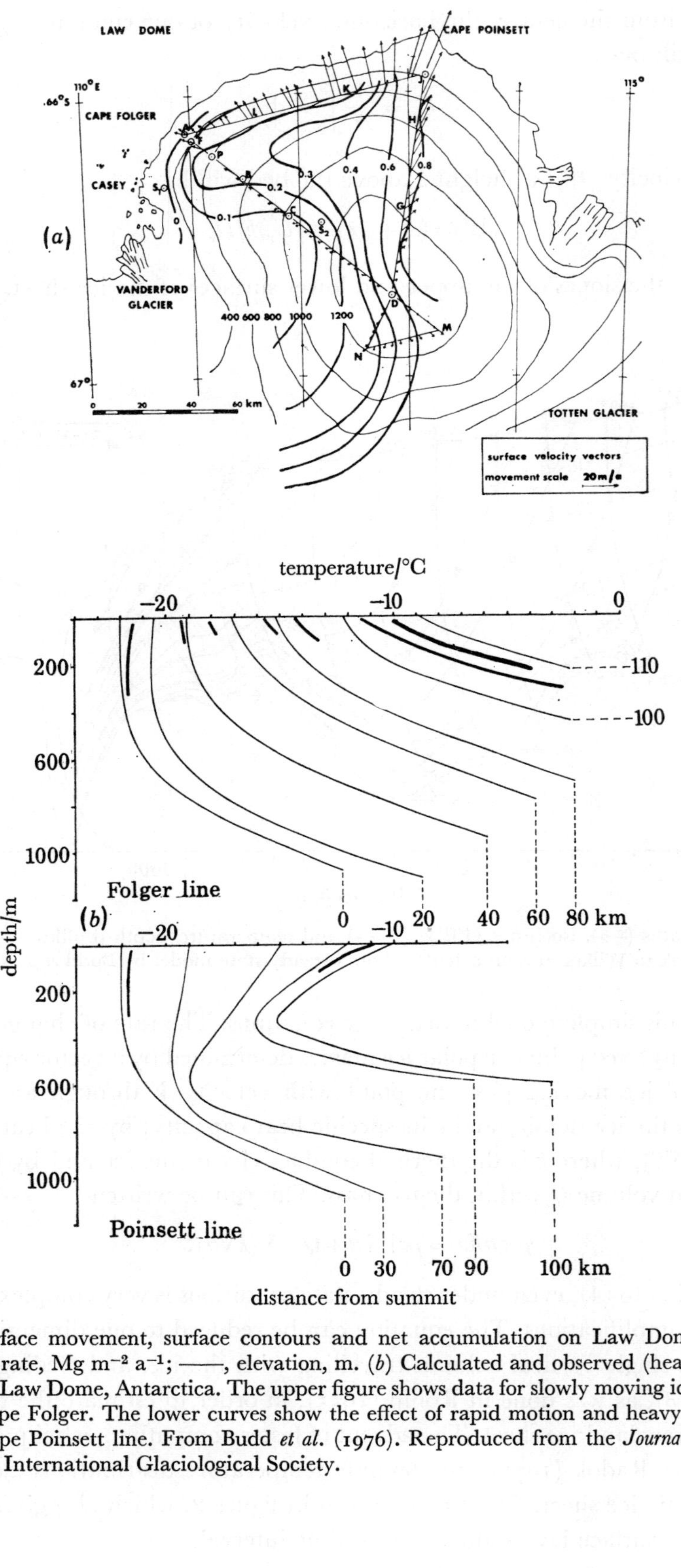

FIGURE 3. (*a*) Surface movement, surface contours and net accumulation on Law Dome, Antarctica. ——, accumulation rate, Mg m^{-2} a^{-1}; ——, elevation, m. (*b*) Calculated and observed (heavy lines) temperature profiles in the Law Dome, Antarctica. The upper figure shows data for slowly moving ice on the line from the summit to Cape Folger. The lower curves show the effect of rapid motion and heavy accumulation on the summit to Cape Poinsett line. From Budd *et al.* (1976). Reproduced from the *Journal of Glaciology* by permission of the International Glaciological Society.

These models take into account the geothermal heat flux, frictional heat, surface accumulation rates and surface temperatures. Where accumulation rates are heavy and ice motion rapid, geothermal heat is advected away by ice motion before it reaches the surface, with the surprising result that colder ice from further inland produces temperatures that decrease with increasing depth. The effect is seen most clearly on Law Dome, Antarctica, a dome of ice some 200 km across, whose flow is largely independent of the main ice sheet (Budd, Young & Austin 1976). Heavier accumulation rates on the eastern side of the dome will, on the steady-state model, produce negative temperature gradients, but smaller accumulation on the western half results in positive gradients. The agreement between theory and observation is clearly seen in figures 3*a* and *b*. The authors (Budd *et al.* 1976) conclude that 'The temperature profiles measured in the Law Dome conform closely to the profiles calculated from steady-state under the present regime except for a slight warming in the firn zone near the surface of a few tenths of a degree. This warming could be caused by some combination of climatic warming during the last century or so, or slight lowering of the ice sheet.'

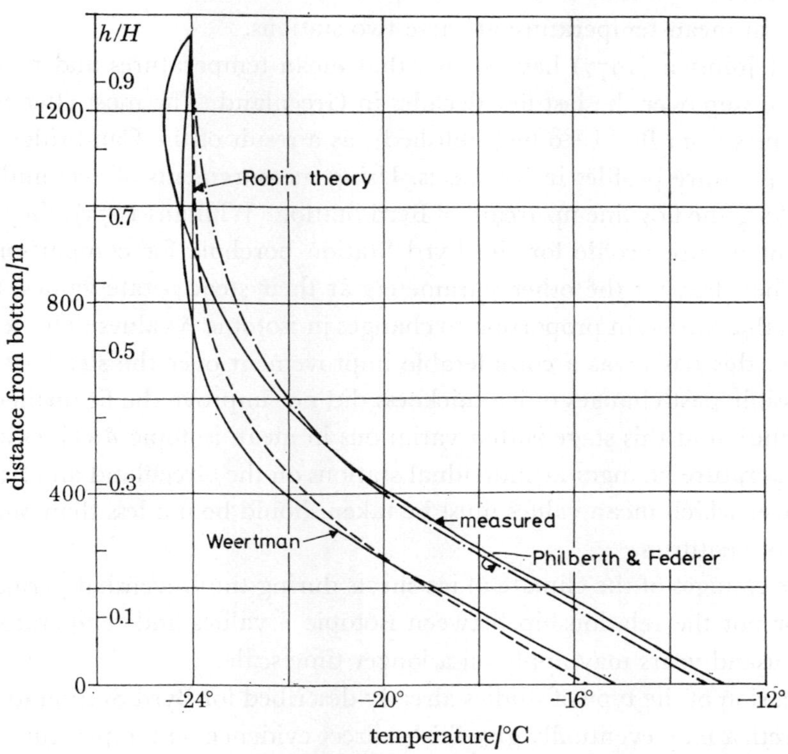

FIGURE 4. Observed and calculated temperature–depth profiles at Camp Century, Greenland (from Philberth & Federer 1971).

When we compare steady-state predictions at Camp Century with observations (figure 4), we find that agreement is limited when using a model involving uniform vertical strain rates (curves for Robin, Weertman) and is considerably improved when using a vertical strain rate that is uniform down to some 400 m or so above the base of the ice, then decreases to zero at bedrock (Philberth & Federer 1971). This type of strain-rate variation is expected when bedrock is frozen. Subsequent studies, shown in figure 5, of the variation with depth of annual layer thickness determined by the isotopic measurements of Johnsen, Dansgaard, Clausen &

Langway (1972) confirm this model of vertical strain rate, provided that accumulation rates have remained constant over the past 8300 years covered by the data.

The temperature models described so far involve input of certain boundary conditions in order to calculate temperature distribution. Budd, Jenssen & Radok (1971*b*) reversed the argument and used the observed temperature distribution, together with the steady-state and uniform vertical strain-rate assumptions to find which boundary conditions gave the best fit to the observed temperatures. The parameters they determined in this way were the accumulation rate, and the rate of change of surface temperature with time. In each of three cases, the best fit was obtained with accumulation rates in good agreement with observed accumulation rates at the present time. This implies that the present-day accumulation rates are typical of those for the last few thousand years. These derived rates of change of temperature with time are plotted in figure 6 against isotopic δ values for Camp Century and Byrd Station. We see good agreement between the trends of temperature and isotopic values over the past 2000–4000 years at these stations, even though the trends are in opposite directions at the two stations. This gives confidence that over this time period, variations of δ values are closely in step with variations in mean temperature at these two stations.

Robin (1976) and Johnsen (1977) have shown that mean temperatures and mean isotopic values have varied in step over the last few decades in Greenland. The most effective illustration of this point comes from Budd (to be published), as a result of the Cambridge Workshop on isotopic and temperature profiles in ice sheets. Using measurements of accumulation rates and temperatures along the flowline upstream of Byrd Station (Whillans 1973), they calculated the steady-state temperature profile for the Byrd Station borehole for comparison with the observed profile. Then leaving the other parameters at their steady-state values, they introduced temperatures that varied in proportion to changes in isotopic δ values. The temperature profile calculated on this basis was a considerable improvement over the steady-state model. Introduction of possible past changes of ice thickness did not improve the fit further.

The general conclusion at this stage is that variations in mean isotopic δ values are proportional to mean temperature changes at individual stations on the Greenland and Antarctic ice sheet. The period over which mean values must be taken should be not less than one year, but preferably decades or centuries.

For a study of the changes of the climate of ice sheets during the Devensian period, we must consider whether or not the relationship between isotopic δ values and temperature derived for the past few thousand years may apply on a longer time scale.

Although an extension of the type of studies already described for Byrd Station to the deeper ice of eastern Antarctica may eventually give fairly direct evidence on temperature changes in the latter part of the Devensian ice age, measurements are not available at present to derive mean surface temperatures from say 10000 to 15000 or 20000 years ago. We must therefore consider the processes involved in isotopic fractionation of water vapour, in order to see if the relationship between isotopic δ values and mean temperatures that has applied during recent millennia on polar ice sheets may also have applied during ice ages.

Figure 7 shows schematically the processes of isotopic differentiation of water vapour. Over the tropics, the δ values of precipitation are close to mean oceanic values. However, as one moves to higher latitudes, evaporation from the oceans falls due to lower temperatures, and there is a net transport of water vapour towards the poles. Since heavier isotopes are precipitated in larger amounts than lighter isotopes, there is a progressive depletion of the heavier

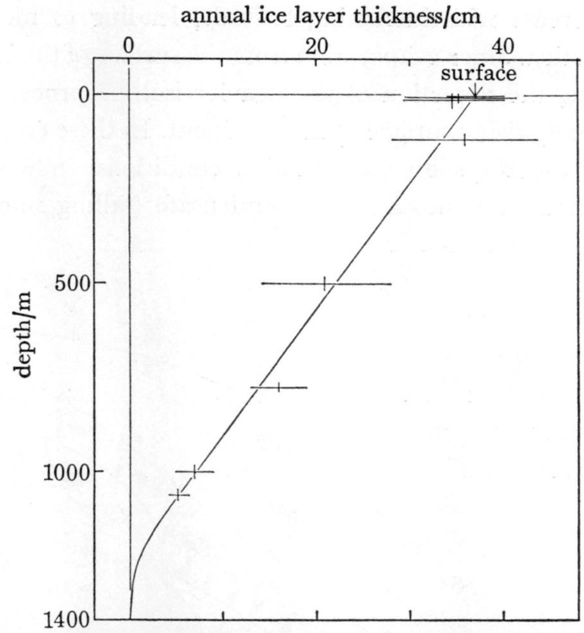

FIGURE 5. Thickness of annual layers determined from isotopic layering studies of the ice core from Camp Century plotted against depth. Data in original form from Johnsen *et al.* (1972) and Dansgaard *et al.* (1973).

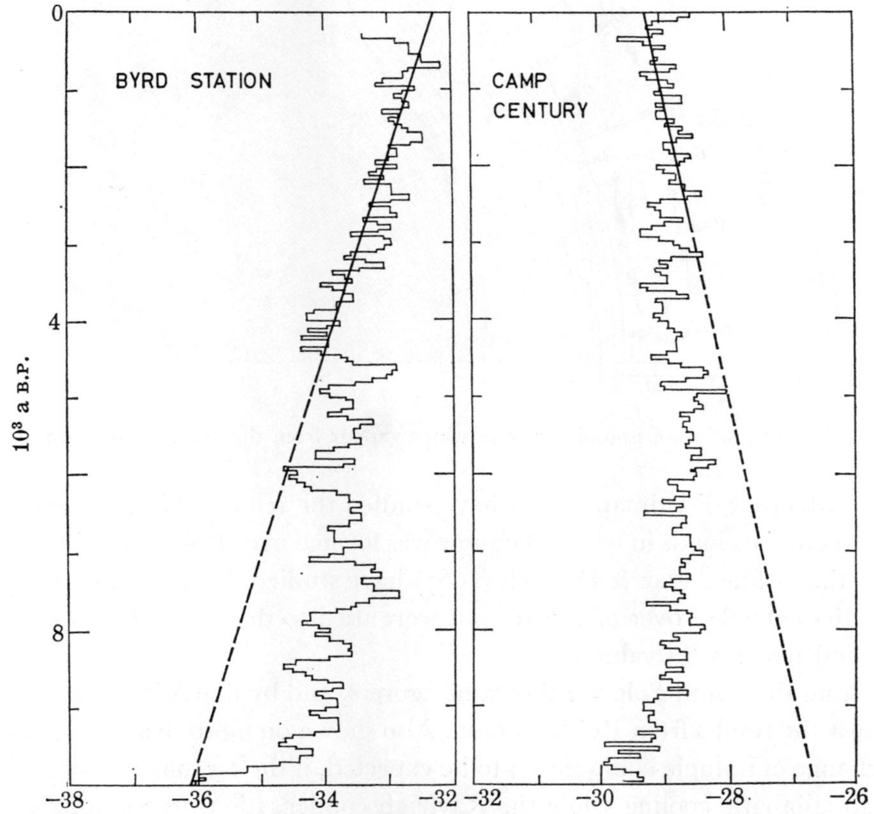

FIGURE 6. Comparison of temperature change with time (straight lines) derived from analysis of temperature–depth profiles at Byrd Station and Camp Century with $\delta^{18}O$ isotopic profiles on a time scale shown by Johnsen *et al.* (1972). From Robin (1976). Reproduced from the *Journal of Glaciology* by permission of the International Glaciological Society.

isotopes, and δ values decrease with increasing latitude, leading to mean δ values around 20 ‰ on the coast of Antarctica. Since evaporation from the surface of the ice sheet is very small owing to low temperatures, the depletion of the heavier isotope proceeds rapidly as an air mass moves over the gradually rising surface of the continent. In these conditions we can apply the equations derived for the Rayleigh condensation conditions, in which we deal with a slow process with the immediate removal of the condensate (falling snow) from the vapour after formation.

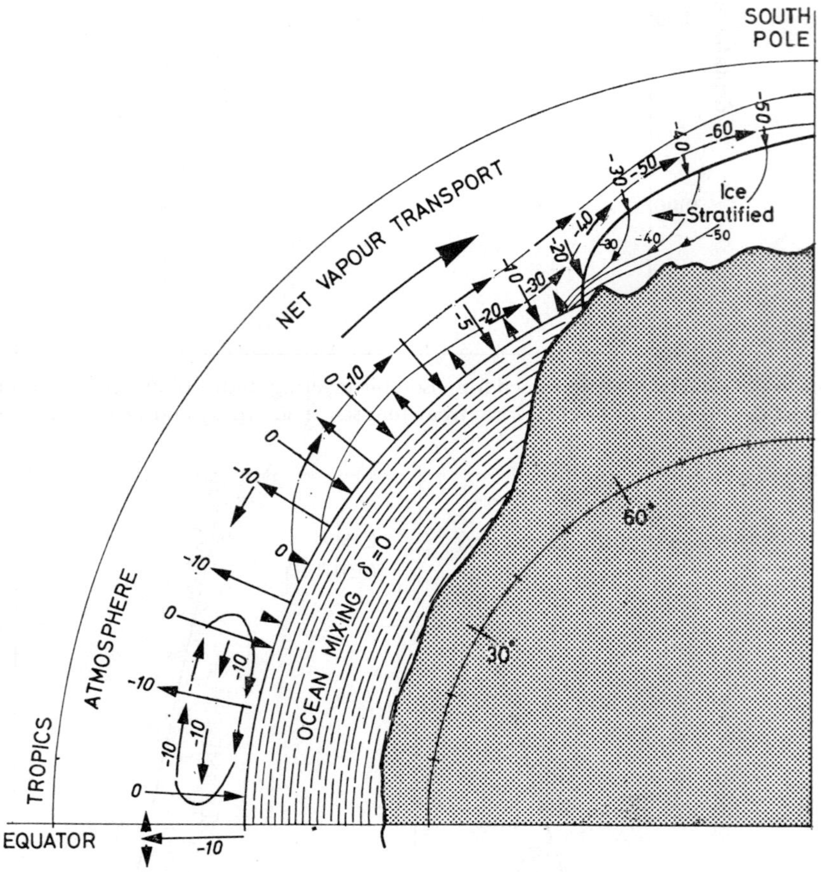

FIGURE 7. Schematic presentation of isotopic cycle of water vapour from the ocean to the Antarctic ice sheet.

Picciotto, De Maere & Friedman (1960) have studied the relationship between $\delta^{18}O$ values and temperatures of the clouds in which the snow was formed over Roi Baudouin Station on the coast of Antarctica, while Aldaz & Deutsch (1967) have studied the same process at the South Pole. Both studies extended over one year, and were used to derive a relation between cloud temperatures and mean $\delta^{18}O$ values.

The results from the South Pole are shown in figure 8, and by line A in figure 9, while line B in figure 9 shows the results from Roi Baudouin. Also shown on figure 9 are two dashed curves showing the change of isotopic composition to be expected, if the air masses over Roi Baudouin were subject to adiabatic cooling while the Rayleigh condensation process applies, as the air mass was forced upwards over the continent. The Rayleigh process gives δ_c the isotopic δ value of the condensate in terms of the remaining fraction of the vapour phase by

$$\delta_c = (\alpha/\alpha_0)\, F_v^{\alpha_m - 1} - 1, \qquad (5)$$

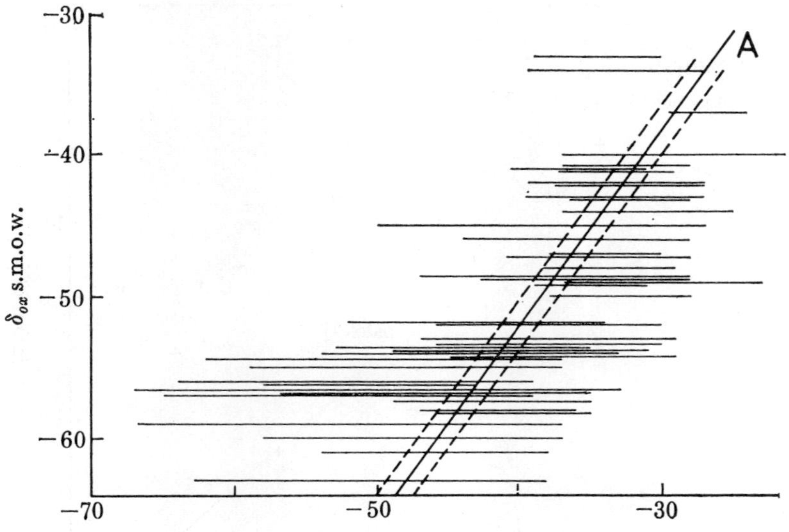

FIGURE 8. Isotopic $\delta^{18}O$ ratios of falling snow at the South Pole plotted against temperatures of clouds from which precipitation fell. Upper and lower temperature of the cloud are shown by ends of the horizontal lines. Based on Fig. 2 from Aldaz & Deutsch (1967).

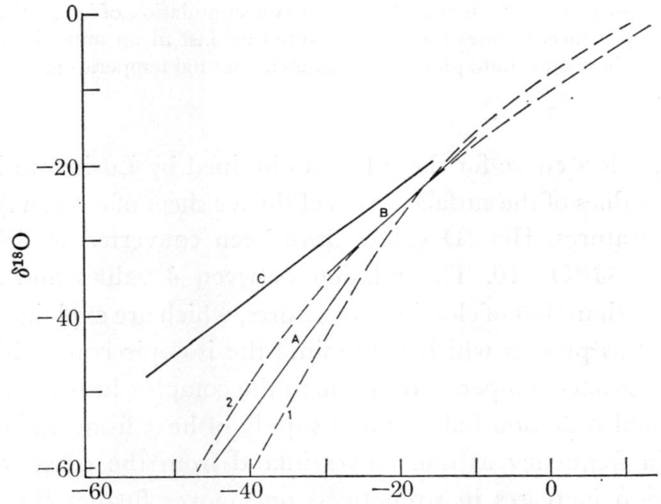

FIGURE 9. Isotopic $\delta^{18}O$–temperature relations. Curve A, clouds at South Pole, as in figure 8. Curve B, clouds at Roi Baudouin Station on the Antarctic coast, from Picciotto *et al.* (1960). Dashed curves 1 and 2, calculated for Rayleigh condensation with adiabatic cooling, based on curve B. Curve 1, α values based on Zhavoronkov *et al.* (1955). Curve 2, α values based on Merlivat & Nief (1967). Curve C, values for isotopic and temperature values for surface layers of firn of eastern Antarctica based on work by Lorius (1975).

where α, α_0 and α_m are the isotopic fractionation factors at condensation temperature t, the initial temperature t_0 and the mean temperature $\frac{1}{2}(t+t_0)$. Since α values have not been satisfactorily determined for ^{18}O at low temperatures, two curves are shown. Curve 1 uses α values obtained by extrapolation of results of Zhavoronkov, Uvarov & Sevryugova (1955) for ^{18}O fractionation at higher temperatures, while curve 2 uses α values measured for deuterium by Merlivat & Nief (1967) but converted to ^{18}O values. The agreement, particularly of observed gradients with curve 2, shows that the explanation of the observations appears satisfactory.

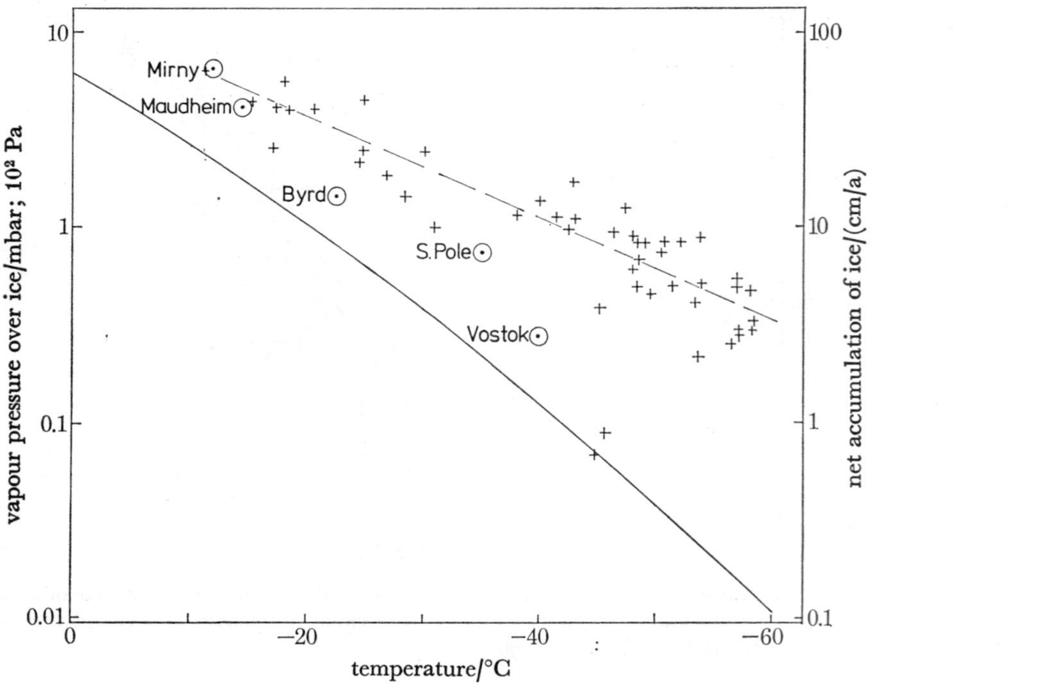

FIGURE 10. Water vapour pressure over ice (line) and annual net accumulation of ice on the Antarctic ice sheet, both plotted against temperature. Crosses show data plotted against mean annual temperature of surface layers of ice sheet. Circles show same data plotted against mean annual temperature of atmosphere above the surface inversion layer.

Also shown on figure 9 is a curve for the relation obtained by Lorius (to be published) for δD values of deuterium values of the surface layers of the ice sheet of eastern Antarctica plotted against mean ice temperatures. His δD values have been converted to δ^{18}O values by the empirical relation $\delta D = 8\delta^{18}O + 10$. The relation between δ values and surface temperatures is more nearly linear than that of cloud temperatures, which are explained by the Rayleigh process. It is, however, this process which determines the isotopic composition of the falling snow, whereas the mean surface temperature is due to the complex heat balance at the surface resulting from the thermal radiation balance and supply of heat from the atmosphere. Since clearer skies increase in frequency as one moves inland from the coast of Antarctica, the resultant surface inversion increases in strength as one moves further inland. Over coastal stations, the mean annual strength of the inversion in the lowest few hundred metres of the atmosphere runs from −0.3 °C at Mirny Station, +3.0 °C at Maudheim, +6 °C at Byrd Station, +13 °C at the South Pole and +15 °C at Vostok Station. Mean temperatures above the inversion are within 4 °C of the effective precipitation temperatures observed in cloud studies. The difference between curves 1 and 2 for cloud temperatures and curve 3 is thus readily explicable in terms of the surface inversion. The linear relation between δ values and mean surface temperatures is thus an empirical approximation to observations, rather than a natural law.

A similar feature can be seen for another parameter which will be useful later. In figure 10 we show by the continuous curve, the relation between the saturation water vapour pressure over ice and temperature. Also shown on the same curve by crosses are the mean annual accumulation at points on the Antarctic ice sheet as a function of mean surface temperature, while the circles show that mean annual accumulation as a function of the mean air temperature above the surface inversion. The trend of the latter shows clearly that the accumulation is primarily governed by the amount of water vapour that can be carried in the air mass above the inversion. The increasing separation between the crosses and circles at low temperatures is due to the increasing strength of the inversion over the cold inland regions – the same effect as we saw in figure 9.

If the whole atmospheric system falls in temperature during an ice age, the numerical relation between δ values and temperature should be similar in value to those observed at the present time. Small changes may arise due to a lower temperature of evaporation of water vapour from the ocean: alternatively, the temperature of evaporation may remain the same, if the main source moves closer to the equator. In the latter case, the amount of isobaric cooling will be greater before air masses reach the Antarctic ice sheet, and adiabatic cooling starts, due to motion over the ice sheets. In either case, once sublimation to snow commences, fractionation dominates the picture and will result in a similar numerical relationship. Dansgaard (1961, 1964) discusses the physics of the processes in detail.

Two other factors that operate in an ice age should be mentioned. The first is a change in the mean isotopic composition of ocean waters, which will increase the $\delta^{18}O$ ratio of the source of water vapour by 1 to 1.5‰. This should be added when temperature changes are estimated from $\delta^{18}O$ ratios during ice ages.

The second factor is changed atmospheric circulation. Such changes are likely to be much greater over Greenland, due to formation of large ice sheets on continents of the northern hemisphere, than over Antarctica, where the geometry of the ice sheet and surrounding areas of pack ice will not change drastically at the end of the ice age. The effect of such changes in Greenland are difficult to estimate.

Variation in size of ice sheets with time

Now that we have seen how much we may deduce about the past temperatures of ice sheets from isotopic δ values, we need to determine how much we can deduce about past climates from our past temperature–isotopic record. Qualitatively, it is broadly probable that the colder the period, the more extensive will be the ice sheets. For quantitative estimates, however, we must determine how much of the changed surface temperatures of ice sheets were due to past changes of ice thickness, and hence surface elevation of ice sheets, and how much was due to broad scale climatic changes at a fixed elevation. An extreme case is that of the highest point of the ice sheet over Canada during major glaciations, which may have been as much as 4000 m in elevation compared with a present-day surface close to sea level. However, for this discussion we need only consider the ice sheets of Greenland and Antarctica.

Surface profiles along flowlines on ice sheets have a certain regularity (Robin 1964) in spite of changes in rates of accumulation and ice temperatures between different locations. In figure 11 we use profiles from two different lines over the ice sheet of western Greenland in showing how the ice thickness along the flowline through Camp Century may have changed during the last ice age. The major effect is due to changes of the outer limit of the ice sheet. In the case of Antarctica, the outer limit is defined by the ocean, either by the line where ice shelves start to float, or by ice cliffs which do not penetrate far beyond the low water line. During ice ages, a fall in sea level of 100–150 m will enable the ice to advance to the new shoreline, but around eastern Antarctica the continental shelf is not wide and such advances will normally not be much more than 100 km. The net effect on surface elevations at the South Pole and Vostok stations will be small, probably not exceeding 100 m. In western Antarctica, the continental shelf is much wider, especially in the Ross Sea and beneath the Ross Ice Shelf. Although the extent of additional grounding is still being studied, it is clear that a rise of a few hundred metres in ice thickness at Byrd Station is likely, together with some change in the flow direction of ice.

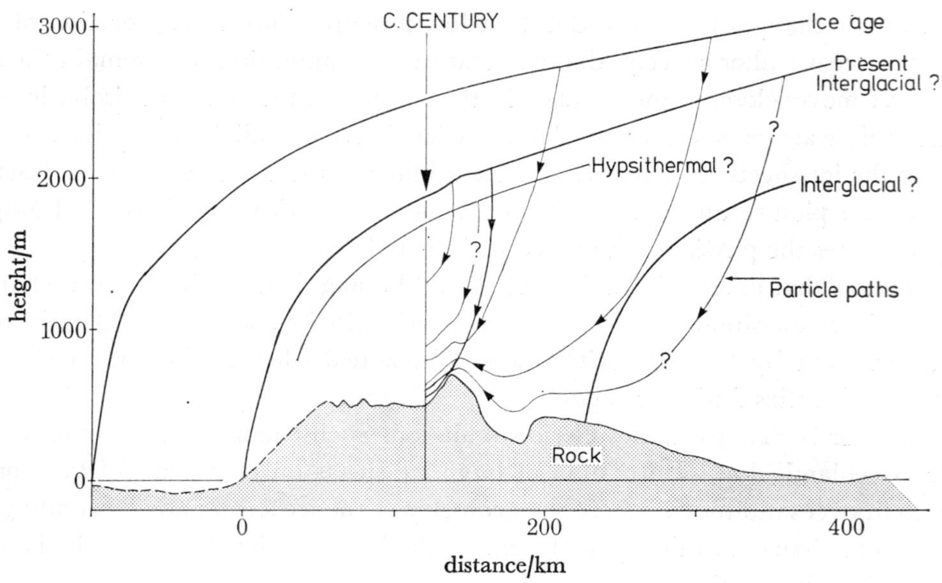

FIGURE 11. Surface and bedrock profiles along the flowline passing through Camp Century, Greenland. See text for discussion of various surface profiles. Lines with arrows show particle paths for various points in the ice core from time ice was deposited on surface to its arrival at Camp Century.

In Greenland, much of the ice now melts before reaching the sea, and effects may be larger than that due to lowering of sea level only. The greatest effect on surface elevation due to the changing size of an ice sheet takes place at the outer boundary, and the effect decreases the further one proceeds up a flowline, as seen in figure 11.

One recently proposed method of studying the elevation at which any piece of ice recovered from an ice core was originally deposited is to determine the amount of air trapped per unit volume of the sample. The volume occupied by air, when the ice ceases to be porous owing to closure of air pockets under increasing loads will be a function of atmospheric pressure and hence elevation, of the temperature at the time of trapping, and of the fractional volume occupied by air at this stage. We can write the fractional volume V of air at s.t.p. at time of closure of pores in terms of the densities of pure ice ρ_1 and of firn at the time of closure ρ_c by

$$V = \left(\frac{1}{\rho_c} - \frac{1}{\rho_1}\right)\frac{T_0}{P_0}\frac{P_c}{T_c} \tag{6}$$

where T_0, P_0 refer to standard temperature and pressure and T_c, P_c refer to temperature and pressure at the time of closure. Now, if the actual volume of voids at the time of closure is a constant and is independent of temperature and other variables, and we can apply appropriate corrections for temperature from our isotopic δ values, we can determine the elevation at which the ice was deposited. A paper by Raynaud & Lorius (1973) shows relevant data for the Camp Century ice core: the authors conclude that the ice from the Wisconsin period was deposited at elevations 1200–1500 m higher than the present surface elevation at that site. Gow & Williamson (1975) in studying the ice core from Byrd Station report little systematic variation in total gas content 'except possibly for slightly reduced values between 1500 and 900 m'. These depths correspond with the culmination and termination of the Wisconsin cold phase in Antarctica. Most recently, a report of studies in Terre Adelie by Lorius (1975) gives values of V for an ice core obtained from a site at 270 m surface elevation, showing that some 'cold' isotopic ice at deeper levels had come from considerably greater elevations. Comparison of the V for shallower layers from this site with the same figures for Camp Century suggests to the present writer that while the concept is useful, the derived numerical values may be too great by a factor of up to two. A similar conclusion comes from comparison of surface profiles in figure 11 with the 1200–1500 m elevation change derived from measurements of V. These different estimates could be due to the assumption of constant density at the time of closure, as opposed to a density at closure that varies slightly with temperature and/or other factors. However, Raynaud (unpublished) and Lorius (private communication), who have carried out the studies on ice cores from Camp Century and Terre Adelie, consider that the evidence favours a constant density of closure. The problem is not fully resolved, but when understood adequately, measurement of V will provide a powerful tool for the study of ice cores.

Dynamic factors that should, if possible, be taken into account when modelling former ice sheets indicate that for a constant basal stress, ice will be slightly thicker for a greater rate of accumulation. On the other hand, lower temperatures will tend to increase ice thicknesses, if these affect the basal layers of ice. Our knowledge is insufficient to develop accurate numerical models for our present needs, and although this should be possible in the future, an empirical approach based on the form of present-day ice sheets is probably the safer method at the present time.

Age–depth relation in ice cores

We have already discussed vertical strain rates and annual layer thicknesses in relation to temperature studies. Provided a continuous ice core is available, and the annual layers can be satisfactorily determined by isotopic methods, annual layers can be counted just as accurately as tree rings or varved clays. This has been done for a period of 740 years at Dye 3 in central Greenland by Dansgaard, Johnsen, Clausen & Gundestrup (1973), and should be a satisfactory technique for up to 10000 years and longer over many parts of Greenland. Barkov, Gordiyenko, Korotkevich & Kotlyakov (1975) have used a similar technique to a depth of 507 m on the Vostok ice core, although their claim to resolve annual layering seems inconsistent with the views of Johnsen *et al.* (1972) on the disappearance of thin isotopic layers due to diffusion. Oeschger, Stauffer, Bucher & Loosli (1977) are developing methods of down-hole extraction of CO_2 from boreholes in polar ice, in order to use ^{14}C dating to provide a time scale, but no results are available at present for the Devensian period, since there are considerable difficulties in working in ice at greater depths. The oldest age determined at Byrd Station is close to 3000 years for ice at 385 m depth.

When dealing with layers over 3000 years old, we are primarily dependent on glacial knowledge of accumulation and deformation rates to date the ice core. If the simple vertical column model is used, the present layer thickness (l) is related to its thickness at the time of deposition (l_0), by Nye's (1963) formula

$$l/l_0 = h/H_0, \tag{5}$$

where h is the present height of the layer above bedrock, and H_0 the ice thickness at the point of deposition. l_0 will normally be the mean annual accumulation at the site of deposition. Integration by numerical or other means will then give the age to any depth. Dansgaard & Johnsen (1969) and Philberth & Federer (1971) have modified equation (5) to take account of the decreasing strain rates at lowest levels shown in figure 5. Dansgaard, Johnsen, Clausen & Langway Jr. (1971) also used preliminary time scales derived by these methods to find periodicities in δ values, which they then used to date still deeper layers in the Camp Century ice cores. However, further evidence is needed before accepting these time scales, and three factors should be considered.

The first is the likelihood of decreased rates of accumulation during glacial periods, due to the inability of the atmosphere to transport water vapour. We see from figure 10 that a general fall of about 7 °C in world temperature is likely to halve the rate of accumulation on the Antarctic ice sheet. While such factors are usually appreciated, they have not been applied in calculating age–depth relations. Nor have ice depths at the points of deposition been used in many cases, since the information is often not available.

The second factor is to determine whether our models of ice flow are satisfactory in various circumstances. Radio-echo methods of sounding polar ice sheets have shown the presence of internally reflecting layers in the ice mass, and in Robin, Evans & Bailey (1969) one such layer, traced over 40 km, was shown to correspond to the surface layer about 1000 years ago. More recently, Whillans (1976) has compared his studies of surface strain and movement along the flowline upstream of Byrd Station with radio-echo soundings along that line. His calculated depths of surfaces of 2500, 5500 and 30000 years of age correspond quite closely to internal layers producing reflexions. Since it is difficult to produce any other hypothesis than that of some depositional effect to explain these layers (see Paren & Robin 1975), we appear justified in regarding these as relic surfaces, or 'isochrons' as shown in our figure 2 from Budd *et al.* (1971*a*).

Improved equipment designed in the Technical University of Denmark, and the use of differentiation of the received signal (Gudmandsen 1975), show great detail in the latest soundings obtained by the joint SPRI-TUD-NSF programme in Antarctica (figures 12 and 13). Figure 12 which is along a flowline about 130 km from Vostok Station, shows that the height of an internal layer above bedrock can vary considerably in the deeper layers, although in the upper part of the ice sheet, layering tends to be parallel to the surface. In the vicinity of Dome C (75° S 123° E) where bedrock is generally level, layering is almost parallel to upper and lower surfaces over at least 80 % of the ice depth (figure 13). The radio-echo technique thus offers clear advantages in selecting sites for any drilling in future. It also provides evidence

Description of plate 1

Figure 12. Radio-echo sounding record along flowline with similar conditions to Vostok but 130 km north, showing how internal layering heights may vary considerably from that expected from equation (5). Smooth bottom echo on left of diagram is the start of a large sub-ice lake.

Figure 13. Suitable layering conditions for interpretation of deep drilling near Dome C in eastern Antarctica.

Figure 12

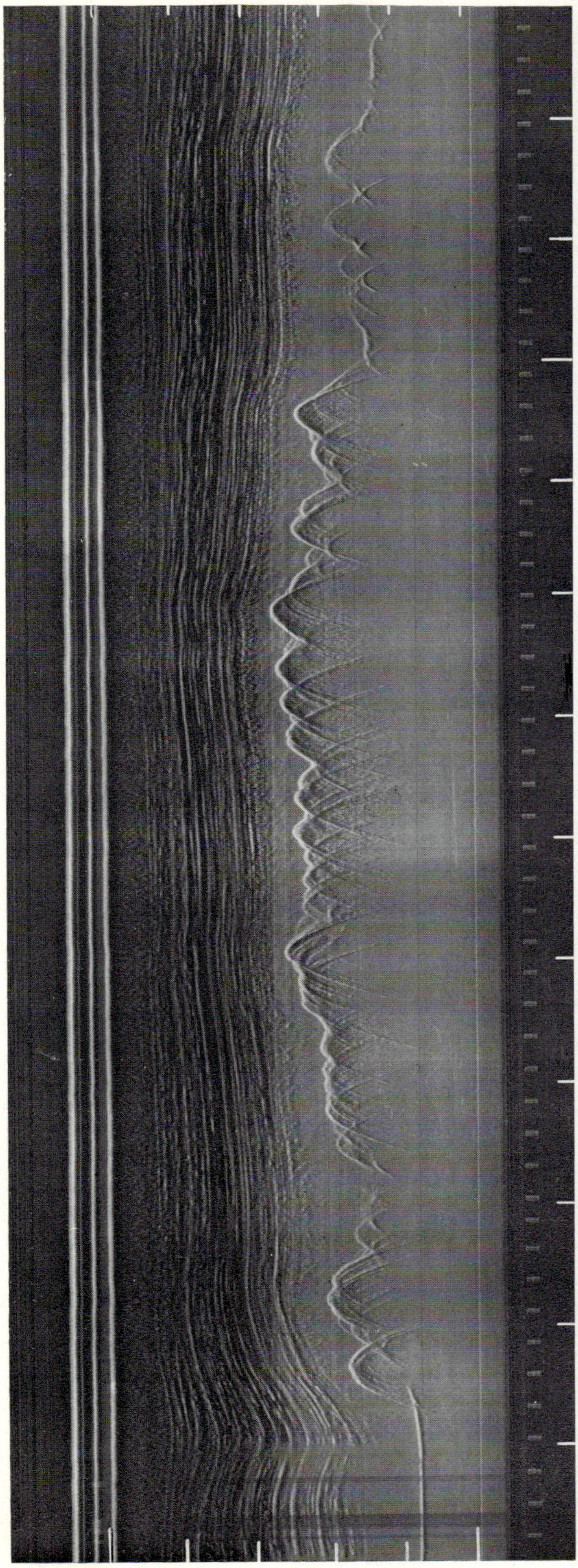

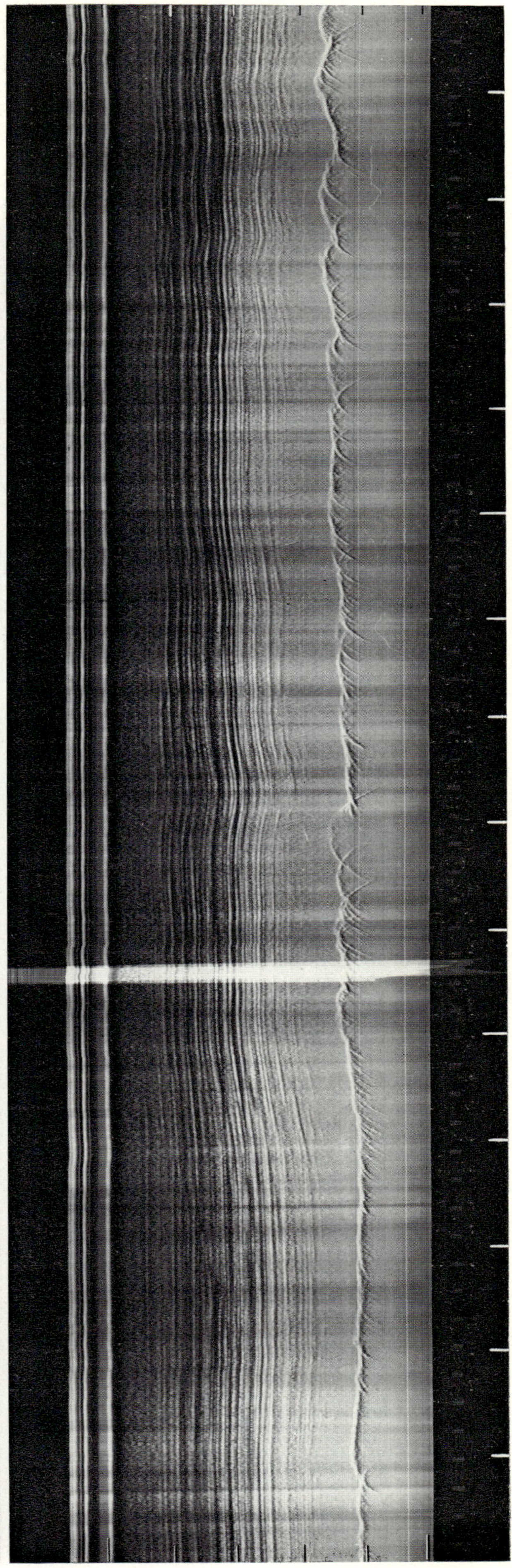

depth/m

Figures 12 and 13. For description see opposite.

(*Facing p.* 156)

that dating of deep ice layers by conventional glaciological models may be inaccurate over uneven terrain.

The third factor that has received little attention is the question of continuity of an ice core. There is no question of the continuity of precipitation on the upper surface of the ice sheet on the time scales we consider, but where significant changes of ice flow have occurred over the last 100000 years in Greenland or Antarctica, the possibility of discontinuities arises. Ice from near the base of the ice cores from Camp Century and Byrd Stations shows isotopic values similar to those of ice deposited in these areas during the last 10000 years, in spite of the intervening cold periods. If the ice is of relatively local origin, it raises a question of continuity that is most readily discussed in terms of conditions around Camp Century.

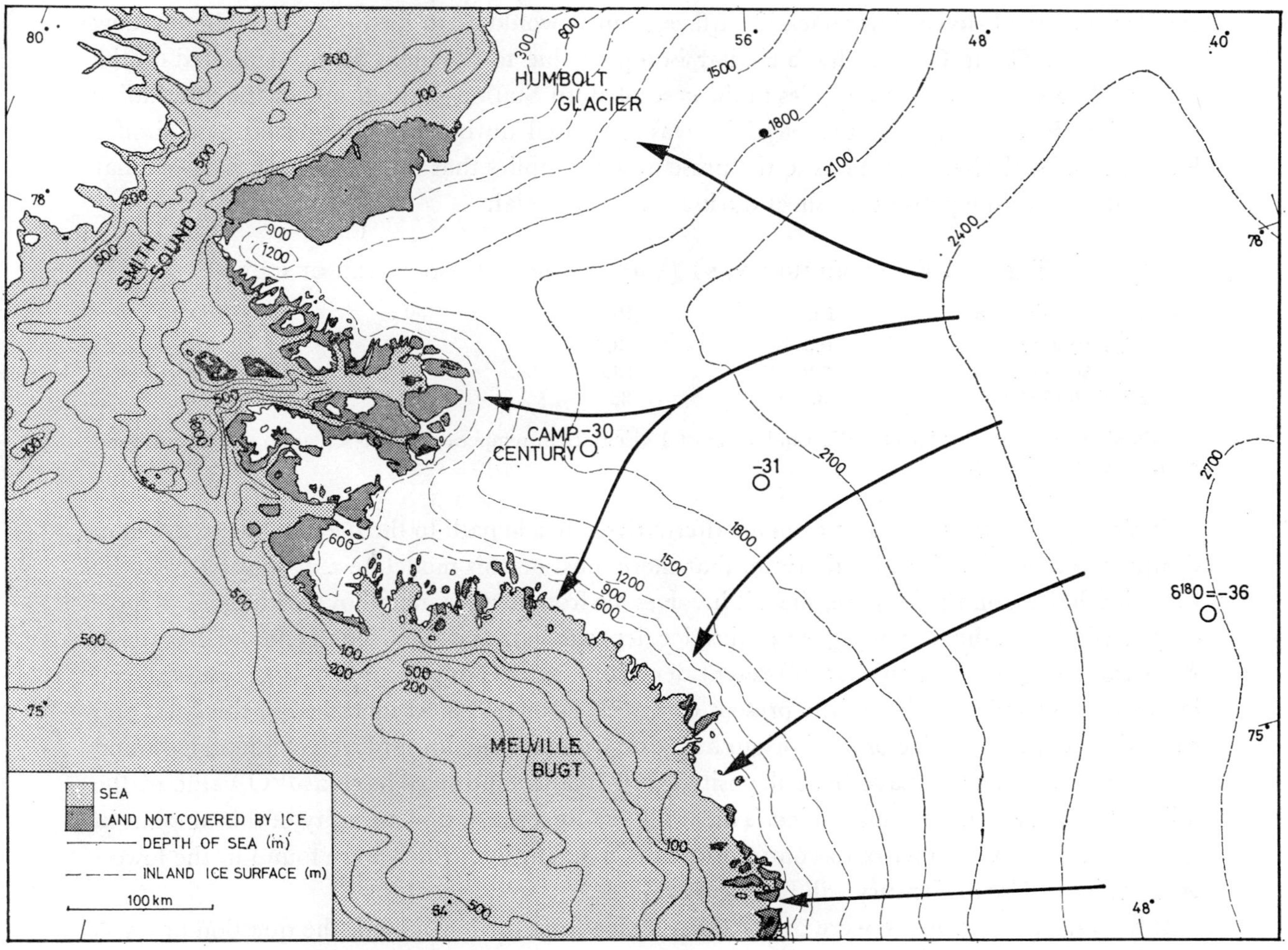

FIGURE 14. Surface contours of the ice sheet of northwestern Greenland from map by Weideck, showing probable flowlines of ice.

In figure 14 we show present-day contours of the ice sheet of northwestern Greenland taken from the map of Weideck (1971). Arrows have been added to show the main directions of ice flow, which follow the generalized contours. Upstream from Camp Century, there appears to be a tendency for the ice flow to be diverted either to the north or east of Camp Century, which leads to the low velocity of flow at Camp Century, which from various pieces of unpublished data appears to be from 5 to 12 m a^{-1}. During a glacial period it seems certain that the ice sheet will expand – at least to the new sea level more than 100 m below the present level – and if ice shelves are formed it would have grounded at depths from 300 to 600 m below the present sea level. In this case, the surface height at Camp Century would be considerably greater, and ice from inland is likely to have moved across the site of Camp Century with less diversion of flow. Figure 11 shows some profiles of possible ice surfaces at different periods that take account of isotopic, temperature, total gas content and flow factors already discussed. Also shown are particle paths to the present ice core from their point of deposition. Although the scales in figure 11 are open to question, and it could be argued that the horizontal scale could be doubled, qualitatively the ideas shown are sound. If no discontinuities are present in the ice column, then the ice sheet during the last interglacial period must have been much smaller (figure 11, lower interglacial? surface), but it would seem fortuitous that the ice now at the base at Camp Century has a similar isotopic value to that now being deposited on the surface. The same argument applies in the case of Byrd Station – which makes the alternative more plausible, namely that the basal ice was deposited during the interglacial at a similar location under similar conditions to the present. This implies that the ice sheets were of similar size and form to the present ice sheet during the interglacial.

TABLE 1. FORWARD MOVEMENT (km) AT TOP OF BASAL LAYER OF ICE†

basal layer thickness	1 m	10 m	50 m	100 m
after 10000 a	0.3	3.2	15.8	32
after 50000 a	7.9	15.8	79	158
after 100000 a	3.2	32	158	316

† Basal shear stress 10^5 Pa (1 bar), random crystal orientation, temperature −9 °C, strain rates from Budd *et al.* (1971 *a*, Table 3).

It also follows, as can be seen from an alternative particle path to the deepest layers at Camp Century, but coming from local origin, that there must on this model be a discontinuity in the time-depth scale on the ice core, since it involves ice from further inland over-riding ice of local origin. We see on the map in figure 14 that ice deposited around the 2700 m contour has mean $\delta^{18}O$ value 6‰ lower than that of present deposition near Camp Century. If the interglacial ice sheet was similar to that of the present day, then as the ice sheet increased in thickness and carried inland ice over ice of local origin at Camp Century, the lowest levels of inland ice found in the core are likely to have been deposited well inland, and may have a $\delta^{18}O$ value of the order of 6‰ lower than the local ice. This would cause a sharp discontinuity of this amount in the ice column. A number of discontinuities of this magnitude which were found in the lowest levels of ice at Camp Century will be discussed later.

If the lowest layers of ice were over-ridden by ice from further inland, the question arises as to whether such basal ice could remain in place over a period of 10^4–10^5 years. This would not be so on our simple model described by equation (2), but we have seen from figure 5 that strain rates decrease rapidly towards bedrock when the basal temperature is well below pressure melting point.

In table 1 we show the forward movement of ice at different heights above the glacier bed under a shear stress of 1 bar, using constants for polycrystalline ice and the present basal temperature. The present basal stress, based on data from Mock (1968), is about 20 kPa (0.2 bar), but it may well have approached 100 kPa (1 bar) when inland ice was moving over the site. Figures in table 1 should be divided by two if the stress falls from 100 to 60 kPa (1.0–0.6 bar), or by five for a fall from 100 to 30 kPa (1.0–0.3 bar). Although our model is imperfect, and we have neglected the possibility of recrystallization of ice in an easy glide direction, it does appear that the observed 'isotopically warmer' layer around 50 m thick at the bottom of the ice core could be a remnant from the ice before the onset of the last ice age which did not move more than a few tens of kilometres in the intervening time.

When the basal ice is at melting point, we expect some basal sliding, which experience elsewhere indicates may be of the order of half the velocity at the surface of a glacier. However, if sliding of only 2 m a^{-1} were present, basal ice would have moved 100 km in 50000 years. At Byrd Station, the basal sliding may be three times this amount, and one must expect the lowest levels of ice to have originated 100 km or more from the present site of the borehole. The chances of finding basal ice that has moved very little while being over-ridden by ice from further inland are very much less in the Byrd ice core than in the case of Camp Century. This may explain the absence of sharp discontinuities in the Byrd core.

We are now in a position to review isotopic data from the three major ice cores, starting with the most straightforward. We have plotted basic data in figures 15, 16, and 18 on a linear depth scale, and also show the time scale derived by those using the data.

(*a*) *Vostok*

From a climatological point of view this is the simplest to interpret. The surface elevation of the ice sheet is unlikely to have changed by more than 100 m during the last ice age. Surface contours (Scott Polar Research Institute, Map 1974) indicate that the deepest ice beneath Vostok will originate near ice 'Dome B', about 300 km upstream from Vostok with a surface elevation around 300 m higher. The ice core results now available come from the top 26 % of the ice sheet, so corrections for deformation are not large. Present-day motion of the ice sheet at Vostok, from astronomical measurements (Liebert & Leonhardt 1974), is about 3 m a^{-1}, which is close to the steady-state figure we obtain by use of equation (3). The isotopic δ value change with depth due to steady-state flow is estimated at $-1.5‰$ at 1000 m, as indicated by the heavy line in figure 15. Uncertainties due to changing elevation should not exceed $\pm 1‰$.

The major difficulties in interpreting isotopic data in figure 15, kindly supplied by V. M. Kotlyakov of the Institute of Geography, Academy of Sciences, Moscow, are to assess the accuracy of the time scale, and the significance of the rapid fluctuations of $\delta^{18}O$ values with depth. The large fluctuations in $\delta^{18}O$ values could result either from sampling techniques, or perhaps from some irregularities in deposition of ice. Barkov *et al.* (1975), in discussing the study of the top 507 m, report that sample thicknesses varied from 3 to 10 cm, and the spacing between samples ranged from 0.1 to 5 m. They then grouped samples into 20 m intervals for palaeoclimatic interpretation. We have grouped the data in figure 15 into 50 m intervals for this purpose, and have added the heavy line showing the mean $\delta^{18}O$ value over each 50 m interval. A shaded zone around this line shows the standard deviation of the mean as an indication of the error of the mean.

The time scale used in Barkov *et al.* (1975) for analysis of the top 507 m of the core was calculated by taking an accumulation rate of 2.4 cm ice a^{-1}, and applying corrections for vertical strain given in equation (5). This uses the best value of local accumulation rate available. Temperature analyses by Budd, Jenssen & Young (1973) gave a best fit accumulation rate of 2.8 cm ice a^{-1}, which indicates the figure used has applied reasonably well for several thousand

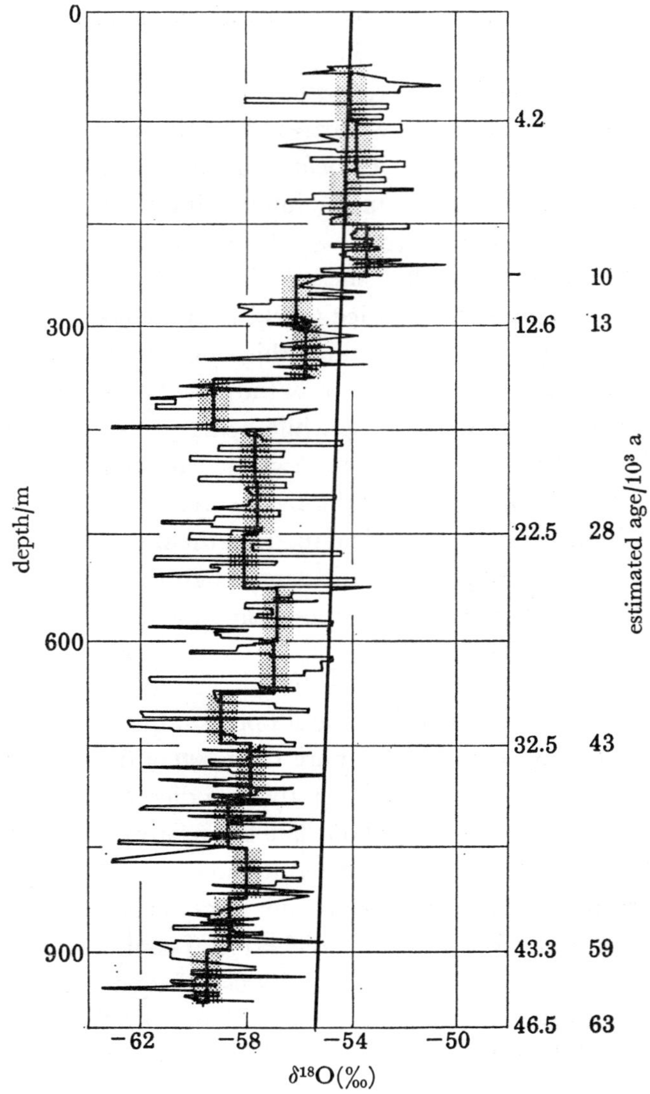

FIGURE 15. Isotopic $\delta^{18}O$ profile from Vostok, Antarctica, plotted on a linear depth scale, supplied by V. M. Kotlyakov and to be published by Barkov *et al.* (in press). Mean values over 50 m depth intervals have been added together with shading to indicate the standard deviation of the mean. The heavy sloping line shows estimated corrections for changing surface altitudes, if the climate had otherwise remained constant. Estimated age scales on the right are those calculated by Barkov and others (inner figures), and estimated in this paper (outer figures).

years. Studies of isotopic layering reported by Barkov *et al.* (1975) at five depths from 48 to 391 m depth (see earlier discussion) showed annual layer thicknesses that decreased from 3.2 to 1.7 cm of ice with a mean value of 2.5 cm ice. They point out that at 391 m depth, a layer of 3.2 cm thickness would decrease to 2.8 cm thickness, which is 40 % greater than that observed. 'This could be explained by lower rates of accumulation, but the illegibility of seasonal variations of $\delta^{18}O$ at horizons 260 and 391 m do not allow us to accept these data with confidence.' They also point out that an average of one year in five or six tends to be missing in any single location around Vostok: hence their figure of 2.4 cm a^{-1} used for determining age. Inspection of figure 15 shows that a figure close to this amount has been used to date ice to 950 m depth (Barkov, Gordiyenko, Korotkevich & Kotlyakov, in press). From our discussion of transport of water vapour in the atmosphere (figure 10), it would appear very likely that a lower rate of accumulation was typical of colder periods, in which case the ages shown on the right hand side of figure 15 should be increased by about 66 % for the ice column below 250 m depth. Inspection of radio-echo records along the flowline upstream from Vostok shows an internal layer at depth around 1700 m from 90 to 20 km upstream of Vostok, increasing in depth to 1900 m over an interval from 20 to 8 km upstream of Vostok. This would involve a layer originally at 850 m depth increasing to 950 m depth between 6000 and 2400 a B.P. When this deformation is also taken into account, we find that the age scale below 100 m should be decreased by 9 % due to this factor. Below 250 m depth this involves a net increase of 50 % in the scale.

When all preceding corrections are taken into account, we conclude that isotopic δ values reached their present-day levels after the ice age 10000 a B.P. with an error of ± 2000 years estimated. The coldest conditions occurred from 17.5 to 20.5×10^3 a B.P. (15.5–17.5 on Barkov *et al.* scale), with an isotopic lowering of 4.5 ‰ to which we should add a correction for changing oceanic composition, giving a probable temperature lowering of 6–8 °C. Before that, temperatures were a little warmer, with regional isotopic values 2 ‰ (*ca.* 4 °C) lower than present values from around $32–40 \times 10^3$ a B.P. (or $25–30 \times 10^3$ a B on the time scale of Barkov *et al.*). Colder periods (-4 ‰ and *ca.* -6 °C) also occurred around 39–43 and $60–63 \times 10^3$ a B.P. (30–32.5 and 43–45 a B.P. Barkov *et al.*).

The Soviet Antarctic Expedition are developing techniques to obtain deeper ice cores from Vostok. The results will be awaited with great interest by all climatologists and glaciologists.

(*b*) *Camp Century*

This ice core has been discussed in more detail than any other, especially by Dansgaard *et al.* (1971) and Dansgaard *et al.* (1973). Earlier discussion in this paper has shown that it is the most difficult to interpret prior to 10000 a B.P., in terms of changing thickness and flow (figure 11), and consequently in dating. Interpretation of the top 25 % of the core deposited since 8300 a B.P. is straightforward, since strain rates appear to have been uniform over this period (figure 5).

The first dating of the ice core used by Dansgaard & Johnsen (1969) was based on a modification of equation (5) to allow for the decrease in strain rate at the lowest levels, but using the present-day accumulation rate and local ice thickness. They found an apparent periodicity in $\delta^{18}O$ values around 2000 years back to 45000 a B.P. (*ca.* 1330 m depth), but before this date the period appeared to decrease to 4000 years. They have assumed that the 2000 year period was present at an earlier date, and used this as a control of earlier dates to determine their time scale.

The above treatment depends on glaciological parameters to 45000 a B.P., but it is not consistent with the model of changes of the ice sheet shown in figure 11. We use this model to estimate $\delta^{18}O$ values, which are plotted against depth in figure 16, as a semi-quantitative estimate of the effect of changing thickness and flow on $\delta^{18}O$ values and on the age scale. The latter assumes an accumulation rate during the last ice age to be two-thirds of the present, and makes a further change by a ratio of 1.7:1.0 to allow for greater ice thicknesses at the point of deposition of ice further inland. The estimates of effects of flow on $\delta^{18}O$ values below 1320 m, shown as approximately -34 ‰, take account of ice that was deposited during the interglacial

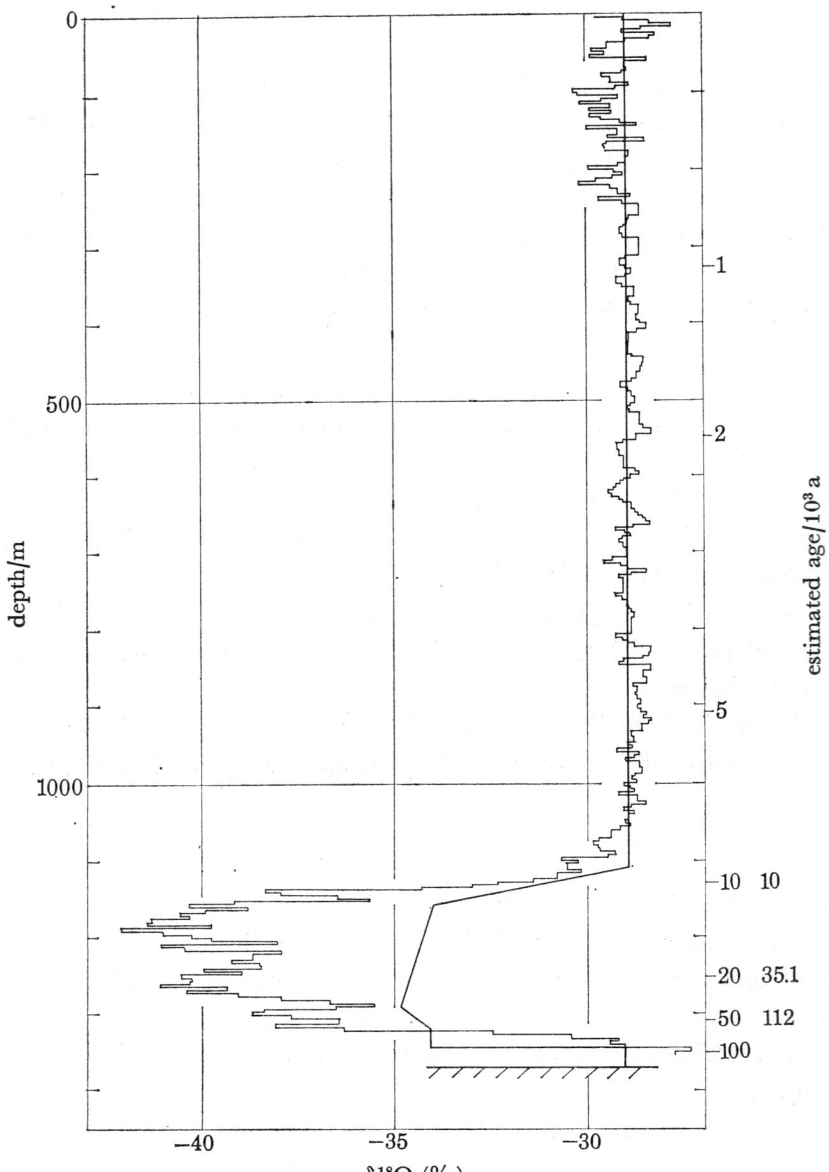

FIGURE 16. Isotopic $\delta^{18}O$ profile from Camp Century, Greenland, plotted on a linear depth scale, supplied by Johnsen. The heavy line shows estimated corrections for changing surface altitudes as in figure 15. Age scales on the right are those of Dansgaard *et al.* (1971) (inner figures) and those estimated in this paper based on figure 11 (outer figures).

period that is assumed to have over-ridden ice deposited locally during the interglacial. This in turn assumes that the ice sheet was then of similar size and of similar climate to the present ice sheet. The effect is shown more dramatically in figure 17 which has a linear time scale as in Dansgaard *et al.* (1971). If much of the lower ice column had originated from the inland ice, the period during which ice of −34‰ was deposited will be uncertain. One would, however, expect this to be some thousands of years. Instead, in the profile of Dansgaard *et al.* we find a few sharp isolated peaks of about −34‰, which they interpret as short periods of intense cold before the onset of the ice age, with longer periods of deposition of ice of −27 to −29‰. An alternative explanation is that the lowest 50 m of the ice column does not represent a continuous

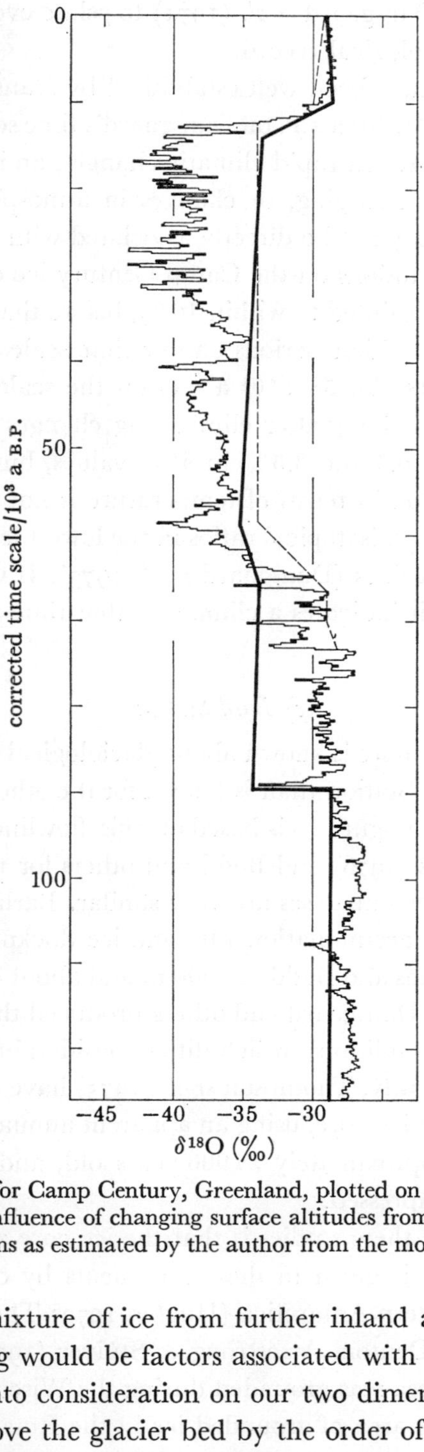

FIGURE 17. Isotopic $\delta^{18}O$ profile for Camp Century, Greenland, plotted on a linear time scale, with dashed lines showing corrections for the influence of changing surface altitudes from Dansgaard *et al.* (1973). The heavy line shows the same corrections as estimated by the author from the model shown in figure 11.

stratigraphy, but shows a mixture of ice from further inland and locally deposited ice. One possible cause of such mixing would be factors associated with the flow of ice in three dimensions which are not taken into consideration on our two dimensional model. Such effects are likely if sharp peaks rise above the glacier bed by the order of a hundred metres or more, in which case ice at lowest levels may tend to flow around a peak, and this may disrupt the stratigraphy when meeting ice flowing over the peak.

The net effect of our discussion so far is to cast doubt on the dating of ice cores by glaciological methods, in cases where the ice was deposited more than 10000 a B.P. If we neglect the glaciological approach, and accept dates for major climatic changes from other sources, then we can calibrate major isotopic changes in the ice core by these dates. A modification of these approaches has been used by Dansgaard *et al.* (1971) to relate events on the isotopic record to known events in the glacial geological record.

Since sampling techniques have been well established by Dansgaard and others, the rapid variations of $\delta^{18}O$ from 15 to 32×10^3 a B.P. on Dansgaard's time scale appear real. It is difficult to determine whether these represent rapid climatic changes, an irregular flow of the ice sheet such as may be associated with surging, or changes in atmospheric circulation and in the sources of water vapour that may not be directly associated with temperature changes.

We can summarize our conclusions on the Camp Century ice core by saying that although information since 10000 a B.P. is dated to within 10 %, before that time dating errors could be as high as a factor of 2.5. The coldest periods on our time scale were centred around 20, 25, 45 and 100×10^3 a B.P. (17, 28, 40, 58×10^3 a B.P. on the scale of Dansgaard *et al.* 1971)). The general climatic effects remaining after eliminating changes of flow and thickness of the ice sheet were maxima of 8, 7, 6.5 and 3.5‰ in $\delta^{18}O$ values, but owing to changes in circulation patterns, an interpretation in terms of temperature is not proposed.

Prior to the onset of the ice age, isotopic δ ratios in the lowest levels of ice were 1–2‰ higher than the present near-surface values (Dansgaard *et al.* 1973). If we accept the argument that this ice was deposited locally, it indicates a climate at that time that was about 2 °C warmer than at present.

(*c*) *Byrd Station*

Of the three major ice cores, more is known about glaciological conditions along the present-day flowline upstream of Byrd Station than is known for the other sites (Whillans 1973, 1976). Primary dating of the ice core in figure 18 is based on this flowline data, and has been carried out independently by Whillans (1976) and Budd and others for the Cambridge workshop (to be published). Results of the two analyses are very similar. Earlier studies by Epstein, Sharp & Gow (1970) used the local accumulation rate and ice thickness to determine ages which were similar to those of Whillans and Budd at 1000 m and about 10000 years older at 2000 m depth. Later work by Johnsen, Dansgaard and others produced three further time scales based on different assumed flowline conditions, which differ considerably from the conditions shown in Whillans (1973). More recently, Thompson *et al.* (1975) have dated the ice column on the basis of the dust content of the ice core, using an apparent annual cycle in dust content. They suggest that the basal ice is approximately 27000 years old, and that the time–depth factor below 1100 m is much too compressed.

The overall impression from these studies is that the ice core at Byrd Station is difficult to interpret and to date. Weight is given to these arguments by current controversy over the stability of the ice sheet of western Antarctica (Hughes 1973; Thomas 1976; Weertman 1976, and others) and by claims of Denton, Armstrong & Stuiver (1971) that the whole of the ice of the present Ross Ice Shelf area was grounded during the Wisconsin period.

There is little doubt that the area of grounded ice in the Ross Sea must have been greater during the last glacial period as a result of a sea level lowering of 100–130 m, and this must in turn have had some effect on the ice thickness and elevation at the site of Byrd Station. In turn, the direction of ice flow may have changed, but we have no evidence of this. Indeed the agreement of the form of the 30000 year isochrons calculated by Whillans (1976), and the form of the internal reflexions at that depth, suggests that the flow near Byrd Station was similar to the present during the last ice age. Unlike Camp Century, Gow & Williamson (1975) report little systematic variation in the total gas content of ice in the Byrd core, 'except possibly for slightly reduced values between 1500 and 900 m depth'. However, the fact that an ice column of about 1200 m in length is available for analysis of conditions prior to the end of the last ice age, compared to the lowest 250 m of the Camp Century ice core, increases the importance of the Byrd ice core. There is no irregularity in the isotopic record (figure 18) to indicate that ice from great inland elevations is present in the ice core. We therefore show a curve for the

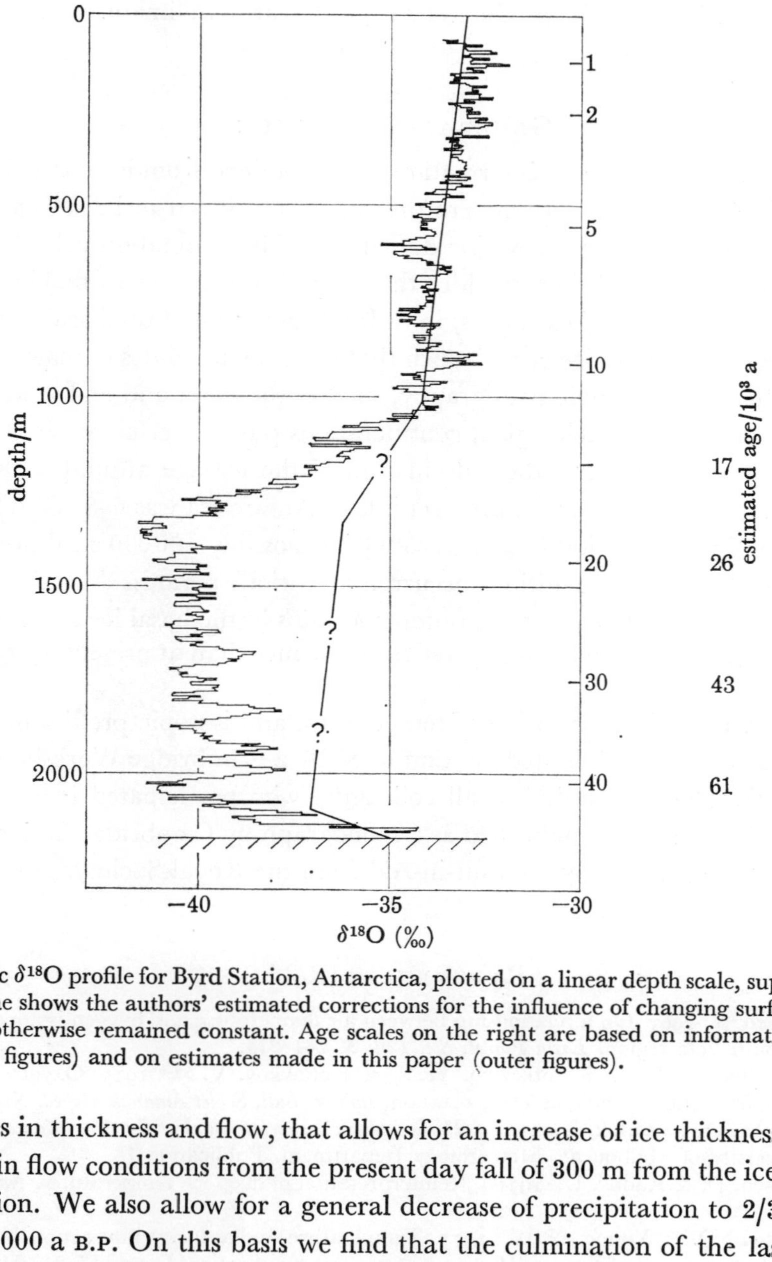

FIGURE 18. Isotopic $\delta^{18}O$ profile for Byrd Station, Antarctica, plotted on a linear depth scale, supplied by Johnsen. The heavy line shows the authors' estimated corrections for the influence of changing surface altitude, if the climate had otherwise remained constant. Age scales on the right are based on information from Whillans (1973) (inner figures) and on estimates made in this paper (outer figures).

effect of changes in thickness and flow, that allows for an increase of ice thickness of 300 m only, and no change in flow conditions from the present day fall of 300 m from the ice divide 150 km from Byrd Station. We also allow for a general decrease of precipitation to 2/3 of the present level, before 10000 a B.P. On this basis we find that the culmination of the last ice age took place at about $22-24 \times 10^3$ a B.P. ($17.5-19 \times 10^3$ a B.P. by Whillans etc.), with a climatic lowering of $\delta^{18}O$ values around 5.3‰, corresponding to a temperature lowering of about 7–8 °C. Although this agrees with the figures for Vostok, the uncertainties of the effect of changing surface levels of ice are much greater. Fluctuations of 2‰ in $\delta^{18}O$ are present at earlier dates, but no prominent colder periods appear until about 65000 a B.P. (41000 a B.P. on Whillans's scale).

Basal ice from before the start of the Devensian ice age shows isotopic δ ratios close to the surface layers of the ice sheet (Epstein *et al.* 1970; Johnsen *et al.* 1972). Since this ice must have been transported by ice movement of 100 km or more from a higher part of the ice sheet, it would appear from the present geometry of the ice that either pre-Devensian climate was about 2–3 °C warmer than the present, or the ice sheet surface was generally lower. The latter condition could also be the product of a slightly warmer climate.

General conclusions

There is a great deal of climatic information to be obtained from ice cores. Use of a relation between isotopic ratios and mean temperatures has been shown to be satisfactory, especially over the past few thousand years. Major difficulties of interpretation arise in determination of the age–depth relation in ice cores older than 10000 a B.P., and in making corrections for changes in the size and flow of ice sheets, in order to determine broad scale climatic changes. Nevertheless the continuity of ice cores is such that one may use dates of major climatic sources to date major changes in isotopic stratigraphy, such as the onset and end of an ice age.

Dates given from purely glaciological considerations put the termination of the last ice age around 10000 ± 1000 a B.P., and the culmination of the ice age around 20000 ± 3000 a B.P. At this time, the lowering of temperature over inland Antarctica was 6–8 °C. Ice age conditions existed continuously for a period of at least 50000 a, possibly 100000 a, although warming of about 3 °C from the coldest conditions occurred at various periods.

Before the onset of this glacial period, isotopic δ ratios in the basal ice at Camp Century and Byrd Station suggest that the climate was slightly warmer than at present – possibly by 2–3 °C.

Many of the ideas on interpretation of temperature and isotopic profiles in polar ice sheets expressed in this paper were initiated or stimulated by a Cambridge Workshop on these topics (see Robin 1973). Thanks are due to all colleagues who participated in the Workshop, the proceedings of which are to be published as a monograph by Cambridge University Press. The Workshop was made possible by a Grant-in-Aid from the Royal Society.

References (Robin)

Aldaz, L. & Deutsch, S. 1967 On a relationship between air temperature and oxygen isotope ratio of snow and firn in the South Pole region. *Earth Planet. Sci. Lett.* **3**, 267–274.

Barkov, N. I., Gordiyenko, F. G., Korotkevich, Ye. S. & Kotlaykov, V. M. 1975 Oxygen isotope study of a 500 m ice core from the borehole at Vostok station. *Inform. Bull. Soviet Antarctic Exped.* No. 90, 39–49.

Budd, W. F., Jenssen, D. & Radok, U. 1971*a* *Derived physical characteristics of the Antarctic ice sheet. Mark I.* Melbourne: University of Melbourne, Meteorology Department. Publication 18.

Budd, W. F., Jenssen, D. & Radok, U. 1971*b* Reinterpretation of deep ice temperatures. *Nature, Phys. Sci.* **232**, 84–85.

Budd, W. F., Jenssen, D. & Young, N. W. 1973 Temperature and velocity interaction in the motion of ice sheets. *First Australasian Conference on Heat and Mass Transfer*, Section I, pp. 17–24. Melbourne: Monash University.

Budd, W. F., Young, N. W. & Austin, C. R. 1976 Measured and computed temperature distributions in the Law Dome Ice Cap, Antarctica. *J. Glaciol.* **16**, 99–109.

Dansgaard, W. 1961 The isotopic composition of natural waters with special reference to the Greenland Ice Cap. *Meddr Grønland,* **165** (2), 1–120.

Dansgaard, W. 1964 Stable isotopes in precipitation. *Tellus* **16**, 436–468.
Dansgaard, W. & Johnsen, S. J. 1969 A flow model and a time scale for the ice core from Camp Century, Greenland. *J. Glaciol.* **8**, 215–223.
Dansgaard, W., Johnsen, S. J., Clausen, H. B. & Langway Jr, C. C. 1971 Climatic record revealed by the Camp Century ice core. In *Late Cenozoic glacial ages* (ed. K. K. Turekian), pp. 37–56. New Haven and London: Yale University Press.
Dansgaard, W., Johnsen, S. J., Clausen, H. B. & Grundestrup, N. 1973 Stable isotope glaciology. *Meddr. Grønland* **197** (2), 1–53.
Denton, G. H., Armstrong, R. L. & Stuiver, M. 1971 The late Cenozoic glacial history of Antarctica. In *Late Cenozoic glacial ages* (ed. K. K. Turekian), pp. 267–306. New Haven and London: Yale University Press.
Epstein, S., Sharp, R. P. & Gow, A. J. 1970 Antarctic ice sheet: stable isotope analyses of Byrd station cores and interhemispheric climatic implications. *Science, N.Y.* **168**, 1570–1572.
Gow, A. J. & Williamson, T. 1971 Volcanic ash in the Antarctic ice sheet and its possible climatic implications. *Earth Planet. Sci. Lett.* **13**, 210–218.
Gow, A. J. & Williamson, T. 1975 Gas inclusions in the Antarctic ice sheet and their glaciological significance. *J. Geophys. Res.* **80**, 5101–5108.
Gudmandsen, P. 1975 Layer echoes in polar ice sheets. *J. Glaciol.* **15**, 95–101.
Hughes, T. 1973 Is the West Antarctic Ice Sheet disintegrating? *J. geophys. Res.* **78**, 7884–7910.
Johnsen, S. J. 1977 Stable isotope profiles compared with temperature profiles in firn and with historical temperature records. Paper presented at International Symposium on Isotopes and Impurities in Snow and Ice, Grenoble, 1975. (To be published by International Association of Scientific Hydrology.) (In the press.)
Johnsen, S. J., Dansgaard, W., Clausen, H. B. & Langway Jr, C. C. 1972 Oxygen isotope profiles through the Antarctic and Greenland ice sheets. *Nature, Lond.* **235**, 429–434.
Liebert, J. & Leonhardt, G. 1974 Astronomical observations aimed at determining ice sheet motion in the Vostok station area. *Inform. Bull. Soviet Antarctic Exped.* No. 88, pp. 68–70.
Lorius, C. 1975 *Rapport d'Activité, 1975*. Grenoble: Centre National de la Recherche Scientifique, Laboratoire de Glaciologie.
Merlivat, L. & Nief, G. 1967 Fractionnement isotopique lors des changements d'état solide-vapeur de l'eau à des températures inférieures à 0 °C. *Tellus* **19**, 122–127.
Mock, S. J. 1968 Snow accumulation studies on the Thule Peninsula, Greenland. *J. Glaciol.* **7**, 59–76.
Nye, J. W. 1963 Correction factor for accumulation measured by the thickness of the annual layers in an ice sheet. *J. Glaciol.* **4**, 785–788.
Oeschger, H., Stauffer, B., Bucher, P. & Loosli, H. H. 1977 Down bore hole extraction of gases and dissolved and particulate matter in ice. Paper presented at the International Symposium on Isotopes and Impurities in Snow and Ice, Grenoble, 1975. (To be published by International Association of Scientific Hydrology.) (In the press.)
Paren, J. G. & Robin, G. de Q. 1975 Internal reflections in ice sheets. *J. Glaciol.* **14**, 251–259.
Philberth, K. & Federer, B. 1971 On the temperature profile and the age profile in the central part of cold ice sheets. *J. Glaciol.* **10**, 3–14.
Picciotto, E., De Maere, X. & Friedman, I. 1960 Isotopic composition and temperature of formation of Antarctic snows. *Nature, Lond.* **187**, 857–859.
Ragone, S. E., Finelli, R. V., Leung, S. & Wolf, C. 1972 Cationic analysis of the Camp Century, Greenland, ice core. *U.S. Army C.R.R.E.L. Special Report* 178. 15 pp.
Raynaud, D. 1976 Unpublished. Thèse de Doctorat d'Etat. University of Grenoble. (Centre National de la Recherche Scientifique, Laboratoire de Glaciologie.)
Raynaud, D. & Lorius, C. 1973 Climatic implications of total gas content in ice at Camp Century. *Nature, Lond.* **243**, 283–284.
Robin, G. de Q. 1955 Ice movement and temperature distribution in glaciers and ice sheets. *J. Glaciol.* **2**, 523–532.
Robin, G. de Q. 1964 Glaciology. *Endeavour* **23**, 102–107.
Robin, G. de Q. 1973 Cambridge Workshop on temperature and isotopic profiles in polar ice sheets, 1973. *Polar Rec.* **16**, 902–907.
Robin, G. de Q. 1976 Temperature-depth and oxygen-isotope profiles in polar ice sheets. *J. Glaciol.* **16**, 9–22.
Robin, G. de Q., Evans, S. & Bailey, J. T. 1969 Interpretation of radio echo sounding in polar ice sheets. *Phil. Trans. R. Soc. Lond. A* **265**, 437–505.
Scott Polar Research Institute, Cambridge. 1974 Map: Antarctica, Radio echo sounding map series A. 1:5000000. Sheet 3. Ice sheet surface and sub-ice relief. 90° E–180°.
Thomas, R. H. 1976 Thickening of the Ross Ice Shelf and equilibrium state of the West Antarctic ice sheet. *Nature, Lond.* **259**, 180–183.
Thompson, L. G., Hamilton, W. L. & Bull, C. 1975 Climatological implications of microparticle concentrations in the ice core from 'Byrd' station, Western Antarctica. *J. Glaciol.* **14**, 433–444.
Weertman, J. 1976 Glaciology's grand unsolved problem. *Nature, Lond.* **260**, 284–286.

Weideck, A. 1971 Quaternary map of Greenland. 1:2500000. Geological Survey of Greenland, Copenhagen. (Undated: received 1971.)

Whillans, I. M. 1973 State of equilibrium of the West Antarctic inland ice sheet. *Science, N.Y.* **182**, 476–479.

Whillans, I. M. 1976 Radio-echo layers and the recent stability of the West Antarctic ice sheet. *Nature, Lond.* **264**, 152–155.

Zhavoronkov, N. M., Uvarov, O. V. & Sevryugova, N. N. 1955 Some physico-chemical constants of heavy oxygen water. In *Primeneniye mechenykh atomov v analiticheskoy khimii* (ed. A. P. Vinogradov), pp. 223–233. Moscow: Akademiya Nauk SSSR (Inst. geokhimii i analit. khimii).

Phil. Trans. R. Soc. Lond. B. **280**, 169–182 (1977) [169]
Printed in Great Britain

The oxygen isotope stratigraphic record of the Late Pleistocene†

By N. J. Shackleton
CLIMAP International Corresponding Member
Sub-department of Quaternary Research, University of Cambridge and Lamont–Doherty Geological Observatory, Columbia University, U.S.A.

Oxygen isotope measurements have been made in foraminifera from over 60 deep-sea sediment cores. Taken together with the oxygen isotope measurements published by Emiliani from Caribbean and Equatorial Atlantic cores, this comprises a unique body of stratigraphic data covering most of the important areas of calcareous sediment over the whole world ocean.

The oxygen isotopic composition of foraminifera from cores of Late Pleistocene sediment varies in a similar manner in nearly all areas; the variations reflect changes in the oxygen isotopic composition of the ocean. The oceans are mixed in about 1 ka so that ocean isotopic changes, resulting from fluctuations in the quantity of ice stored on the continents, must have occurred almost synchronously in all regions. Thus the oxygen isotope record provides an excellent means of stratigraphic correlation.

Cores accumulated at rates of over about 5 cm/ka provide records of oxygen isotopic composition change that are almost unaffected by post-depositional mixing of the sediment. Thus they preserve a detailed record of the advance and retreat of the ice masses in the northern hemisphere, and provide a unique source of information for the study of ice-sheet dynamics.

Introduction

Over the past few years the writer has made oxygen isotope measurements in over 60 deep-sea sediment cores widely distributed over the oceans. Figure 1 shows the distribution of those cores analysed with the intention of studying the last glacial stage (see also table 1). The oxygen isotope data are given in tables 2–65.‡

Although it is not possible to discuss all the data in a single paper, the results constitute a unique synthesis of the stratigraphy of the Late Quaternary of the ocean floor, and as such deserve to be made available.

There are three aspects of the data that will be discussed here. First, a major objective of oxygen isotope studies is to obtain an accurate record of the changing oxygen isotopic composition of the oceans as a function of time. For this purpose it is evident that some cores are more suitable than others. Second, the recognition of this record (albeit in a modified form) in each core permits that core to be correlated with another or with a synthetic 'ideal' sequence. Third, differences between individual records provide valuable insight into sea-floor processes.

† Contribution No. 2435 of the Lamont–Doherty Geological Observatory.
‡ Tables 2–65 appear in the accompanying microfiche.

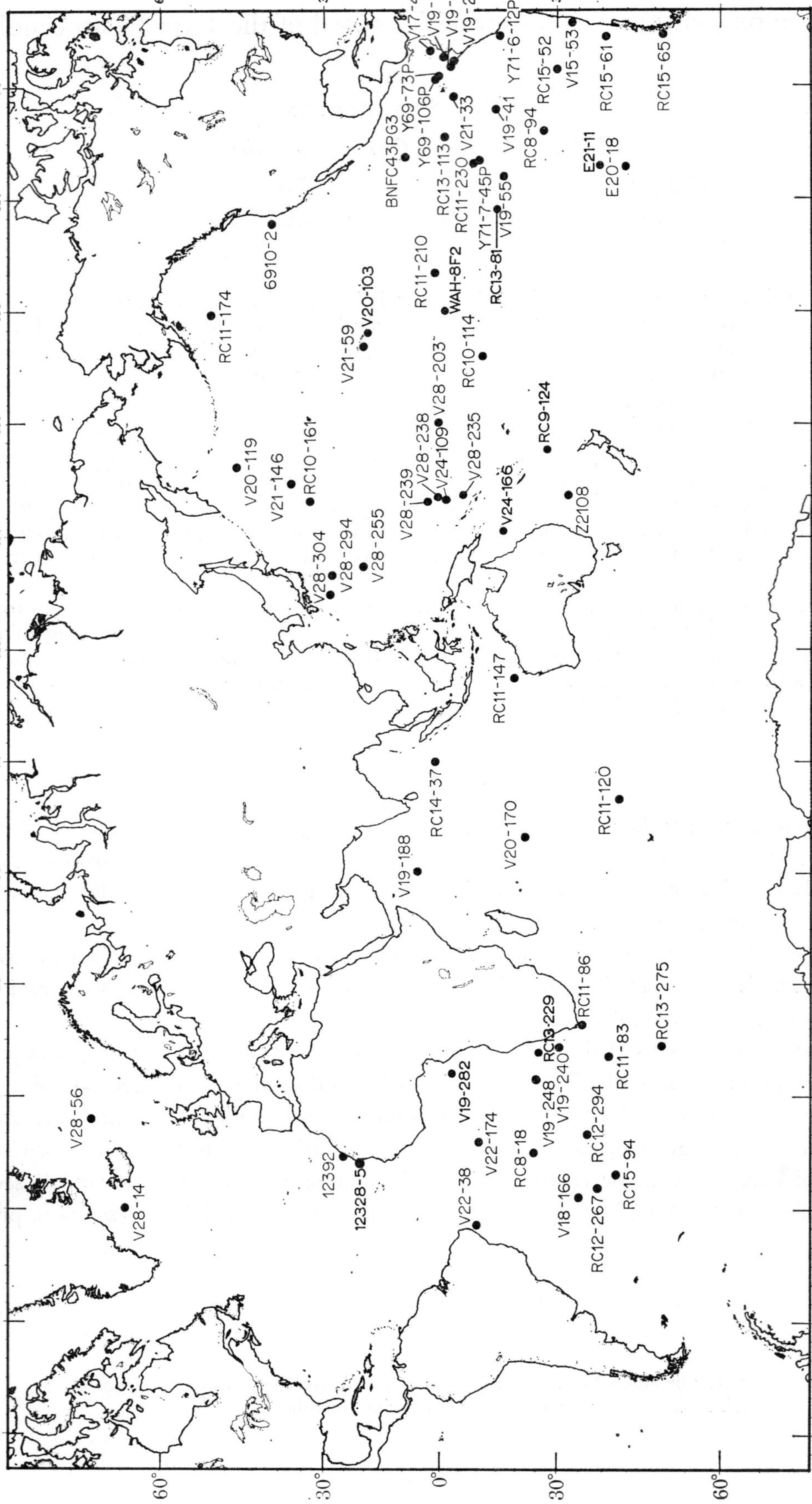

FIGURE 1. Distribution of the cores from which oxygen isotope data are presented (see also table 1).

TABLE 1. LOCATION OF CORES

core	latitude	longitude	depth	table
North Atlantic				
V28-14	64° 47′ N	29° 34′ W	1855	2
V28-56	68° 02′ N	06° 07′ W	2914	3
12328-5	21° 08.7′ N	18° 34.4′ W	2812	4
South Atlantic				
RC8-18	24° 04′ S	15° 07′ W	3977	5
RC11-86	35° 47′ S	18° 27′ E	2829	6
RC12-267	38° 41′ S	25° 47′ W	4144	7
RC12-294	37° 15.6′ S	10° 05.8′ W	3308	8
RC13-229	25° 29.6′ S	11° 18.6′ E	4191	9
V19-240	30° 35′ S	13° 17′ E	3103	10
V19-248	24° 34′ S	04° 50′ E	3321	11
V19-282	02° 45′ S	04° 35′ E	4356	12
V22-38	09° 33′ S	34° 15′ W	3797	13
V22-174	10° 04′ S	12° 49′ W	2630	14
Antarctic				
RC11-83	41° 36.0′ S	09° 43.0′ E	4718	15
RC11-120	43° 31.0′ S	79° 52.0′ E	3913	16
RC13-275	50° 43.0′ S	13° 26.0′ E	1984	17
RC15-94	42° 53.9′ S	20° 51.3′ W	3762	18
Indian				
V19-188	06° 52′ N	60° 40′ E	3356	19
V20-170	21° 48′ S	69° 14′ E	2479	20
RC11-147	19° 03.5′ S	112° 45′ E	1953	21
RC14-37	01° 28′ N	90° 10′ E	2226	22
Pacific				
6910-2	41° 16′ N	126° 24′ W	2743	23
BNFC 43 PG3	10° 29.5′ N	109° 01.7′ W	2874	24
E20-18	44° 33′ S	111° 20′ W	2867	25
E21-11	39° 58′ S	112° 09′ W	2798	26
RC8-94	27° 17′ S	102° 05′ W	3074	27
RC9-124	28° 44.7′ S	172° 35.6′ E	2540	28
RC10-114	11° 11′ S	162° 55′ W	2791	29
RC10-161	33° 05′ N	158° 00′ E	3587	30
RC11-174	52° 34.6′ N	151° 21′ W	1618	31
RC11-210	01° 49′ N	140° 03′ W	4420	32
RC11-230	08° 48′ S	110° 48′ W	3259	33
RC13-81	19° 01.1′ S	124° 13.5′ W	3751	34
RC13-113	01° 39′ S	103° 38′ W	2195	35
RC15-52	29° 14.3′ S	85° 59.1′ W	3952	36
RC15-61	45° 36.6′ S	77° 12.1′ W	3771	37
RC15-65	53° 04′ S	78° 57.1′ W	3200	38
V15-53	32° 27′ S	73° 40′ W	3915	39
V17-42	03° 32′ N	81° 11′ W	1814	40
V19-27	00° 28′ S	82° 04′ W	1373	41
V19-28	10° 22′ S	84° 39′ W	2720	42
V19-29	03° 35′ S	83° 13′ W	3157	43
V19-41	14° 06′ S	96° 12′ W	3248	44
V19-55	17° 00′ S	114° 11′ W	3177	45
V20-103	33° 59′ N	177° 50′ W	3442	46
V20-119	47° 57′ N	168° 47′ E	2736	47
V21-33	03° 48′ S	92° 05′ W	3726	48
V21-59	20° 55′ N	158° 06′ W	2992	49
V21-146	37° 41′ N	163° 02′ E	3968	50
V24-109	00° 26′ N	158° 48′ E	2367	51
V24-166	16° 31′ S	150° 47′ E	781	52
V28-203	00° 57′ N	179° 25′ W	3243	53

TABLE 1 (*cont.*)

V28-235	05° 27′ S	160° 29′ E	1746	54
V28-238	01° 01′ N	160° 29′ E	3120	55
V28-239	03° 15′ N	159° 11′ E	3490	56
V28-255	20° 06′ N	142° 27′ E	3261	57
V28-294	28° 26′ N	139° 58′ E	2308	58
V28-304	28° 32′ N	134° 08′ E	2942	59
WAH-8F2	00° 01.5′ N	147° 59′ W	4308	60
Y69-73P	01° 27′ N	87° 56′ W	2707	61
Y69-106P	02° 59′ N	86° 33′ W	2870	62
Y71-6-12	16° 26.6′ S	77° 33.8′ W	2734	63
Y71-7-45P	11° 04.7′ S	110° 62′ W	3096	64
Z-2108	33° 23′ S	161° 37′ W	1448	65

Notes: Cores are curated in the following institutions from which routine lithological descriptions of the cores are available.

prefix	V, RC	Lamont–Doherty Geological Observatory, Palisades, New York, U.S.A.
(no prefix)	12392, 12328-5	University of Kiel, West Germany
(no prefix) / prefix	6910-2 / Y	Oregon State University, School of Oceanography, Corvallis, Oregon, U.S.A.
prefix	BNFC, WAH	Scripps Institute of Oceanography, La Jolla, California, U.S.A.
prefix	E	Florida State University, Talahassee, Florida, U.S.A.
prefix	Z	New Zealand Oceanographic Institute, Wellington, New Zealand

PREVIOUS WORK

The most important body of Late Pleistocene oxygen isotope data other than that contained here has been obtained by Cesare Emiliani (1955*a*, *b*, 1958, 1964, 1966, 1972). The cores that he analysed cover the Caribbean (several cores), the Equatorial Atlantic, and one core each in the Eastern Mediterranean and the North Atlantic. Additional work has been done by others in the Mediterranean, the Gulf of Mexico and the Red Sea; these areas present special problems because of their relative isolation from the ocean water masses, and will not be discussed further here. Duplessy, Chenouard & Reyss (1974) analysed an Equatorial Atlantic core and Duplessy, Chenouard & Vila (1975) analysed one from the Norwegian Sea, which is another area which presents some unusual problems. In the Indian Ocean, Oba (1969) made oxygen isotope measurements in two cores from south of Ceylon, and more recently M. Sommer (Brown University) and J.-C. Duplessy (Gif-sur-Yvette, France) have analysed several cores as a part of the CLIMAP project.

In addition to the measurements reported here the writer has recently published detailed data for core 12392, illustrated in figure 2 (Shackleton 1977). A few other cores have been analysed but are not included here because work on them is not complete.

A careful programme of inter-laboratory standardization operates between the Cambridge laboratory and that of Dr J.-C. Duplessy at Gif-sur-Yvette, and has been extended to that of Dr M. Sommer at Brown University. This calibration in terms of the P.D.B. reference standard is probably concordant with that used by Emiliani while working at Chicago (Shackleton 1974). All measurements reported here are in terms of that standard using the δ notation:

$$\delta = 10^3\left(\frac{R_{\text{sample}}}{R_{\text{standard}}} - 1\right),$$

where R is ^{18}O–^{16}O ratio (see also Craig 1957).

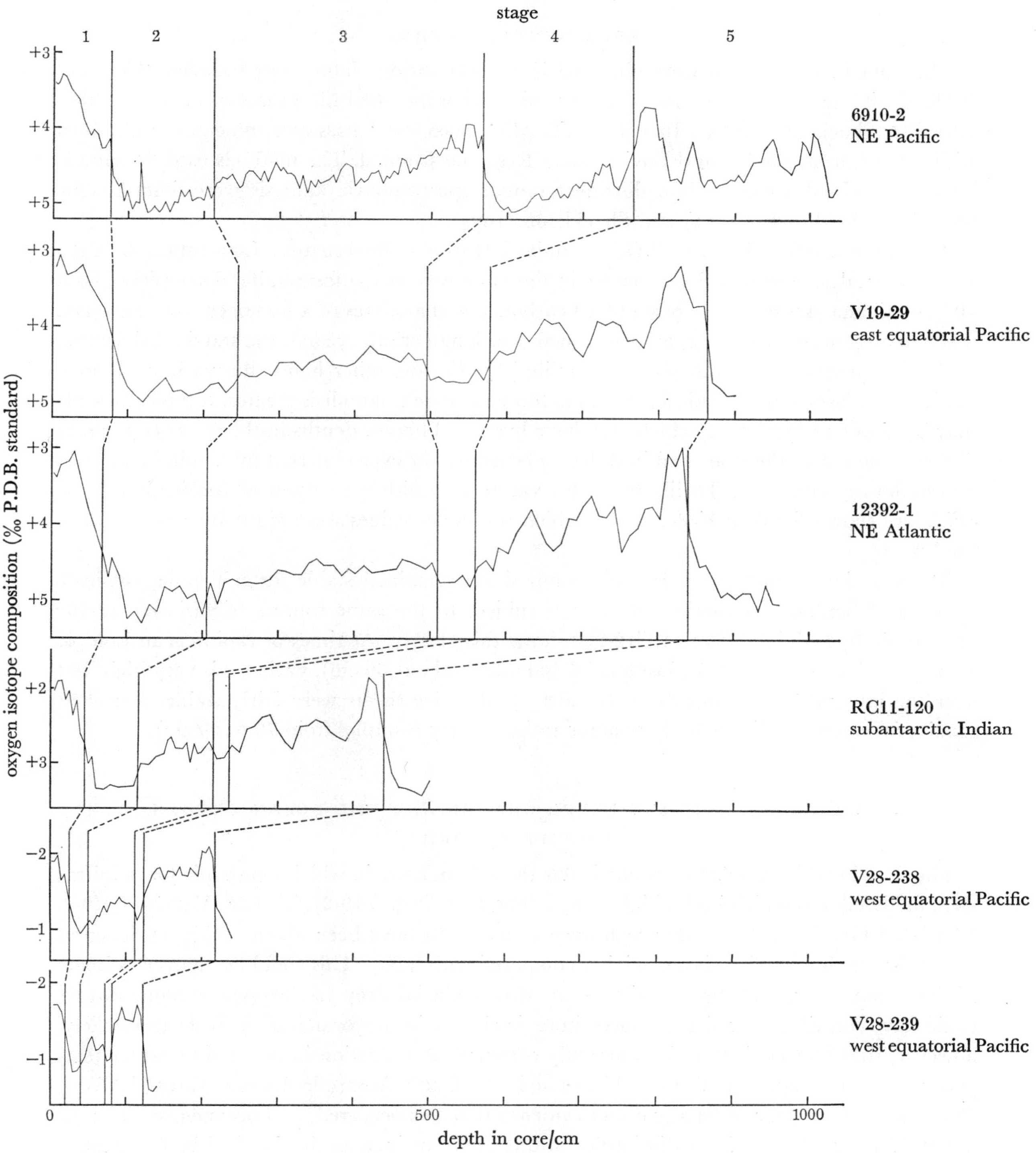

FIGURE 2. Oxygen isotope records in six cores from different areas characterizing sediment accumulation rates from about 10 cm/ka (top) to about 1 cm/ka (bottom). Core 12392-1 was taken at 23° 08.4′ N 17° 44.7′ W in 2066 m water depth; the record illustrated is based on analyses of several species of benthonic foraminifera published by Shackleton (1977). The locations of the remainder are given in table 1, and the data in tables 23, 43, 16, 55 and 56. Stages after Emiliani (1955) and Shackleton & Opdyke (1973).

Experimental methods

The data in tables 2–65 were obtained in the Cambridge laboratory between 1972 and, 1976. Until late 1973 measurements were made with a modified MS3 mass spectrometer, described by Shackleton (1965). In 1973 a VG Micromass 602C mass spectrometer was obtained with a grant from the Natural Environment Research Council. The methods used for sample handling and release of carbon dioxide for mass spectrometric analysis were described by Shackleton & Opdyke (1973) and Shackleton (1974).

Analyses are referred to the P.D.B. standard (Epstein, Buchsbaum, Lowenstam & Urey 1953) by analysing standard carbonates in the same way as the foraminiferal samples. About 10 % of all analyses made are of standard carbonates. If analyses of a homogeneous carbonate taken in aliquots of *ca.* 0.3 mg, are performed over long periods (years), the standard deviation of the measurements is about ± 0.07 per mille (‰). On the other hand, the variance among replicate analyses of planktonic foraminifera from a sediment sample is greater. A typical sample may be about 12 specimens, which may have lived at different depths in the water column, at different seasons of the year, and in different centuries (or even different millennia in a slowly accumulating sediment). To illustrate this variance, multiple analyses of *Globigerinoides sacculifer* are shown for core V28-238 (for which the mean values were given by Shackleton & Opdyke 1973).

At those times when ocean isotopic composition remained stable for millennia, replicate analyses of benthonic foraminifera are not subject to the same sources of variation as the planktonic foraminifera. Core RC12-229 shows this clearly. At times of rapid ocean isotopic change, such as the end of the last glacial (sample at about 50 cm), variance is very high, but over the interval from about 24 to 14 ka B.P., when ice sheets were fairly stable near their maximum extent, replicate analyses agree rather closely (samples from 70 to 90 cm).

Glacio-eustatic sea-level change and the ocean oxygen isotope record

The data in tables 2–65, and in particular those from cores in which benthonic foraminifera were analysed and which had a high rate of deposition (e.g. V19-27, V19-28, V19-29, 6910-2, 12392-1, RC12-267), show that ocean deep waters must have been about 1.65‰ enriched in ^{18}O at the maximum of the last glacial as compared with today. This could be an over-estimate of the ocean isotopic change, to the extent that a glacial drop in deep-water temperatures could have contributed to the measured isotopic change in the foraminifera. It seems unlikely, however, that any significant drop actually occurred. A reduction in ocean deep-water temperatures would require a substantial increase in the flux of Antarctic Bottom Water (AABW). There does not appear to be any direct evidence that this occurred, and oceanographic arguments (Weyl 1968) predict a diminution rather than an increase in the AABW flux during glacial periods.

An average oxygen isotopic composition of -30‰ for the excess ice stored during glacials has been estimated by Dansgaard & Tauber (1969) and by Shackleton (1967). Taking this figure, we can calculate that the ocean would have been enriched in ^{18}O by about 0.1‰ for every 10 m drop in sea level (neglecting isostatic effects). Hence, an ^{18}O enrichment of 1.65‰ would represent an ice excess equivalent to a sea level lowering of about 165 m. Although this figure is somewhat greater than most current estimates (for example, those of Donn, Farrand & Ewing 1962), recent work by members of the CLIMAP project working at the University of Maine suggests that the larger estimate is probably correct (Denton 1976, personal communication). If so, this greatly strengthens the case for regarding the oxygen isotope record as an ideal monitor of glacio-eustatic sea-level change, subject only to the reservations discussed below concerning ocean mixing times and bioturbation.

Discussion: the 'ideal' record of ocean isotopic change

In figure 2 six oxygen isotope records are illustrated with the same depth and isotopic difference scales. These clearly illustrate the extent to which the details of the record become lost as a result of bioturbation in slowly accumulating sediments. A comparison of the records in core V19-29 (Ninkovich & Shackleton 1975) from the south side of the Panama Basin, and Meteor core 12392-1 (Shackleton 1977) from off the coast of Morocco, shows a high degree of similarity in two areas of rather rapid sedimentation (average about 7 cm/ka). The similarity is enhanced by the fact that in both areas the high sedimentation rate is the result of wind-driven coastal upwelling; more intense glacial winds have given rise to a higher accumulation rate in glacial than in interglacial times in both areas.

Core 6910-2 (Oregon State University; discussed by Heusser, Shackleton, Moore & Balsam 1975) shows an even higher accumulation rate and perhaps less variability in rate. The isotopic peak at about 550 cm is believed to represent the base of stage 4, equivalent to about 470 cm in core V19-29. If this is correct, its age would be about 60 ka; it probably represents the deglacial event marked by the stand of sea level dated in Barbados (James, Mountjoy & Omura 1971), Japan (Konishi, Omura & Nakamichi 1974) and New Guinea (Bloom *et al.* 1974), at which time the sea is estimated to have reached within 28 m of its present level (Bloom *et al.* 1974). If the isotopic event at 550 cm in core 6910-2 is correctly identified, its duration may be estimated as about 1 ka on the basis of a uniform sedimentation rate. Since the interglacial accumulation rate for core V19-29 is about 5 cm/ka, an event as short as this would probably not be recorded in that sediment.

Bloom *et al.* (1974) have inferred sea level stands at −38 m at a date near 40 ka ago, and at −41 m about 28 ka ago. These events do not appear to be recorded in core 6910-2 and may have been of even shorter duration than the 60 ka event. If so, this is unfortunate, in that the mixing time of the ocean may be as much as 1 ka (Gordon 1975), so that events which lasted only a few centuries will have given rise to transient isotopic differences between water masses and will not have been recorded uniformly even in the most detailed sedimentary records.

Kennett & Shackleton (1975) have already documented large transient isotopic changes in the sediments of the Gulf of Mexico, associated with the flow of glacial meltwater into that sea at a fast rate compared with the rate of ocean mixing. Thus the interpretation of a record of ocean isotopic composition with a resolution below about 1 ka will require a careful consideration of ocean mixing, particularly since the ocean mixing rate will itself have been affected by the transient changes in North Atlantic surface salinity associated with glacial advance and retreat.

Discussion: stratigraphic correlation

As was discussed by Shackleton & Opdyke (1973, 1976), the oxygen isotope record in deep-sea sediments provides a tool for stratigraphic correlation that is generally limited only by the stratigraphic resolution of the sediments themselves. The reconstruction of the distribution of ocean surface temperatures at 18 ka B.P. undertaken by the CLIMAP project (CLIMAP 1976) used the oxygen isotope record as a primary means of long-distance and inter-oceanic correlation. Within each geographically limited area many other stratigraphic tools were available. The cores discussed in this paper form the basis for this stratigraphic framework.

In areas with a high rate of sediment accumulation, several samples, taken from the interval representing *ca.* 24–14 ka B.P., typically yield oxygen isotope values that are statistically indistinguishable (for example, from 100 to 200 cm in core V19-29, figure 2), because over this interval the major ice sheets apparently remained relatively stable. In this case the mean sedimentation rate must be used to judge the sample most likely to represent 18 ka B.P. In core V19-29 the 150 cm sample probably represents a real, though brief, interstadial event, possibly the Erie Interstade (Mörner & Dreimanis 1973).

In areas of lower accumulation rate the most isotopically positive value may or may not be statistically different from its neighbours; in either case it is the sample likely to contain the highest proportion of glacial-maximum material. This statement recognizes the fact that if it were possible to separate from that sample only the specimens that lived (for example) 18 ka ago, they would be found to be isotopically heavier than the intermingled specimens that may be several thousand years older or younger. Thus the selection of the 29 cm level in core V28-239 (figure 2) as representing 18 ka attempts a best choice in an area in which any sediment sample must represent the imperfectly homogenized record of events covering several thousand years.

A few of the cores analysed (e.g. V28-294, V15-94, RC11-174) yield isotopic values for the top of the core, which indicate it is of glacial age rather than Holocene. Others (e.g. V20-103, V24-166) yielded an isotopic record that was not recognized in terms of the pattern that emerges from the majority of cores. This may be a function of the inefficiency of a burrowing animal as a homogenizer of sediment (i.e. an isolated, visible burrow in the sediment may contain material of quite different age and character from its surroundings), or it may be an indication that an unconformity is present. In the latter case other biostratigraphic tools are available for assessing the approximate age of the sediment. Thus oxygen isotope stratigraphy is a valuable means of checking that a core does in fact record events of the Late Pleistocene.

Discussion: sedimentation processes

There is considerable interest in the study of the process of bioturbation as it affects the record in deep-sea sediment (Berger & Heath 1968; Guinasso & Schink 1975). In the past the effect has been studied in detail mainly by examining the vertical distribution in the sediment of material deposited over a geologically brief interval (for example, a volcanic ash horizon). Another approach is to study the decay profile of radiocarbon or other radioisotopes (Peng, Broecker, Kipphut & Shackleton 1977). A simple approach, using the oxygen isotope data, is to examine the extent to which the isotopic change between the most recent sediment and that representing the last glacial maximum has been reduced by mixing. As discussed above, the ocean water masses probably underwent an isotopic change of about 1.65‰, yet the range observed in sediments accumulated at about 1 cm/ka is only a little over half the figure (e.g. core V28-239 in figure 2). This approach was used by Peng *et al.* (1977).

A second approach to the sediment mixing problem using oxygen isotope data is to examine the range of isotopic composition among individual benthonic foraminiferal tests in a single sediment sample, which (as discussed above) is a function of the amount of time averaged in that piece of sediment.

Since the oxygen isotope records may be expressed as a set of continuous records representing isotopic composition as a function of time, a third approach is to compare these records either at selected points (Shackleton & Opdyke 1976, table 3) or statistically, using all the data. The oxygen isotope records of core RC11-120 (illustrated in figure 2) and of another core nearby, both show spectral energy at a frequency corresponding to a period of about 20 ka (Hays, Imbrie & Shackleton 1976), while that of core V28-238 (also illustrated in figure 2) shows insignificant energy at this frequency (unpublished study by J. Imbrie). This suggests that in core V28-238 (accumulation rate 1.7 cm/ka) the signal with frequency 1 cycle per 20 ka (i.e. wavelength in the sediment 34 cm) has been attenuated by bioturbation to within the stratigraphic noise level. This is illustrated by the fact that the five substages in stage 5 cannot be seen in this latter core (figure 2).

Conclusions

In view of the fact that sediment accumulation rates below 2 cm/ka characterize a great proportion of the ocean area, the oxygen isotope record in most sediment cores is not an accurate representation of the changes in oxygen isotopic composition in the ocean water masses. For the same reason, the temporal resolution that may be obtained by the use of the oxygen isotope record as a tool for stratigraphic correlation is limited, and is probably inversely proportional to the accumulation rate. However, several regions are already known in which accumulation rates are sufficiently high that it is apparently possible to obtain oxygen isotope records whose time resolution is less than the mixing time for ocean water masses. In these areas there is the potential for the extraction of a great deal more information than has been obtained to date. Close attention to analytical accuracy, to optimal sampling frequency and to inter-core comparisons both over small and inter-oceanic distances, will provide unique insights on two important processes. The first is the dynamic behaviour of large and perhaps unstable ice sheets. The second is the mixing of the oceans under the flux boundary conditions associated with the removal of water by evaporation from, and the addition of meltwater to, the ocean surface layer. The importance of calibration accuracy must be particularly stressed. From the analytical point of view it may become possible to measure horizontal and vertical isotopic gradients in the deep waters through much of Late Pleistocene time with an accuracy of ±0.03‰ or less, equivalent to a palaeotemperature accuracy of about 0.1 °C; it is not at present possible to approach this accuracy in inter-laboratory comparison.

Oxygen isotope analysis has been supported by N.E.R.C. under grants GR3/768 and GR3/1762. Mr M. A. Hall has been in charge of the operation of the mass spectrometer throughout the period during which the measurements were made. The great majority of the cores discussed were taken and curated by the Lamont–Doherty Geological Observatory under grants

DES 72-01568 and ONR N00014-75-c-0210. The possibility of making this comprehensive survey of oxygen isotope variations in sediments of the world's oceans would not have arisen without the chance to work with the members of the CLIMAP project, which has been supported under the International Decade of Ocean Exploration program by the National Science Foundation of the U.S.A. through grants ID071-04204, GX28672, GX28673 and GX39773.

I am deeply indebted to a large number of my friends and colleagues in CLIMAP for the free exchange of exciting ideas which this project has engendered, and in particular I would like to thank T. C. Moore and J. D. Hays for giving me such a geographically wide perspective on palaeo-oceanographic events in the Late Pleistocene, through their responsibility for the Pacific and Antarctic interests in CLIMAP. This work would not have been possible, and certainly would not have been undertaken in Britain, but for the support and encouragement of Sir Harry Godwin, F.R.S., and later of Richard West, F.R.S., and the Sub-department of Quaternary Research in the University of Cambridge.

I am especially grateful for the opportunity to work closely with my CLIMAP colleagues, and to have immediate access to the enormous library of deep-sea sediment cores at the Lamont–Doherty Geological Observatory, which was provided by the receipt of the Senior Visiting Research Fellowship at that institution in 1974–5.

References (Shackleton)

Berger, W. H. & Heath, G. R. 1968 Vertical mixing in pelagic sediments. *J. Mar. Res.* **26**, 134–143.

Bloom, A. L., Broecker, W. S., Chappell, J. M. A., Matthews, R. K. & Mesolella, K. J. 1974 Quaternary sea level fluctuations on a tectonic coast: new $^{230}Th/^{234}U$ dates from the Huon Peninsula, New Guinea. *Quat. Res.* **4**, 185–205.

CLIMAP 1976 The surface of the ice-age earth. *Science, N.Y.* **191**, 1131–1137.

Craig, H. 1957 Isotopic standards for carbon and oxygen and correction factors for mass-spectrometric analysis of carbon dioxide. *Geochim. Cosmochim. Acta* **12**, 133–149.

Dansgaard, W. & Tauber, H. 1969 Glacier oxygen-18 content and Pleistocene ocean temperatures. *Science, N.Y.* **166**, 499–502.

Donn, W. L., Farrand, W. R. & Ewing, M. 1962 Pleistocene ice volumes and sea-level lowering. *J. Geol.* **70**, 206–214.

Duplessy, J. C., Chenouard, L. & Reyss, J. L. 1974 Paléotempératures isotopiques de l'Atlantique Equatorial. *Colloques Internationaux du Centre National de la Recherche Scientifique* **219**, 251–258.

Duplessy, J. C., Chenouard, L. & Vila, F. 1975 Weyl's theory of glaciation supported by isotopic study of Norwegian core K11. *Science, N.Y.* **188**, 1208–1209.

Emiliani, C. 1955*a* Pleistocene temperatures. *J. Geol.* **63**, 538–578.

Emiliani, C. 1955*b* Pleistocene temperature variations in the Mediterranean. *Quaternaria* **2**, 87–98.

Emiliani, C. 1958 Palaeotemperature analysis of core 280 and Pleistocene correlations. *Geol.* **66**, 264–275.

Emiliani, C. 1964 Palaeotemperature analysis of the Caribbean cores A 254-BR-C and CP-28. *Geol. Soc. Am. Bull.* **75**, 129–144.

Emiliani, C. 1966 Palaeotemperature analysis of Caribbean cores P 6304-8 and P 6304-9 and a generalised temperature curve for the last 425,000 years. *J. Geol.* **74**, 109–126.

Emiliani, C. 1972 Quaternary paleotemperatures and the duration of the high-temperature intervals. *Science, N.Y.* **178**, 398–401.

Epstein, S., Buchsbaum, R., Lowenstam, H. A. & Urey, H. C. 1953 Revised carbonate–water isotopic temperature scale. *Geol. Soc. Am. Bull.* **64**, 1315–1326.

Gordon, A. L. 1975 General ocean circulation. In *Numerical models of ocean circulation*, pp. 39–53. Washington, D.C.: National Academy of Science.

Guinasso, N. L. Jr. & Schink, D. R. 1975 Quantitative estimates of biological mixing rates in abyssal sediments. *J. geophys. Res.* **80**, 3032–3043.

Hays, J. D., Imbrie, J. & Shackleton, N. J. 1976 Variations in the earth's orbit: pacemaker of the ice ages. *Science, N.Y.* **194**, 1121–1132.

Heusser, L. E., Shackleton, N. J., Moore, T. C. & Balsam, W. L. 1975 Land and marine records in the Pacific Northwest during the last glacial interval. *Geol. Soc. Am. Abstracts*, 1975 Annual Meetings.

James, N. P., Mountjoy, E. W. & Omura, A. 1971 An early Wisconsin reef terrace at Barbados, West Indies, and its climatic implications. *Geol. Soc. Am. Bull.* **82**, 2011–2018.

Kennett, J. & Shackleton, N. J. 1975 Laurentide ice sheet meltwater recorded in Gulf of Mexico deep sea cores. *Science, N.Y.* **188**, 147–150.

Konishi, K., Omura, A. & Nakamichi, O. 1974 Radiometric coral ages and sea level records from the Late Quaternary reef complexes of the Ryukyu Islands. *Proc. Second Int. Coral Reef Symp.* **2**, 595–613.

Mörner, N.-A. & Dreimanis, A. 1973 The Erie Interstade. *Geol. Soc. Am. Mem.* **136**, 107–134.

Ninkovich, D. & Shackleton, N. J. 1975 Distribution, stratigraphic position and age of ash layer 'L', in the Panama Basin region. *Earth Planet. Sci. Lett.* **27**, 20–34.

Oba, T. 1969 Biostratigraphy and isotopic paleotemperature of some deep-sea cores from the Indian Ocean. *Sci. Reps. Tohoku Univ. (Geol.)* **41**, 129–195.

Peng, T.-H., Broecker, W. S., Kipphut, G. & Shackleton, N. J. 1977 The relation of sediment mixing to the distortion of climatic records in the deep sea sediments. In *The fate of fossil fuel* CO_2 *in the oceans* (eds N. R. Andersen & A. Malahoff). New York: Plenum.

Shackleton, N. J. 1965 The high-precision isotopic analysis of oxygen and carbon in carbon dioxide. *J. scient. Instrum.* **42**, 689–692.

Shackleton, N. J. 1967 *The measurement of palaeotemperatures in the Quaternary era*. Ph.D. Thesis, University of Cambridge.

Shackleton, N. J. 1974 Attainment of isotopic equilibrium between ocean water and the benthonic foraminifera genus *Uvigerina*: isotopic changes in the ocean during the last glacial. *Colloques Int. Centre National de la Recherche Scient.* **219**, 203–210.

Shackleton, N. J. 1977 Carbon-13 in *Uvigerina*: tropical rainforest history and the Equatorial Pacific carbonate dissolution cycles. In *The fate of fossil fuel* CO_2 *in the oceans* (eds N. R. Andersen & A. Malahoff). New York: Plenum.

Shackleton, N. J. & Opdyke, N. D. 1973 Oxygen isotope and palaeomagnetic stratigraphy of Equatorial Pacific core V28–238: oxygen isotope temperatures and ice volumes on a 10^5 year and 10^6 year scale. *Quat. Res.* **3**, 39–55.

Shackleton, N. J. & Opdyke, N. D. 1976 Oxygen isotope and palaeomagnetic stratigraphy of Equatorial Pacific core V28-239, Late Pliocene to Latest Pleistocene. In *Investigation of Late Quaternary paleoceanography and paleoclimatology* (eds R. M. Cline & J. D. Hays), pp. 449–464. *Geol. Soc. Am. Mem.* **145**.

Weyl, P. K. 1968 The role of the oceans in climatic change: a theory of the Ice Ages. *Meterol. Monogr.* **8**, 37–62.

Discussion

H. H. LAMB (*Climatic Research Unit, University of East Anglia, Norwich NR4 7TJ*). Can one yet compare the land and ocean oxygen isotope records and say that the times when the ice sheets were forming or growing most rapidly were marked by much less depression of temperature than later developed when the ice cover was most extensive?

N. J. SHACKLETON. At present, we know very little about sea surface temperatures at the time of inception, or of most rapid growth, of the Northern hemisphere ice sheets.

B. C. WORSSAM (*Institute of Geological Sciences, 5 Princes Gate, London SW7 1QN*). The succession of warm and cold peaks shown by isotope stage 5 seemed to have a bearing on the question of the definition of the base of the Devensian. Are substages 5b and 5d likely to have been of such intensity and duration as to have allowed ice sheets to invade the British Isles?

J. M. HODGSON (*Soil Survey of England and Wales, Rothamsted Experimental Station, Harpenden, Herts*). The marked climatic fluctuations in stage 5 shown by Dr Shackleton prompt me to ask if there is any evidence from his work or from that of Professor McIntyre suggesting that there could be sea ice in the English Channel at this period.

N. J. SHACKLETON. So long as the base of the Devensian Stage remains defined by a till, I believe that it is unlikely that it should be correlated with any part of stage 5. However, climate during substages 5b and 5d was certainly of glacial character; at these times the size of the

Laurentide ice sheet was probably sufficient to have lowered sea level to a point where the English Channel was no longer marine.

A. R. LORD (*Department of Geology, University College London, Gower Street, WC1E 6BT*). The factors affecting the precision of ^{14}C dating have been discussed earlier today by Professor Shotton for material of generally terrestrial origin. Would Dr Shackleton assess the relative reliability of ^{14}C dates based on material from deep-sea cores, for example in relation to diagenetic change in the foraminiferid tests used, as this is clearly crucial when comparing continental events with those in the ocean basins.

F. W. SHOTTON, F.R.S. (*Dept. of Geological Sciences, University of Birmingham, Birmingham B15 2TT*)) In core V19-29, which has rather more rapid sedimentation rates than most, Dr Shackleton shows three well-marked peaks in stage 5 and suggests that the earliest and warmest, 5e, might mark the Ipswichian. If that is so, the subsequent record accords excellently with the principal interstadials of the Devensian, putting 5c as Amersfoort, 5a as Chelford and the steep change from stage 4 to 3 being the very rapid amelioration which ushered in the Upton Warren Interstadial. An age of about 125000 years had been suggested for the centre of the Ipswichian (Eemian). If this were so, one would have to think of dates of the order of 105–110000 for 5c, 80000 for 5a and 65000 for the 4/3 junction, and these figures disagreed seriously with the radiocarbon dates for the land-based interstadials. While the 'enriched' ^{14}C dates for Amersfoort (68000) and Brørup/Chelford (64000/60000) might still have to be regarded as only minima, there was no reason to question the validity of the several Upton Warren Interstadial dates around 43000.

What methods have been used to put dates on the ocean cores and do they have a degree of uncertainty sufficient to reconcile them with ^{14}C dates as discrepant as those cited above?

N. J. SHACKLETON. The accuracy of ^{14}C age determinations in marine sediments from the end of the last glacial is limited by bioturbation and related processes, rather than by diagenesis, while for older samples laboratory contamination is a more serious hazard than geological diagenesis. Fortunately, the chronology available is based on several other techniques as well as on ^{14}C. It is extremely unlikely that these techniques could all be so far in error that the true age of the stage 4–3 boundary could be as young as 43000 B.P.

D. J. SCHOVE (*St David's College, Beckenham, Kent*)

Temperature turning points

Temperature curves show characteristic turning points which need to be fixed independently in ^{14}C, lake-core, ice-core and deep-sea chronology.

The writer would be interested to know how far the climatic turning points can be identified in ice-cores and deep-sea cores. It is clear from Dr Shackleton's paper that the ^{14}C chronology is proving too short when compared with K-Ar dating, and it would be helpful to know how the magnitude of the discrepancy varies. If the cold period hitherto dated 53000 B.P. is really 70000 B.P. should we stretch the ^{14}C scale in the Devensian not merely by the 3 % needed for the right half-life but by some 30 % throughout?

Chronology of the Bölling (Zone Ib)

Minor temperature fluctuations, like the major changes demonstrated by Professor McIntyre and Dr Shackleton, are in the same sense on both sides of the Atlantic. This has been true in the last 200 years.

Varve counts on both sides of the Atlantic have thus been teleconnected using cross-correlation of non-random features. Curves of glacial recession, derived separately for the U.S.A. and for the Baltic and based on the isochrones of bottom varves, show parallel trends when superimposed at the positions proposed (see D. J. Schove (1971), Varve teleconnection across the Baltic, *Geografiska Annaler, Stockholm* **53**, 214–234).

In the Bölling (Pollen Zone Ib in northern Europe) for instance the abnormal warmth implied by Dr Coope's beetles is matched not only by a phase of rapid recession in southern Sweden but also by a synchronous recession in the U.S.A. In Vermont the rate of 335 m/year was described by Antevs (cf. his varves numbered 6950/7125) as 'the fastest rate observed in New England (cf. D. J. Schove (1969), A varve teleconnection project, *VIIIe Congres INQUA, Paris*, p. 932). In Europe the Alleröd recession of Zone II was slower if more prolonged; in the U.S.A. there is again a synchronous slower but more prolonged recession corresponding to varves 7280/8050.

The warmth of Bölling summers in the British Isles has been stressed at this conference. We should, nevertheless, distinguish between the Bölling in the new broad sense of 13000/12000 B.P. and in the narrow sense of the hot phase which lasted less than two centuries (12400/12200 B.P.) and was, surprisingly, punctuated in S. Sweden by a brief glacial readvance! In pollen diagrams the Bölling has often been included wrongly with the Alleröd in an extended but too broad sense of Zone II.

From the analysis of varves in SE Sweden an important result was announced by Tarling & Noel (in progress) at the Meeting of European Geological Societies in September 1975 that 'the Laschamp geomagnetic event can be dated as lasting 200 years with a reversal lasting for 30 years', that is between 12077 and 12103 varve-dates B.P. On the Nilsson scale this makes the reversal 10127/10153 N which I date 12400 B.P. $\pm$ 100 (cf. 1971, *Geografiska Annaler*, **53**, 230, Table 5). This geomagnetic event would have hindered modulation of the cosmic ray flux and this would

(*a*) have increased the ^{14}C content in the atmosphere, so making some radiocarbon dates that ought to relate to the mid-Bölling appear to be much too young;

(*b*) be partly responsible for the abnormal warmth of the Late Bölling.

L. H. N. Cooper (2 *Queens Gate Villas, Lipson, Plymouth*). There are two distinct oceanic boundaries of interest during the advance and retreat of Quaternary ice. One is the southern limit of polar water becoming precisely defined by the work of the CLIMAP group and reaching as far south as 43° N on the Iberian shores. This is confirmed by evidence on British Admiralty charts. An oceanic area of seasonal ice separates this polar front from the second boundary, i.e. the edge of the permanent ice. North of this permanent ice edge the parameters which control climate cease to be maritime, equable and moist, but tend towards continentality with wider diurnal and seasonal ranges of temperature, lower absolute humidity and precipitation largely limited to dry snow which may readily drift.

The evidence available for defining this limit of permanent ice is much like that drawn from facial features which enable us to identify people but is hard to put down in words. The extreme southern limit seems to be set along the western and southern edges of the Celtic Sea, retaining permanent ice on the bed of the English Channel and Celtic Sea down to 48° N lat. but continuing as a belt of shelf ice down the Biscayan coast of France. Kellaway and his associates (1975) seem to be alone in sharing this view.

The descriptive terms applied to present-day air masses break down during glacial maxima unless arbitrarily the permanent oceanic ice is treated without reservation as 'continental'. Even so redefinition is needed. Air masses of continental polar and continental arctic origin approached from directions which are now considered as sources of maritime polar air.

Though today maritime tropical and maritime polar air masses and the occluded weather systems to which they may give rise contribute most of our precipitation, this is mostly as rain. Such snowfall as they bring, though sometimes heavy, is usually 'wet' and 'warm'. Severe weather in the British Isles arises from continental polar and continental arctic air streams approaching from between southeast and northeast. The snow deposited as small crystals is 'dry' and 'cold' and readily drifts. The moisture to be precipitated here as snow has largely been acquired by very cold and dry Siberian air masses during passage over the present-day open waters of the North Sea. Air masses approaching over an ice filled North Sea would have arrived much colder and with less moisture and have been able to deposit less snow. Climate would have been sunny and arid.

When conditions at the end of an interglacial deteriorated, oceanic permanent ice would have crossed the Greenland–Scotland Ridge into the North Atlantic, properly so called. On my view at its maximum the permanent ice edge reached to 48° N lat., though there may have been an embayment in the ice edge west of the Celtic Sea and over the deeper water of the Porcupine Forebight. At this ice edge westerly winds would have tended to produce upwelling not of cooler water as is commonplace today but of warmer more saline water. Under such conditions a (pseudo-) continental air mass from the ice-bound northwestern Atlantic would have blown over the Porcupine Forebight to have its moisture content increased as over the North Sea today. A decision as to whether and when this embayment in the permanent oceanic ice occurred is of importance for understanding Quaternary climates in the British Isles.

Thus as the permanent oceanic ice moved south there would have been suppression, probably complete, of maritime tropical air masses and then replacement of maritime polar by continental polar and continental arctic air masses. The greatest volume of snowfall was likely when maritime polar air from the southwest was frequently occluded over continental polar air from west or northwest. But as the ice front advanced southwards prevalence of much continental arctic air should have led to long periods of very cold sunny weather with infrequent falls of 'cold' and 'dry' drifting snow. This leads to difficulties in defining the terms 'periglacial' and 'glacial'.

Reference

Kellaway, G. A., Redding, J. H., Shephard-Thorn & Destombes, J.-P. 1975 The Quaternary history of the English Channel. *Phil. Trans. R. Soc. Lond.* A **279**, 189–218.

Phil. Trans. R. Soc. Lond. B. **280**, 183–198 (1977) [183]
Printed in Great Britain

THE PERIGLACIAL ENVIRONMENT

The periglacial environment of Great Britain during the Devensian

By E. Watson

Department of Geography, University College of Wales, Aberystwyth, Dyfed SY23 3DB

The main question considered is the extent to which extra-glacial Britain had a permafrost environment in the Devensian, so the three features used in identifying contemporary permafrost are discussed. These are ice wedge polygons, pingos and thaw lakes. As their environments have been studied estimates of temperatures may be made when they are found in a Devensian context. A theoretical case can be made for regarding some involutions as representing the former active layer of permafrost but they have not been related to a contemporary environment. They cannot therefore give a measurement of Devensian conditions.

Fossil ice wedges are very widespread in Great Britain, being found from the English Channel to the Scottish Highlands, and associated with late Early, Middle and Late Devensian deposits. Today ice wedges grow in continuous permafrost where the mean annual air temperature is −6 to −8 °C or lower. Remains of pingos of the open system type which today develop in discontinuous permafrost with a mean annual air temperature of −3 to −6 °C, are found in Wales and East Anglia. These continued to form until the end of the Younger Dryas, and indicate that discontinuous permafrost in England and Wales thawed completely only with the beginning of the Flandrian. The pingos must have begun to form after the transition from continuous to discontinuous permafrost, which probably took place with the rise in temperatures ahead of the Allerød.

The area of continuous permafrost and ice wedge formation in Alaska is almost limited to the tundra. Apart from the coastal zone, discontinuous permafrost carries boreal forest with mean July air temperatures of around 15 °C, so that periods of tree growth in the Devensian do not necessarily imply the temporary disappearance of permafrost.

Continuous permafrost in southern England implies a minimum fall in mean annual air temperatures of 16–17 °C, and a fall in the July mean of 4–5 °C. During the maximum ice advance the fall in mean annual air temperatures is estimated to be as much as 25 °C, with a fall in the July mean of 10 °C.

1. Introduction

The term periglacial was introduced to describe phenomena produced by a severe frost climate in Poland south of the Baltic Ice Cap during the last glaciation, and it has often been used in western and central Europe in this sense of being around or marginal to an ice sheet. Investigation of frozen ground phenomena on a world scale has shown that some of the more widespread and obvious categories such as patterned ground and solifluxion deposits are formed whether the frozen ground cycle is diurnal, as in the high mountain areas of the arid tropics, or seasonal, as in mountain areas of the temperate zone. In neither are they formed in a surface layer which is underlain by perennially frozen ground. In the permafrost regions of high latitudes, the freezing of the surface layer is seasonal, but it takes place over perennially frozen ground. Some authorities use periglacial to include all three environments (cf. Troll 1958; Cailleux & Taylor

1954; Tricart 1970), so that although the original 'periglacial' area in Poland proves to have been part of a permafrost zone surrounding the Baltic Ice Cap, *periglacial* cannot be equated with *permafrost*, as Péwé reluctantly admits (1969, pp. 2–4). Since many periglacial phenomena occur in a number of environments it is often difficult if not impossible to draw precise conclusions from their presence. The aim of the author is, therefore, to concentrate on permafrost phenomena, and to consider how far Great Britain beyond the limits of glacier ice had a permafrost environment during the Devensian Cold Stage.

2. Indicators of contemporary permafrost

Contemporary permafrost is usually divided into two zones, that of continuous permafrost and that of discontinuous permafrost. In the former which is found in the more polar regions, high ground and low are perennially frozen below a shallow surface zone, the active layer, the only permafrost-free areas being below large rivers and lakes. In the discontinuous permafrost zone of the subpolar regions the permafrost is increasingly interrupted towards its outer limit, by areas subject only to surface seasonal freezing. These breaks are related to local factors such as aspect and permeability. South-facing slopes become permafrost-free while north-facing slopes retain it; coarse deposits and bedrock thaw out while fine sediments remain frozen. Formerly North American workers listed a zone of sporadic permafrost, where only isolated pockets of permafrost occur, but this is now regarded as the outer fringe of the discontinuous zone (Brown & Péwé 1973, p. 72).

In the mapping of contemporary permafrost three reliable indicators have generally been accepted, ice wedge polygons, pingos and thaw lakes. Though these have surface expression, they form within the permafrost. Other surface features such as solifluxion terraces, sheets and lobes, sorted patterned ground such as stone polygons, stone circles and stone stripes which form in the active layer, also occur in non-permafrost areas and can not be used as independent evidence (Hopkins, Karlstrom *et al.* 1955).

(*a*) *Ice wedge polygons*

Aerial photography has shown that large areas of the Arctic coastlands and islands are covered by a net of polygonal markings, which usually reflect surface differences in vegetation and microrelief. They are especially clear in the tundra, from which they are sometimes called tundra polygons, but they also occur on bare polar desert as in Antarctica and Greenland, and on sloping as well as on flat land. Where investigated, these polygonal nets have been found to overlie a lattice of vertical ice sheets, each of which tapers downwards so that it is wedge-shaped in vertical section. In the more arid areas, the wedges may be partly filled by ice and partly by blown sand or entirely filled by sand, apparently depending on the degree of aridity (Black 1973, p. 193).

An ice wedge consists of an accretion of thin vertical ice layers of variable length due to the freezing of water which seeps down a thermal contraction crack each spring. As the ice wedge widens, the expansion of the permafrost in summer increasingly deforms the sediments along its margins. When the ice subsequently melts, it is replaced by material falling in from above and from the sides. In well drained but cohesive material such as angular gravels, the infill comes largely if not entirely from above and a pseudomorph of the ice wedge is formed. The marginal deformation is preserved, though the infill differs to some degree from the material on either side. A fairly accurate measurement of the original wedge can be made. In loose materials such as clean river gravels, collapse from the sides tends to destroy the marginal deformation; if bedded, the bedding appears to sag into the wedge and only approximate measurements of the width of the wedge may be made. Modern ice wedges may be forked, bent or hooked in profile and a study of the structures underlying a typical polygonal net in outwash gravels in central Alaska shows that the form of the sedimentary replacement may differ greatly from the idealized geometrical wedge form (Péwé, Church & Andresen 1969). Because they are part of a polygonal net, fossil ice wedges should occur in groups, though in a small exposure only one may be seen. Larger exposures may show several, usually irregularly spaced, as the polygons are irregular (cf. Morgan 1973, p. 286), and it may be possible to show that the wedges are linear in plan or that they form a net (cf. Worsley 1966*a*).

From a study of the distribution of contemporary ice wedges in Alaska, Péwé (1966) concluded that they are growing in the continuous permafrost. 'In the discontinuous-permafrost zone the wedges are essentially inactive' (Brown & Péwé 1973, p. 82), and survive only in silt, the ice having already been replaced by a sedimentary fill in gravels. The claim that fossil ice wedges may indicate former discontinuous permafrost (Wright 1961, p. 942) needs to be qualified if they do not develop there; in such cases they indicate former continuous permafrost which has subsequently been degraded.

(*b*) *Pingos*

A pingo is an ice-cored mound formed within the permafrost; the permafrost table is found to be close to the surface so that almost the whole of the pingo is below the active layer. Borings and natural exposures suggest that the cores of many pingos are of clear ice, but others consist of stratified ice and sediments. A study of existing forms suggests that a pingo goes through a cycle of growth and decay which is unrelated to climatic change (Washburn 1973). In its growth stage it forms a mound, which as it grows larger, splits open at the summit to expose the ice core. The exposed core melts in summer to form a crater lake (figure 1). The heat of the water causes the lake to sink deeper into the core and to get wider as it melts the ice on its shore, undercutting the crater walls and causing the continued slumping of material down their inner face. These processes result in the surface of the crater lake being steadily lowered, and its area increased, until only the lowest part of the mineral skin of the pingo remains forming a rampart surrounding the lake. The field evidence on this point seems unequivocal; the skin collapses inwards into the central crater. This is a point of prime importance in the recognition of former pingos. The form of the pingo core is destroyed during its evolution; as the ice support for the surrounding skin or shell melts, the shell collapses. It is difficult to see a Pleistocene pingo core being replaced by a pseudomorph, by a lens-shaped basin of sediments with overhanging walls. The final form in the evolution of the pingo which is the initial form of the fossil pingo, consists of a basin, enclosed in whole or in part by a rampart. The basin has received a large part of the skin which forms its basal deposit, and in the central part fine material in suspension will have settled to form muds. Both have been laid down on the ice core, and must be subject to settling and deformation as the core thaws. Undisturbed stratification will start only when the core has entirely melted.

Contemporary pingos are usually classified in the two categories proposed by Müller (1959), the closed system and the open system. The former type results from the freezing of a pocket of unfrozen sediments beneath a lake in continuous permafrost, the excess water being expelled into the surface zone where it freezes to form a pingo. This is a 'non-repetitive process' which tends to produce a 'solitary hill' (Müller 1959, p. 101). Open system pingos, on the other hand, form in the discontinuous permafrost zone at the foot of valley sides, where the uplands are subject to seasonal freezing only but the valley floor deposits are perennially frozen (figure 2). Meteoric water percolates into the ground in the uplands, moves beneath the permafrost

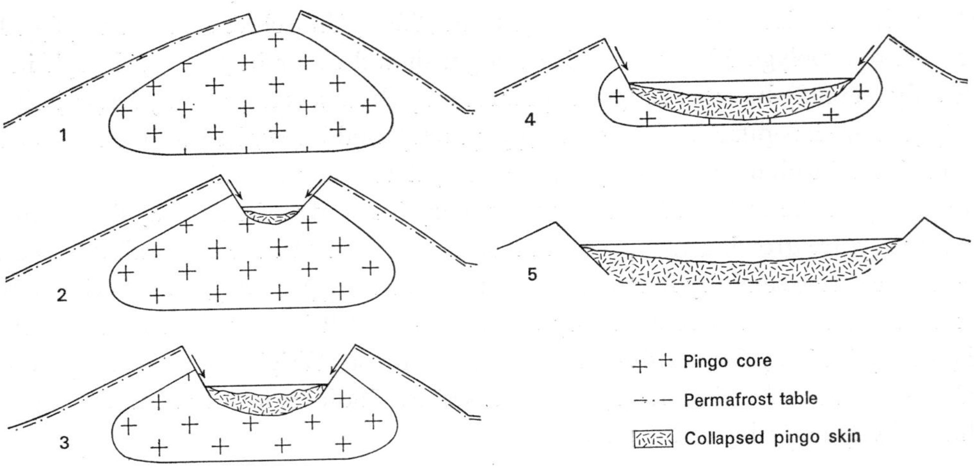

FIGURE 1. Collapse of a pingo. 1, the ice core is exposed by the rupture of the summit; 2–4, a pond develops on the exposed core and is enlarged by the melting back of the ice on its shores, which leads to the collapse of the overlying sediments into the crater; 5, the core has thawed and only the lower flanks of the mound remain, forming a low rampart around the pond.

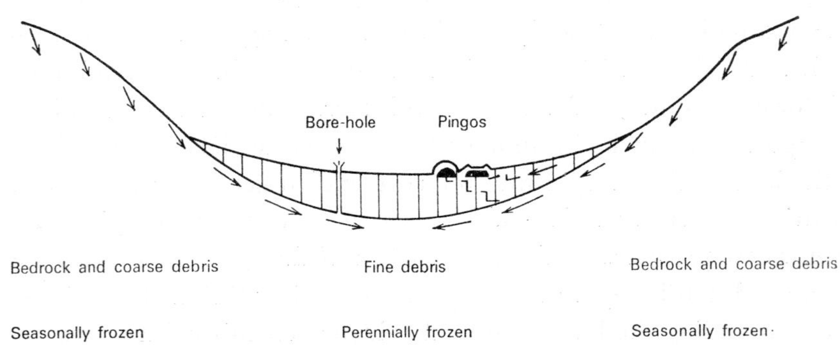

FIGURE 2. Hydrology of open system pingos in central Alaska. The permafrost (shown by vertical lines) is typically 30–45 m thick beneath these pingos. It is confined to the finer sediments of the valley bottom and is impervious except for any coarse beds which will be thawed. Water, entering through the upper permafrost-free slopes, is under pressure beneath the permafrost and rises to the surface in bores. It may reach the surface naturally to form a spring but where it freezes before reaching the surface it forms a pingo.

of the lower slopes and is under pressure beneath the valley floor (Hopkins *et al.* 1955). Unlike the closed system type, open system pingos continue to grow and decay in a favourable area so long as these hydrological conditions persist, so that 'clusters of mutually interfering pingos of different ages' are produced (Holmes, Hopkins & Foster 1968, p. 15).

As the two types of pingo represent two different environments, it is important to determine to which type fossil pingos belong. In brief, closed system pingos occur on a flat terrain, almost entirely on the sites of shallow lakes so that their skins are largely of lacustrine sediments. They tend to occur singly and their final form is an oval or circular rampart which completely surrounds the basin. Open system pingos occur in hilly country, are sited at the foot of valley slopes, involving slope deposits or valley floor deposits. They tend to occur in clusters. On slopes the ramparts may survive as a horse-shoe or crescent on the downslope side of the basin (Holmes *et al.* 1968, p. 27).

From their siting and clustered character the rampart forms which have been reported so far undoubtedly belong to the open system type. Those in southwest Wales in particular resemble closely contemporary pingos in central Alaska and Yukon Territory, rather than those of east Greenland (Watson & Watson 1974, p. 222). Most of the Welsh ramparts are on sloping sites; they are interrupted on the upslope side, and the enclosed basin is asymmetrical. On flat sites on the valley bottom the rampart surrounds the basin whose profile is symmetrical (figure 3).

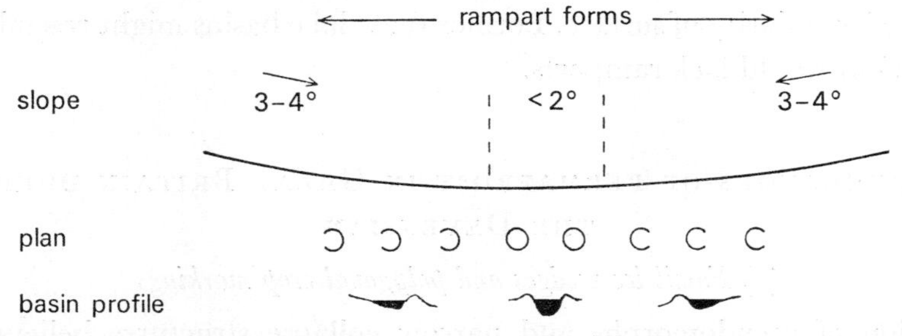

FIGURE 3. Relations between slope, rampart form and basin form in pingo remains in Wales.

Only general information is available on the diameters of contemporary open system pingos; in central Alaska the range is given as 15–440 m (Holmes *et al.* 1968, p. 30). In Wales typical basins have diameters between 30 and 60 m, while the external diameter of the ramparts is frequently 50–120 m. The diameter of the original mound would lie between these two measurements, for although marginal collapse into the pond as the core thaws would enlarge the basin, the movement of rampart material outwards by slope processes would increase the external diameter. The diameter of the former mounds in Wales may be estimated as typically between 45 and 90 m. This is comparable with the values given for contemporary *closed* system pingos in the Yakutsk basin in Siberia of 30–80 m for the higher steep-sided mounds and 50–100 m for low flat domes (Czudek & Demek 1973, pp. 53–54). None appear to be as large as the largest contemporary pingos. The largest basin measured in Wales has a diameter of 130 m, the largest in East Anglia, 120 m (Sparks, Williams & Bell 1972, p. 329). Available evidence, however, suggests that very large mounds form only a small percentage of contemporary pingos (Stager 1956).

Again little information exists on the depth to which pingo cores extend below the surface though Russian workers believe it is generally between 5 and 10 m (quoted in Maarleveld 1965, p. 8). In southwest Wales, the deepest basins which have been augered, exceed 10 m in depth.

(*c*) *Thaw lakes*

The upper layers of the permafrost, 'where fine grained materials are present, commonly contain 30 to 90 % of ice by volume' (Black 1969, p. 133; see also Mackay 1971). Many flat areas in both the continuous and discontinuous permafrost zones are pitted by lakes due to localized thawing of this ice. 'Simple lakes are round, oval or roughly rectangular in outline, but groups of lakes commonly coalesce to form compound water bodies with scalloped outlines' (Hopkins *et al.* 1955, p. 140). These thaw lakes rarely exceed 300 m in diameter (Black, 1969, p. 138), and are generally less than 3 m deep.

Lakes 'similar in outline are common in areas in Alaska where little or no permafrost exists today' and they 'probably persist in recognizable form in areas where no frozen ground remains' (Hopkins *et al.* 1955, p. 140). But no systematic investigation of the possibility of thaw lake basins in the British Isles has been published so far, so that, although important in the identification of contemporary permafrost areas, they have not contributed to our knowledge of the extent of permafrost in Devensian Britain.

The author suggests that a thaw lake identification might be considered for small roundish lake basins in flat areas of silty or clay sediments including subdued (probably older) till. Kettle lake basins due to buried glacier ice are likely to be associated with gravelly (dead ice) deposits and some irregularity of surface. Former thaw lake basins might resemble pingo basins in shape and size but would lack ramparts.

3. Indicators of permafrost in Great Britain during the Devensian

(a) *Fossil ice wedges and polygonal crop markings*

The distribution of pseudomorphs and narrow collapse structures believed to represent former ice wedges together with polygonal crop markings is shown on figure 4. The latter resemble the polygonal nets overlying contemporary ice wedges (cf. Shotton 1960; Péwé *et al.* 1969), and in some cases have been shown to overlie fossil ice wedges (e.g. Gruhn & Bryan 1969; Morgan 1973). It should be emphasized that the map shows localities and not individuals; at this stage an attempt to show densities would be misleading and tend to overemphasize centres of active recording.

The outstanding feature of the distribution is that it extends from the south coast to the limit of the Loch Lomond readvance, some of the Scottish examples being as late as Pollen Zone III (Galloway 1961, p. 190). In many cases in east Scotland and the Midland Valley the wedges pre-date the last ice advance in the locality, and Galloway (1961) concluded that the majority of Scottish examples were of Pollen zone I age. In the northeast of England the same is probably true as some of the recorded examples are covered by later deposits. In general the evidence suggests that, inside the limit of the maximum extension of the Late Devensian ice, the area of ice wedge formation extended northwards as the ice cap diminished.

In mid-Wales, where recorded fossil wedges are much more numerous than figure 4 indicates, those on sloping sites are usually covered by 1–2 m of head which accumulated after the wedge ice had completely melted, since the overlying head has not been disturbed. The majority is probably of similar age to those of northern Britain.

A common location for fossil wedges is flat gravel terraces where they may extend up to the present soil. In these cases they may be of any age younger than the gravels they pierce. If the gravels are Late Devensian, one may estimate the age of the wedges to within a fairly small time error, but if they are Early Devensian, the possible time range is so great that the fossil wedges have little meaning in terms of chronology. Where the complete (or truncated) fossil wedge is within the body of the gravels, an attempt at dating can only be made after a careful study of its relation with the beds it intrudes. Such data exist only in a relatively small number of cases, and it is possible to discuss the significance of much of the wedge and crop pattern distributions only in very general terms. Collapsed wedge structures and casts have been found in association with datable deposits of Early Devensian age, as at Wretton (West *et al.* 1974) and Chelford (Worsley 1966*a*), and of Middle Devensian age, as at Four Ashes (Morgan 1973) and Syston (Bell, Coope, Rice & Riley 1972). It seems probable, therefore, that permafrost made its appearance as early as the latter part of the Early Devensian.

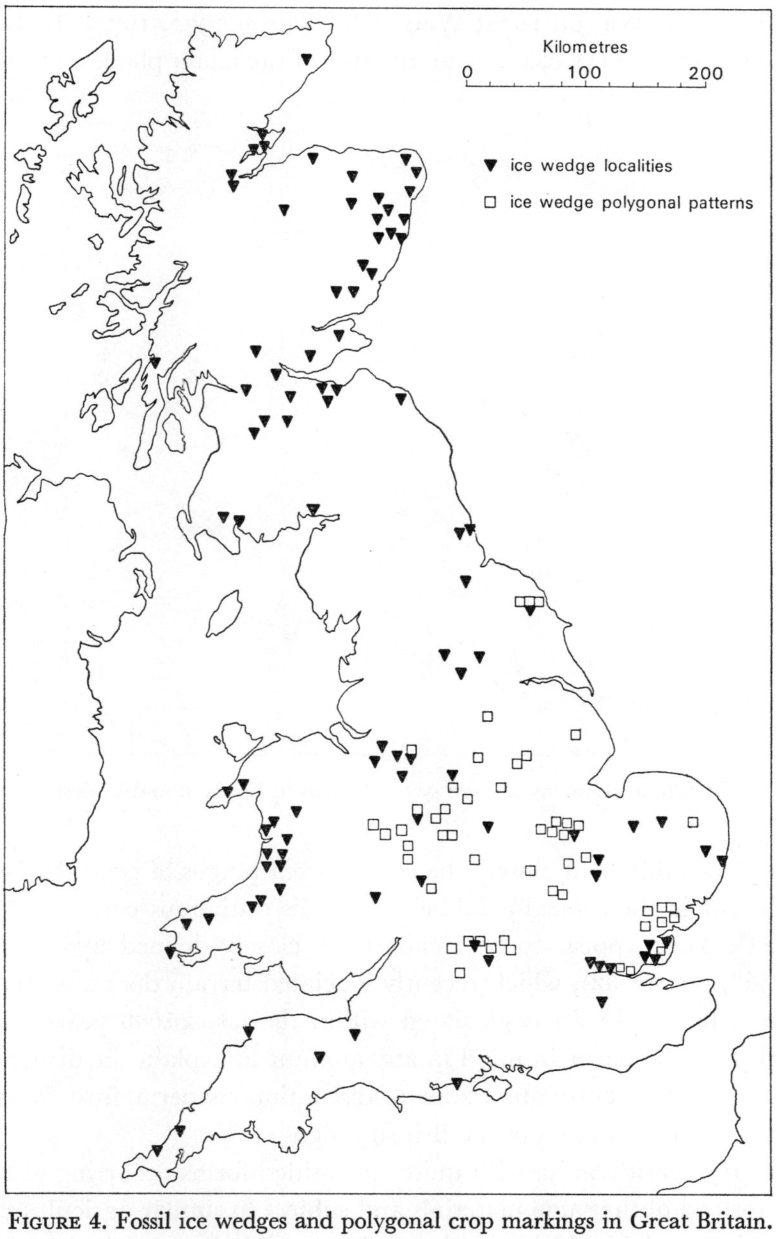

FIGURE 4. Fossil ice wedges and polygonal crop markings in Great Britain.

The map must be regarded as an interim compilation, and no doubt some of the blank areas will be eliminated as more sites are recorded. Not all will disappear, for ice wedge polygons are not found in all parts of the contemporary continuous permafrost zone, but are replaced by other types of patterned ground. Such appears to be the case in the blank representing the chalk areas (cf. figure 4 with Fig. 4.8 in Sparks & West 1972, p. 115). In the present state of knowledge, fossil ice wedges add much more to our understanding of the Late Devensian than in earlier periods where the information they give tends to be of a general character.

(*b*) *Fossil pingos*

Rampart forms representing Devensian pingos have been found in two main areas, Wales and East Anglia (figure 5). In southwest Wales they occur in groups in minor valleys scattered over a considerable area (Watson 1971; Watson & Watson 1972, 1974). In East Anglia they tend to be located in the valleys of the western edge of the chalk plateau (Sparks *et al.* 1972).

FIGURE 5. Remains of open system pingos in England and Wales.

In both areas they resemble most closely the open system pingos of central Alaska and Yukon Territory which occur in the central axial belt of the discontinuous permafrost zone (Brown & Péwé 1973, p. 81). They appear to be localized by closely defined hydrological conditions (Holmes *et al.* 1968, pp. 30–32), which recently glaciated terrain does not readily provide, as 'pingos are extremely rare in areas glaciated within the last 25000 years' (Brown & Péwé 1973, p. 81). This must be borne in mind in any attempt to explain the distribution of former pingos. One may, however, postulate a zone of discontinuous permafrost from Wales to East Anglia at the time of the formation of the British pingos.

The ramparts vary considerably, with quite steep-sided forms occurring alongside low subdued ones, both formed of the same materials and subject to similar agricultural hazards. The pingo basins contain a variable thickness of grey clay overlain by organic deposits. In southwest Wales the base of these organic deposits has been sampled for radiocarbon dating in three basins in the Cledlyn valley and one in the Cletwr valley (Shotton & Williams 1973, pp. 460–461; Shotton, Williams & Johnson 1974, p. 287, and 1975, pp. 257–258). All gave dates between 9500 and 10000 B.P. The same deposits were also sampled for pollen analysis, and show that organic sedimentation began prior to the 'initial expansion of *Juniperus* at the opening of the Flandrian'

(Handa & Moore 1976, p. 205). As the deposits include those surrounded by the steepest freshest rampart, it may be concluded that all the pingo ice had thawed by this time. Similar results from pollen analysis have been obtained for pingo basins at Llangurig in east Wales (Trotman 1963) and in East Anglia (Sparks *et al.* 1972), indicating that the discontinuous permafrost of the pingo belt ended with the beginning of the Flandrian.

It is more difficult to date the beginning of the phase of pingo development. A group of ramparts does not represent a group of mounds which grew together, but a series of mounds which grew individually over a period time. Thus the form of the ramparts indicates successive growth (Watson 1971, p. 386), and the fresher, more obvious ramparts stand on the ruins of older. The older pingos must generally have collapsed completely before the later pingos grew, otherwise the later ramparts must have suffered deformation in the overlap area. Though there are no measurements concerning rates of growth and collapse of open system pingos, it is believed that in central Alaska they have been 'forming continuously for several thousand years' (Holmes *et al.* 1968, p. 37). It is highly probable that pingo development in England and Wales lasted longer than the conventional Pollen Zone III, and that the transition from continuous to discontinuous permafrost took place with the climatic amelioration that ushered in the Allerød.

(*c*) *Involutions*

In the continuous permafrost zone today, considerable areas do not have ice wedge polygons, but have instead various types of sorted and unsorted patterned ground. The active layer or surface zone in which these patterns develop thaws and freezes annually. Its base, the top of the permafrost, is known as the permafrost table.

Since the 1920s, 'involutions' have been regarded by many as representing the fossil active layer. These consist of deformations of former sedimentary structures often as a series of pockets, but sometimes highly contorted and at times resembling load structures. As such convolutions of bedding and load structures are widespread in sedimentary deposits, they must be carefully considered in the context of their deposition. If they were formed in an active layer they must relate to a former surface and to a limited horizon. Discontinuous and random structures are suspect, as are those in deposits likely to be lacustrine (cf. Fries, Wright & Ruben 1961).

In many cases the involution layer lies just below the present surface which slopes at $< 2°$. The depth of the deformation varies within narrow limits, and in a large exposure often appears remarkably constant. The base truncates dipping beds, and may reasonably be interpreted as a fossil permafrost table. Even where the pockets consist of sand and gravel sunk in clay or silt, if the base of the deformation is parallel to the surface and can be followed for a considerable distance, it probably represents a former active layer. But very often the question of load structures does not arise; the pockets are of silt or clay sunk in sandy or gravelly material. These pockets in gravel, called festoons, are usually accompanied by erected pebbles or vertical stones where elongated pebbles have their long axis vertical (Watson & Watson 1971). Figure 6 is an example from the west coast of Wales, south of Aberystwyth, where the thickness of the deformed layer varies between 2 and 3 m over a distance of about 200 m. Below its base the till is undisturbed, as is shown by its constant preferred stone orientation. Elsewhere on the coast, festoons are developed in coarse fluvial gravels and in fluvio-glacial gravels, with the imbrication in the former, and the bedding in the latter, preserved undisturbed below the base.

The context of these systematically developed involution horizons is periglacial, in that they are developed on flat surfaces where surrounding slopes have developed periglacial slope deposits. In a few cases where the head at its outer limit is thin, it has been involved in festoons (cf. Watson & Watson 1971, pp. 111). The distribution of involutions in Britain seems to be limited to south England and Wales, and they may well be related to the maximum of the last glaciation, as many workers in central and western Europe believe (cf. Washburn 1973, p. 247). Poser's map (1948, p. 59, Fig. 2 – reproduced in Washburn 1973, p. 249) shows they indicate a depth of thaw of < 1.5 m in East Germany and western Poland, 1.5 to 2.0 m in West Germany and > 2 m in the low countries and in the area on both sides of the English Channel. In Cardigan Bay it reaches 3 m. Such a thickening active layer with decreasing continentality might appear anomalous, but it probably reflects the fact that 'heat is distributed through earth materials more rapidly by circulating water than by direct conduction' (Hopkins *et al.* 1955, p. 117).

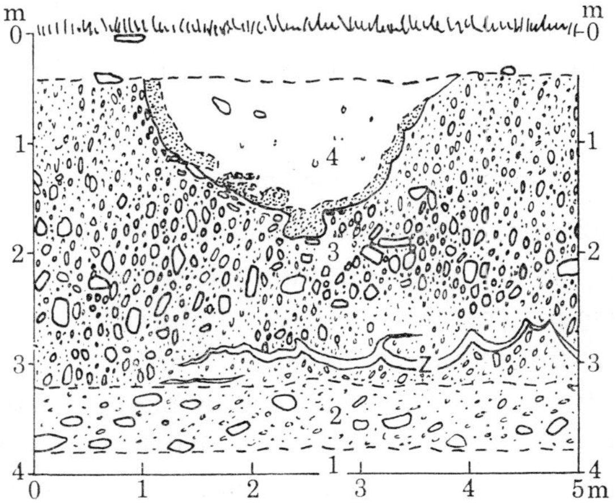

FIGURE 6. Festoon in vertical stones south of the River Clydan, near Aberystwyth. 1, modern beach; 2, undisturbed local till; 3, cryoturbated local till; 4, festoon of decalcified Irish Sea till surrounded by a deformed layer of gravelly coarse sand; z, deformed lens of silty clay and sand.

In Britain, there also appears to be a second involution layer which is shallower, 0.8 m thick, and is not consistently developed but patchy in distribution. It is found north of the deeper type considered above, extending to Scotland (Galloway 1961). South of Aberystwyth, both types may be seen together in cliff exposures, the shallower being developed in the surface zone of upper alluvial gravels which overlie the festoons of the deeper type.

Some of the processes which have played a part in involution layers also operate in the active layer of the continuous permafrost zone; the churning of the material beneath unsorted circles (Hopkins & Sigafoos 1951), the tendency for elongated stones to migrate vertically towards the surface (Washburn 1973, pp. 66–80). The case for permafrost remains hypothetical however, as it has yet to be demonstrated that structures comparable with festoons or other involutions are forming in permafrost regions today. It is therefore not possible to define the conditions under which they form, as one can for ice wedges (cf. Jahn 1975, p. 134), nor to use them with confidence for an environmental reconstruction of the Devensian.

4. The periglacial environment of Devensian Britain

(a) *The contemporary permafrost environment*

In an attempt to interpret the environment of Great Britain during the Devensian, Alaska, with its west coast situation, is probably the most instructive of contemporary permafrost regions. Furthermore, it has also been the scene of many studies of the relationship of permafrost phenomena to the natural environment in the last 20 years so that considerable data are available.

Table 1. Features of the permafrost area of the northwest of North America

permafrost	ground ice features		mean air temperatures °C: annual	mean air temperatures °C: July	vegetation
continuous	active ice wedge polygons in fine and coarse sediments	closed system pingos	−13/−15	4/6	polar desert .... tundra
	....		−6/−8	10/11	....
discontinuous	inactive ice wedge polygons in fine sediments	open system pingos			coastal tundra : interior boreal forest
			0/−1	13/17	

It may be urged that the Bering Sea and Arctic Ocean are no counterpart for the North Atlantic, but the evidence indicates that the latter had a much reduced climatic influence on the British Isles during the Devensian. There is consistent evidence for a great expansion westwards of the present Siberian permafrost region as far as the British Isles. This can only have taken place with a great increase in continentality of climate and very cold winters, while the considerable drop in the surface temperatures of the North Atlantic would have resulted in cool cloudy summers (Velitchko 1974).

On the northwestern coast of North America the continuous permafrost zone falls into two main vegetation zones. The polar desert is found on the Canadian Arctic Archipelago north of 68°–72° N, while the tundra occupies a small part of the islands south of this and occurs largely on the mainland, its western part being in Alaska between 66 and 70° N. In both areas mean annual precipitation is low, being generally < 125 mm in the polar desert, and < 250 mm in the tundra, and a high proportion falls in the summer months as rain. Even in the polar desert about 50 % falls in the three months with a mean temperature above 0 °C. The cool summers have a high percentage of cloud cover. Snowfall is relatively light so that the ground is not insulated from frost. Mean January air temperatures are generally < − 20 °C. In the polar desert or High Arctic, the surface is very largely bare rock debris with saline crusts and alkali flats. In the southern marginal zone, the subpolar desert or Middle Arctic, plantless areas are still common and a stony sedge-moss-lichen tundra is characteristic (Brown 1972).

In the discontinuous permafrost zone not only are temperatures generally higher but there are important differences between the coastal area and the interior. There is a decrease in rainfall inland, but its distribution is similar to that in the continuous permafrost zone. In the coastal area the mean January air temperature is around −15 °C, but with cloudy summers the July figure reaches only 10–11 °C. The vegetation for up to 300 km inland is tundra. In the interior, the January air temperature falls to around −24 °C, but in July with a lower average cloud cover, the mean air temperature is around 15 °C. With June, July and August mean temperatures between 12 and 16 °C, the coastal tundra is replaced by a boreal forest of spruce, birch, aspen, balsam poplar, alder and willow except on badly drained areas and uplands.

The environment of the two permafrost indicators used in this discussion for Devensian Britain has been studied in Alaska (see Brown & Péwé 1973). Active ice wedges are limited to continuous permafrost and by a mean annual air temperature of −6 to −8 °C. They are virtually all on the tundra. Open system pingos are found in both the coastal tundra and the interior forest of the discontinuous permafrost, the great majority between isotherms −3 and −6 °C.

(*b*) *Great Britain during the Devensian*

(i) *The Early Devensian, preceding* 50000 B.P.

As yet there is very little evidence for permafrost which can confidently be assigned to this period. The undisturbed birch-pine-spruce woodland horizon at Chelford dated 60800 B.P. has been interpreted as representing boreal forest with a mean July air temperature of about 15 °C (Simpson & West 1958). Intraformational fossil ice wedges in the sands covering the above bed indicate continuous permafrost subsequent to the forest phase (Worsley 1966*b*). An undated horizon with birch, pine and spruce pollen (FG) at Wretton has been tentatively equated with the Chelford organic layer, though the beetle assemblages of the two suggest very different environments (R. G. Coope in West *et al.* 1974, pp. 414–418). Fossil ice wedge horizons above and below FG indicate continuous permafrost, which suggests the probability of at least discontinuous permafrost during the birch-pine-spruce period. In the overlying horizon (H), halophytes are well represented, suggesting conditions approaching polar desert by the end of the sequence, and that it was already becoming as severe as in the Middle Devensian.

(ii) *The Middle Devensian,* 50000–26000 B.P.

Fossil ice wedges are known associated with dated deposits which span most of this period (e.g. Earith, 42000 B.P. and >45000 B.P., Syston, 37420 B.P. and Four Ashes, with radiocarbon dates between 42000 and 30000 B.P.), so that permafrost was present. Pollen and beetle assemblages indicate that the landscape was treeless, and as this is broadly the period of the 'full-glacial floras' which include obligate and facultative halophytes (Bell 1969), conditions were fairly arid with probably subpolar desert at times. This picture of a permafrost environment fluctuating between tundra and subpolar desert is interrupted only in the period between 42530 and 38500 B.P. (Coope, Morgan & Osborne 1971).

(iii) *The Late Devensian* 26000–10000 B.P.

Although no fossil ice wedges are datable to the period of maximum glaciation, they occur in the most recent till at Four Ashes (Morgan 1973) and Chelford (Worsley 1966*b*). As the insect assemblage from Dimlington (18500 B.P.) has few species and these are tolerant of very low summer temperatures (Coope *et al.* 1971, p. 93), it is probable that polar desert with continuous permafrost existed south of the ice margin.

Fossil ice wedges in northern England and southeast and east Scotland suggest that continuous permafrost was established as these areas were uncovered by ice, and some in central Scotland overlain by glacial deposits may indicate permafrost in the period preceding the Aberdeen–Lammermuir Readvance and in that preceding the Perth Readvance. In Wales, it has already been suggested that the fossil ice wedges are broadly Pollen Zone I. In England and Wales polar desert with continuous permafrost probably continued until around 15000 B.P., and the barren silty-clay deposits in lake basins which frequently underlie the lowest late-glacial pollen bearing material probably result from rain and melting snow run-off flowing over a surface with a very reduced vegetation cover.

In the period preceding the development of open birch woodland in the Allerød, the permafrost of England and Wales probably became discontinuous. Mean July air temperatures during the Allerød were probably about 12–13 °C, while on the Alaskan model, Ireland would have been in the coastal zone with slightly lower July temperatures and a tundra vegetation. Though the Younger Dryas saw the replacement of birch woodland by tundra, discontinuous permafrost persisted until the beginning of the Flandrian, when open system pingos which had been forming throughout most of the discontinuous permafrost period, ceased to develop. In Scotland, however, fossil ice wedges suggest continuous permafrost in the Younger Dryas, as in southern Scandinavia (Svensson, 1974). If we assume that the fossil pingos in Wales and East Anglia occupy the same position within the discontinuous permafrost belt as the open system pingos of Alaska today, the southern limit of discontinuous permafrost should have lain in the English Channel area, and the northern limit in the English–Scottish border area, with continuous permafrost and active ice wedges north of this.

(iv) *Devensian temperatures*

The minimum fall in mean annual air temperature compared with the present day, which continuous permafrost in England implies is of the order of 16–17 °C. This is based on the continuous permafrost limit at −7 °C determined for central Alaska. Active ice wedge formation, on which the identification of former continuous permafrost has been based in this paper, probably implies a somewhat lower figure, say −10 °C, giving a fall of around 20 °C.

At the maximum of the Devensian ice advance the fall in temperature was greater than this, for the southern limit of ice wedge formation in the last glaciation lay some 500 km south of the southern English coast (Kaiser 1960, Plate 1, reproduced in Washburn 1973, p. 250). In the northwest of North America today, the northern limit of the tundra is about 500 km north of the continuous permafrost limit, and has a mean annual air temperature of −14 to −12 °C. Full polar desert conditions probably existed in the English Midlands with mean annual air temperatures of at least −16 to −17 °C. There is a scarcity of floristic and faunistic data on this point but what does exist, indicates very severe conditions (cf. Coope *et al.* 1971, pp. 92–93).

On the evidence of the pingos, the mean annual air temperature of southern England and Wales in the Younger Dryas was between −4 and −5 °C, representing a fall of 13–14 °C compared with the present day.

Table 2 summarizes the temperatures deduced for the English Midlands–East Anglia at various points of time. The fall in the July mean temperatures compared with those of today is small. The great difference between temperatures in Britain today and those of permafrost regions is in the winter temperatures.

TABLE 2. ESTIMATED DEVENSIAN TEMPERATURES (°C)

B.P.	mean air temperatures		fall from present	
	annual	July	annual	July
10500	−5	10	15	5
11500	−4	12/13	14	3
13000	−7	10	17	5
18000	−15	5	25	10
55000	−7	11	17	4

5. GENERAL CONCLUSIONS

1. Ice wedge casts remain the most useful indicator of former permafrost, partly because they are the most widespread and partly because more is known about contemporary forms, about recently thawed forms and about their environment.

2. Pingos are more restricted in distribution but their remains may be important in discriminating between continuous and discontinuous permafrost. Considerable work is being done on contemporary forms and their environments, but little has been recorded about the final rampart stage and especially on the internal structure. This is urgently required as identification of fossil pingos at present depends very largely on the survival of their surface form.

3. Boreal forest may occur in permafrost or permafrost-free areas. The mean July air temperatures may have much the same range in both cases.

4. Once continuous permafrost is established the short-term replacement of tundra by boreal forest is likely to be within permafrost conditions rather than to represent a change to permafrost-free conditions.

The author thanks D. B. Smith for unpublished information on ice wedge casts in north-east England.

REFERENCES (Watson)

Bell, F. G. 1969 The occurrence of southern, steppe and halophyte elements in Weichselian (Last Glacial) floras from southern Britain. *New Phytol.* **68**, 913–922.

Bell, F. G., Coope, G. R., Rice, T. J. & Riley, T. H. 1972 Mid-Weichselian fossil bearing deposits at Syston, Leicestershire. *Proc. geol. Ass. Lond.* **83**, 197–211.

Black, R. F. 1969 Thaw depressions and thaw lakes. *Biul. Peryglacj.* **19**, 131–150.

Black, R. F. 1973 Growth of patterned ground in Victoria Land, Antarctica. In *Permafrost, N. American Contribution to 2nd Int. Conf. on Permafrost*, pp. 193–203. Nat. Acad. Sci., Washington, D.C.

Brown, R. J. E. 1972 Permafrost in the Canadian Arctic Archipelago. *Z. Geomorph.* N.F., Suppl. Band **13**, 102–130.

Brown, R. J. E. & Péwé, T. L. 1973 Distribution of permafrost in North America and its relationship to the environment: a review, 1963–1973. In *Permafrost, N. American Contribution to 2nd Int. Conf. on Permafrost*, pp. 71–100. Nat. Acad. Sci., Washington, D.C.

Cailleux, A. & Taylor, G. 1954 *Cryopédologie.* Paris: Hermann.

Coope, G. R., Morgan, A. & Osborne, P. J. 1971 Fossil Coleoptera as indicators of climatic fluctuations during the Last Glaciation in Britain. *Paleogeog. Paleoclimatol. Paleoecol.* **10**, 87–101.

Czudek, T. & Demek, J. 1973 Die Reliefentwicklung während der Dauerfrostbodendegradation. *Rozpravy ČSAV, Řada matemat. a přírod.* **83**, seš. 2.

Fries, M., Wright Jr., H. E. & Rubin, M. 1961 A Late Wisconsin peat at North Branch, Minnesota. *Am. J. Sci.* **259**, 679–693.

Galloway, R. W. 1961 Ice wedges and involutions in Scotland. *Biul. Peryglacj.* **10**, 169–193.

Gruhn, R. & Bryan, A. L. 1969 Fossil ice wedge polygons in southeast Essex. In *The periglacial environment* (ed. T. L. Péwé), pp. 351–363. Montreal: McGill-Queens University Press.

Handa, S. & Moore, P. D. 1976 Studies in the vegetational history of mid-Wales. IV. Pollen analyses of some pingo basins. *New Phytol.* **77**, 205–225.
Holmes, G. W., Hopkins, D. M. & Foster, H. L. 1968 Pingos in central Alaska. *U.S. Geol. Surv. Bull.* 1241-H.
Hopkins, D. M., Karlstrom, T. N. V. *et al.* 1955 Permafrost and groundwater in Alaska. *U.S. Geol. Surv. Prof. Paper* 246-F.
Hopkins, D. M. & Sigafoos, R. S. 1951 Frost action and vegetation patterns on Seward Peninsula, Alaska. *U.S. Geol. Surv. Bull.* 974-C.
Jahn, A. 1975 *Problems of the periglacial zone.* Warsaw: Polish Scientific Publishers.
Maarleveld, G. C. 1965 Frost mounds, a summary of the literature of the past decade. *Mededelingen van de Geologische Stichting*, N.S. **17**.
Mackay, J. R. 1971 The origin of massive icy beds in permafrost, western arctic coast, Canada. *Can. J. Earth Sci.* **8**, 397–422.
Morgan, A. 1973 The Pleistocene geology of the area north and west of Wolverhampton, Staffordshire. *Phil. Trans. R. Soc. Lond.* B **265**, 233–297.
Müller, F. 1959 Beobachtungen über Pingos. *Meddr Grønland* **153**, no. 3.
Péwé, T. L. 1966 Palaeoclimatic significance of fossil ice wedges. *Biul. Peryglacj.* **15**, 65–73.
Péwé, T. L. 1969 The periglacial environment. In *The periglacial environment* (ed. T. L. Péwé), pp. 1–9. Montreal: McGill-Queens University Press.
Péwé, T. L., Church, R. E. & Andresen, M. J. 1969 Origin and palaeoclimatic significance of large scale polygons in the Donnelly Dome area, Alaska. *Geol. Soc. Am. Spec. Pap.* **109**.
Shotton, F. W. 1960 Large scale patterned ground in the valley of the Worcestershire Avon. *Geol. Mag.* **97**, 404–408.
Shotton, F. W. & Williams, R. E. G. 1973 Birmingham University Radiocarbon Dates VII. *Radiocarbon* **15**, 460–461.
Shotton, F. W., Williams, R. E. G. & Johnson, A. S. 1974 Birmingham University Radiocarbon Dates VIII. *Radiocarbon* **16**, 287.
Shotton, F. W., Williams, R. E. G. & Johnson, A. S. 1975 Birmingham University Radiocarbon Dates IX. *Radiocarbon* **17**, 257–258.
Simpson, I. M. & West, R. G. 1958 On the stratigraphy and palaeobotany of a late-Pleistocene organic deposit at Chelford, Cheshire. *New Phytol.* **57**, 239–250.
Sparks, B. W. & West, R. G. 1972 *The Ice Age in Britain.* London: Methuen.
Sparks, B. W., Williams, R. B. G. & Bell, F. G. 1972 Presumed ground-ice depressions in East Anglia. *Proc. R. Soc. Lond.* A **327**, 329–343.
Stager, J. K. 1956 Progress report on the analysis of the characteristics of pingos east of the Mackenzie delta. *Can. Geogr.* **7**, 381–392.
Svensson, H. 1974 Distribution and chronology of relict polygon patterns on the Laholm plain, the Swedish west coast. *Geogr. Ann.* **54**A, 159–175.
Tricart, J. 1970 *Geomorphology of cold environments.* London: Macmillan.
Troll, C. 1958 *Structure soils, solifluction and frost climates of the earth.* U.S. Army SIPRE Trans. **43**.
Trotman, D. M. 1963 Peat deposits within a pingo near Llangurig, Wales. *Z. Geomorph.* N.F. **7**, 168–171.
Velitchko, A. A. 1974 Paragenesis of a cryogenic (periglacial) zone. *Biul. Peryglacj.* **24**, 89–110.
Washburn, A. L. 1973 *Periglacial processes and environments.* London: Arnold.
Watson, E. 1971 Remains of pingos in Wales and the Isle of Man. *Geol. J.* **7**, 381–392.
Watson, E. & Watson, S. 1971 Vertical stones and analogous structures. *Geogr. Ann.* **53**A, 107–114.
Watson, E. & Watson, S. 1972 Investigations of some pingo basins near Aberystwyth, Wales, *Rep. 24th Int. Geol. Congr.* (Montreal), Section 12, 212–233.
Watson, E. & Watson, S. 1974 Remains of pingos in the Cletwr basin, southwest Wales. *Geogr. Ann.* **56**A, 213–225.
West, R. G., Dickson, C. A., Catt, J. A., Weir, A. H. & Sparks, B. W. 1974 Late Pleistocene deposits at Wretton, Norfolk. II. Devensian deposits. *Phil. Trans. R. Soc. Lond.* B **267**, 337–420.
Worsley, P. 1966*a* Fossil frost wedge polygons at Congleton, Cheshire, England. *Geogr. Ann.* **48**A, 211–219.
Worsley, P. 1966*b* Some Weichselian fossil frost wedges from east Cheshire. *Mercian Geologist* **1**, 357–365.
Wright, H. E., Jr 1961 Late Pleistocene Climate of Europe: a review. *Geol. Soc. Am. Bull.* **72**, 933–984.

Discussion

A. V. Morgan (*Department of Earth Sciences, University of Waterloo, Waterloo, Ontario, Canada*). (In reply to a question by Dr Starkel to Dr Watson requesting an estimation of the time elapsed for the formation of periglacial features.)

There are well developed ice-wedge casts and fossil polygonal ground present within the

Late Devensian Irish Sea till in the area north and west of Wolverhampton. Assuming that the maximum ice advance position at the Wolverhampton Line was reached *ca.* 20000–18000 a B.P., then the relict permafrost zone must post-date ice retreat. About 12 km north of Wolverhampton the Late Devensian ice-wedge casts are overlain by glacio-fluvial terrace gravels. The glacio-fluvial gravels have been cut and filled with an organic sequence at Stafford, dated at 13490 ± 375 a B.P. (Birm. 150). Although the fauna associated with this date could co-exist in an environment with permafrost, it seems likely that perennially frozen ground was disappearing about this time. The permafrost regime must, therefore, have existed (in the Late Devensian) from ?18000 a B.P. to about 13500 a B.P. in the west English Midlands. It is possible that it may have reappeared *ca.* 10800–10200 a B.P. when coleopteran faunas indicate a return of severe climatic conditions.

A number of colour slides were shown illustrating the sequence exposed at the Devensian type section at Four Ashes, Staffordshire, between 1967 and 1970. Dr Morgan pointed out that over 50 organic sites had been collected from the gravel sequence which ranged in thickness from 50 cm to over 5 m. All of the organic lenses came from a restricted geographic area *ca.* 400 m × 100 m and all had fragments of Coleoptera preserved within them. Eight lenses were dated and provided ages ranging from 30500 ± 440 a B.P. (Birm. 195) to in excess of 45000 a B.P. (Birm. 44). Because of the fauna and flora within the gravel sequence it was shown to have been deposited in a time period ranging from the Ipswichian Interglacial through the Early Devensian Chelford episode (Brørup equivalent) to the Middle Devensian Interstadial complex, and finally covered by a Late Devensian Irish Sea till. Fossil ice-wedge casts at the base and within the sequence indicated at least two severe climatic deteriorations within the gravel deposit, although, at Four Ashes, there was no evidence of an Early Devensian ice advance. The dates of 30500 and 30655 ± 700 a B.P. (Birm. 25) indicate that the Late Devensian ice crossed the area sometime later, although how much later is uncertain. Following ice-retreat permafrost conditions prevailed resulting in the ice-wedge casts, polygonal ground and cryoturbation affecting the Late Devensian deposits at Four Ashes.

At Stafford, 15 km to the north, investigations have begun on a thick (19.1 m) Late Devensian and Holocene sequence consisting of clay, silts, gyttjas and peats. A sample at the 15.1 m level has provided a date of 13500 a B.P. Well preserved pollen, plant macrofossils, diatoms, cladocera, molluscs, ostracods, beetles and other insects will hopefully provide a detailed picture of biological change, while geochemical analyses of the core sediment and molluscs both for major and trace elements as well as stable isotopes should help provide additional data for this important time period.

Phil. Trans. R. Soc. Lond. B. **280**, 199–209 (1977) [199]
Printed in Great Britain

Periglacial Ireland

By G. F. Mitchell, F.R.S., P.R.I.A.
Department of Geology, Trinity College, Dublin

During the colder parts of the Last Cold Stage at some times there were large ice masses, and at others there were periglacial conditions apparently without ice masses. It is a matter of doubt whether ice masses did form early in the Cold Stage, or whether periglacial conditions were dominant for the greater part of the stage, with ice only appearing at a relatively late date; some evidence from Ireland does not appear to tally with that from England.

The paper presents a hypothetical frost cycle, and suggests that all the periglacial features created during the Last Cold Stage that have so far been recognized in Ireland could have been formed in two such cycles. The first cycle, which was very severe, perhaps took place at the end of the Early Devensian; the second cycle, which was sharp but short, occurred at the end of the Late Devensian, between 11000 and 10000 B.P., and corresponds with Pollen Zone III.

Nearly all the types of periglacial features that are to be seen in Britain also occur in Ireland, and therefore this paper will concentrate on the possibility of drawing climatic inferences from fossil periglacial structures, rather than provide a catalogue of the features themselves. But, as Washburn (1973) has warned, any such attempts, in the present limited extent of knowledge, can only be very tentative and highly subjective.

During the Last Cold Stage three types of environment seem to have developed in Ireland:

(A) Interstadial conditions too warm for snow or frost;

(B) Glacial conditions suitable for snow accumulation, and leading to the development of ice masses; on the areas that were not buried by ice, tundra vegetation of some variety probably continued to grow; and

(C) Periglacial conditions too dry or too cold for snow accumulation, but suitable for frost action, and leading to the formation of frost structures; any vegetation would have been sparse and open.

Condition A gave rise to park-tundra with scattered trees and bushes, sometimes with large mammals. A mammoth bone from the cave at Castlepook, Co. Cork, was dated to 33500 B.P. (Mitchell 1976†), and if the faunal remains from the cave were contemporaneous, then in Middle Devensian time giant deer, reindeer, woolly mammoth and spotted hyena were wandering in the Blackwater valley. Giant deer and reindeer flourished in the Woodgrange Interstadial, between 14000 and 11000 B.P.

In Co. Fermanagh two organic deposits (with arctic vegetable remains, beetles and molluscs) lie between tills, and below a topography of drumlin form. That at Derryvree (Colhoun *et al.* 1972) has been dated to 30500 B.P., and that at Hollymount to greater than 41500 B.P. Two mountains in northwest Ireland, Ben Bulbin (530 m) in Sligo, and Slieve League (600 m) in Donegal, which appear to have stood as nunataks above the Devensian ice masses, today

† Much of the material of this paper is discussed in Mitchell (1976), and constant reference to that book is omitted. Again, where full details of a publication referred to in this paper are given in the references in Mitchell, Penny, Shotton & West (1973), the details are not repeated here.

carry clusters of arctic plants which probably survived the accompanying glacial conditions in or near their present locations. Many of the same plants also characterize the deposits at Derryvree and Hollymount, and perhaps these deposits can do no more than tell us that the plants and animals they record could coexist with ice masses in Ireland; they do not tell us that conditions were interstadial, in the sense of Sparks & West (1972), who require 'a climax dominated by a conifer such as pine or, in the case of a shorter amelioration, a phase of birch woodland between phases of Arctic treeless vegetation' (p. 28).

We have thus only unequivocal evidence for two interstadials in the Last Cold Stage in Ireland, in the Middle Devensian at Castlepook (33500 B.P.), and in the Woodgrange Interstadial (14000–11000 B.P.) in the Late Devensian.

Condition B perhaps occurred three times. As we have seen, at Derryvree the organic deposit (30500 B.P.) lay below Late Devensian till which had been moulded into drumlin form; it rested on a lower till which showed no signs of weathering or of cryoturbation. What appeared to be the same upper till – again in drumlin form – also buried the nearby deposit at Hollymount, where the record (in exposed section and by drilling) was:

0–1000 cm	till moulded in drumlin form
1000–1200 cm	gravel
1200–1665 cm	laminated sandy silt with arctic vegetable debris of age greater than 41500 years
1665–1695 cm	brown silt
1695–1740 cm	black silt with arctic vegetable debris and lamellibranchs
1740–1670 cm	varved clay
at 1760 cm	calcareous till

This record strongly suggests that the lower part of the sequence – till, varved clay, lacustrine deposit with vegetable debris – was deposited without interruption, and that there was no opportunity for the till either to be subaerially weathered, or to be disturbed by frost action, before the arrival of the vegetable debris. Thus unless the vegetable debris is Late Wolstonian in age – and it could hardly have survived the weathering of the Ipswichian Warm Stage – it must be of Devensian age, and if so that underlying till must also be of Devensian age, and formed before 41500 B.P. In Britain there is no accepted record of till so early in the Devensian.

If we accept this interpretation of the Hollymount evidence, Condition B will have occurred for the first time in the Devensian before 41500 B.P. The second occurrence was in the Late Devensian, as drumlins of the type that occur at Derryvree must be of this age, because they overlie material dated to 30500 B.P., while the closed basins between them contain lacustrine deposits of Woodgrange age (14000–11000 B.P.). The third occurrence was only on a very small scale, and confined to ice partly reoccupying older corries. Lough Nahanagan (at 500 m in the Wicklow Mountains) held a corrie glacier in the Devensian, but this melted away in the Woodgrange Interstadial, and a lake occupied the corrie hollow; muds accumulated on the lake floor. Small glaciers then formed on the corrie walls, and ploughed up the lake muds, thus clearly demonstrating their post-Woodgrange age. They thus fall into the final cold snap of the Devensian, formerly called Zone III, but now named in Ireland the Nahanagan Stadial (11000–10000 B.P.). Their moraines are buried by muds of Flandrian age.

It is more difficult to say how often condition C occurred. When the Devensian ice sheets were at their greatest extent in Ireland (figure 4*b*) about one-fifth of the country, chiefly in the south, was not covered by ice. In this area marked periglacial structures are to be seen. It seems fair to assume that if the end of the Wolstonian Cold Stage had left periglacial structures imprinted on the country, these would have been dissipated by the deep weathering of the Ipswichian Warm Stage. Obliteration would have been aided by the deep penetration of tree-roots, and by the wind-throw of trees. At the opening of the Devensian Cold Stage, the soils would have shown no signs of relict features. It can of course be argued that if tree-roots can destroy periglacial soil structures, then the roots of the Flandrian forests should have destroyed Devensian periglacial structures. But the Ipswichian soils and forests ran a full interglacial cycle, whereas in Ireland man began the destruction of the postglacial forests no more than five thousand years after the woodlands had first developed.

We must next ask 'Can in fact "periglacial" features form in the vicinity of an ice mass?' This question at once throws up the modern misuse of the term 'periglacial' which originally referred to the climate and related geological features peripheral to ice sheets. For Tricart (1969) glacial and periglacial conditions coincide. But ice masses, just like the snow from which they derive, blanket the surface of the Earth, and greatly reduce loss of heat by radiation, a loss that is essential, if deep freezing, and the structures it produces, are to develop. Once temperatures fall to the level necessary for deep freezing, the air will be too cold to hold the quantities of water necessary for massive snowfalls. Flint (1971) considers it was possible for the ice sheets to have been built up by precipitation at today's rates, provided only that temperatures were moderately lower, and notes (p. 268) that 'important sectors of former ice sheets seem not to have been fringed with extensive features created by frost action'. Embleton & King (1968) see the formation of snow and ice flourishing in temperatures 5–7 °C cooler than those of today. Condition B, which encourages the accumulation of snow and ice, should perhaps be called 'niveogenic'.

Parts of the periphery of a large ice sheet, which was nourished by distant snowfalls, might advance into an area which was already in the grip of deep frost, or frost might come to dominate an area already occupied by an ice sheet, whose growth would then have to stop, but ice sheets of themselves cannot induce deep freezing. Where frost conditions obtain, ice masses cannot melt away; thus if we find ice-wedge casts in meltwater gravels, the ice wedges must have formed in a phase of cold later than that that produced the ice from whose melting the gravels derived.

Sparks & West (1972) have warned against the use of synthetic jargon, but we might do better to speak of 'cryogenic' or 'cryokinetic', rather than 'periglacial', processes, because the structures that are formed are the result of the energy reactions of water as it freezes and thaws. A wealth of frost structures certainly formed in the Last Cold Stage, and when we try to see how many cycles of frost conditions are involved, we must remember that today it is being shown more and more clearly that an ice sheet in decay can produce a suite of deposits of different facies superimposed on one another. In the past the deposits composing such a suite have often been attributed to more than one ice sheet; a single frost cycle as it grows in intensity and then ebbs away may produce several structures of different facies.

We can construct a diagram (figure 1) to represent a hypothetical frost cycle. As the cycle develops, the winter cold freezes lakes, and produces a relatively thin frozen layer of soil: there may be a phase of heavy snowfall. For a time the frozen layer melts completely in the following summer, and so we have 'seasonally frozen ground'. But as long as a watertight sheet of frost – however thin – survives, it prevents the downward movement of water, and if the slope of the ground is less than 2° an 'active layer', a layer of supersaturated soil, will develop above the frozen horizon. Convectional disturbances in the active layer may bring about the erection of contained stones.

As the cold grows further in intensity, snowfall drops away, the winter freezing-front moves deeper down into the ground; the weaker heat of summer can no longer penetrate so deeply, and the active layer gets shallower. Ultimately very severe winter frost brings the mean annual air temperature (m.a.a.t.) below 0 °C, summer heat can no longer pierce the frozen layer, and permafrost is firmly established, while the active layer becomes negligible. For Péwé (1969) the permafrost zone is the periglacial zone. If m.a.a.t. falls below −7 °C, then ice wedges will develop, arranging themselves in polygonal patterns.

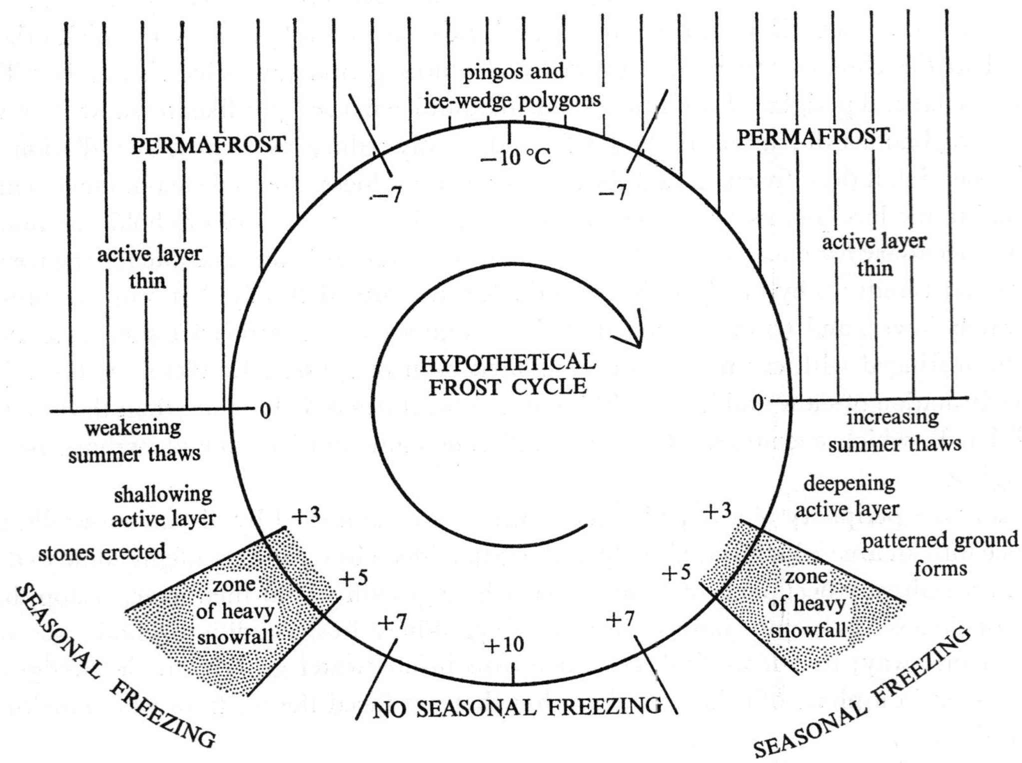

FIGURE 1. A hypothetical frost cycle.

Warmer conditions then begin to return, winter frosts ebb away, and the active layer deepens again. As it becomes deeper and warmer, movements analogous to convection currents in cell-like patterns develop in it, and these sort the contained debris, drawing the coarser material up to form the walls of honeycomb-like cells, and filling the cells with the finer fragments. As the active layer deepens, the polygonal patterns become more prominent and more perfect, until one summer the basal sealing-layer of frost completely melts, the water of the active layer drains away downwards, while its contained polygonal structures become immobilized, and survive to form polygonally patterned ground. It must be noted that patterned ground produced in this manner is quite different to the patterned ground produced by ice-wedge polygons. A phase of heavy snowfall may reoccur.

Thus the one cycle of frost may leave in one place erected stones, in another ice-wedge casts, and in a third polygonally patterned ground, and it may be difficult to interpret the record.

Also, we cannot assume that the frost cycle will always rotate at a constant speed, or complete a full circle. The tendency towards cold can be reversed at any point, and parts of the cycle can be speeded up or retarded, or even short-circuited.

But at one locality in Ireland, Ballymakegogue, Kerry (Mitchell 1970, Figs. 1 and 5), we appear to see a fairly complete cycle in deposits of Wolstonian age. Here the local rock is a brecciated Namurian shale, which is particularly susceptible to frost action.

When seasonal freezing of the ground set in at the beginning of the Wolstonian Cold Stage, the countryside was essentially free from trees, even though some pollen from warmth-demanding trees still lingered in the soil. Frost action broke up the plant cover, and as soil was moved downslope it brought derived pollen with it. Solifluxion then became more vigorous, and coarser debris was moved downslope to form a thick deposit of head. Developing permafrost then checked solifluxion, and as the frost cycle reached its coldest level, ice wedges formed. Material primarily ice-borne in origin, and rich in striated stones, rests on the head, but it is not clear whether this material was introduced by a contemporaneous glacier, or was solifluted down from an older deposit upslope. At any rate solifluxion was resumed, as rising temperature allowed an active layer to establish itself once more, and a further thick layer of head was deposited. As the head grew thicker and thicker, the slope on its upper surface will have got less and less, and cryoturbation will have replaced solifluxion. The upper layers of the head are deeply disturbed by cryoturbation, and involutions are well developed in it. It is very probable that these involutions belong to a later frost cycle of Devensian age, because as we have already seen any structures that were in the upper layers of the head as the Wolstonian Cold Stage ended, would probably have been obliterated by weathering during the following warm stage.

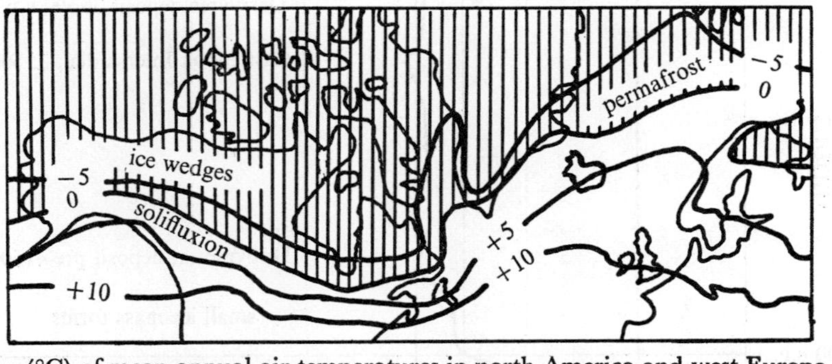

FIGURE 2. Isotherms (°C) of mean annual air temperatures in north America and west Europe, showing zones of solifluxion, permafrost and ice wedges.

We can also get a contemporary view, because the climatic conditions from temperate to extreme cold that develop in the first half of the frost cycle can be paralleled as we move from south to north, particularly in western North America (figure 2). We pass from the frost-free Pacific west coast to the solifluxion lobes of the Seward Peninsula, to the permafrost around Fairbanks, and to the ice wedges and pingos of the coastal plain around Point Barrow. But when we think about the active layer in these areas, we must remember that the long summer days of the arctic north may well bring about a deeper melting than the shorter days further south.

As noted at the opening, we have in Ireland a wide range of Devensian periglacial features on deposits of different ages. How many frost cycles do they represent? If we take an arbitrary horizontal scale, with five levels of mean annual air temperature, we can construct a very tentative curve (figure 3) to suggest how climate may have changed from one level to another during the Devensian. The levels are: +10 °C, the temperature of the British Isles today; +5 °C, the level at which seasonally frozen ground starts to appear, and some cryoturbation and solifluxion will take place; 0 °C, when permafrost becomes established; −5 °C, when the contraction-cracking that leads to the development of ice wedges will start; −10 °C, a temperature at which ice wedges may continue to grow, but little else will happen, as the ground is too solidly anchored by frost for much summer movement to take place. In addition it is accepted that when the temperature falls to below 7 °C cooler than at present – say to +3 °C – the air will be too cold to hold the water that is necessary to provide snow, if ice masses are to build up. As temperature falls from +5 to +3 °C, there may be a phase of heavy snowfall.

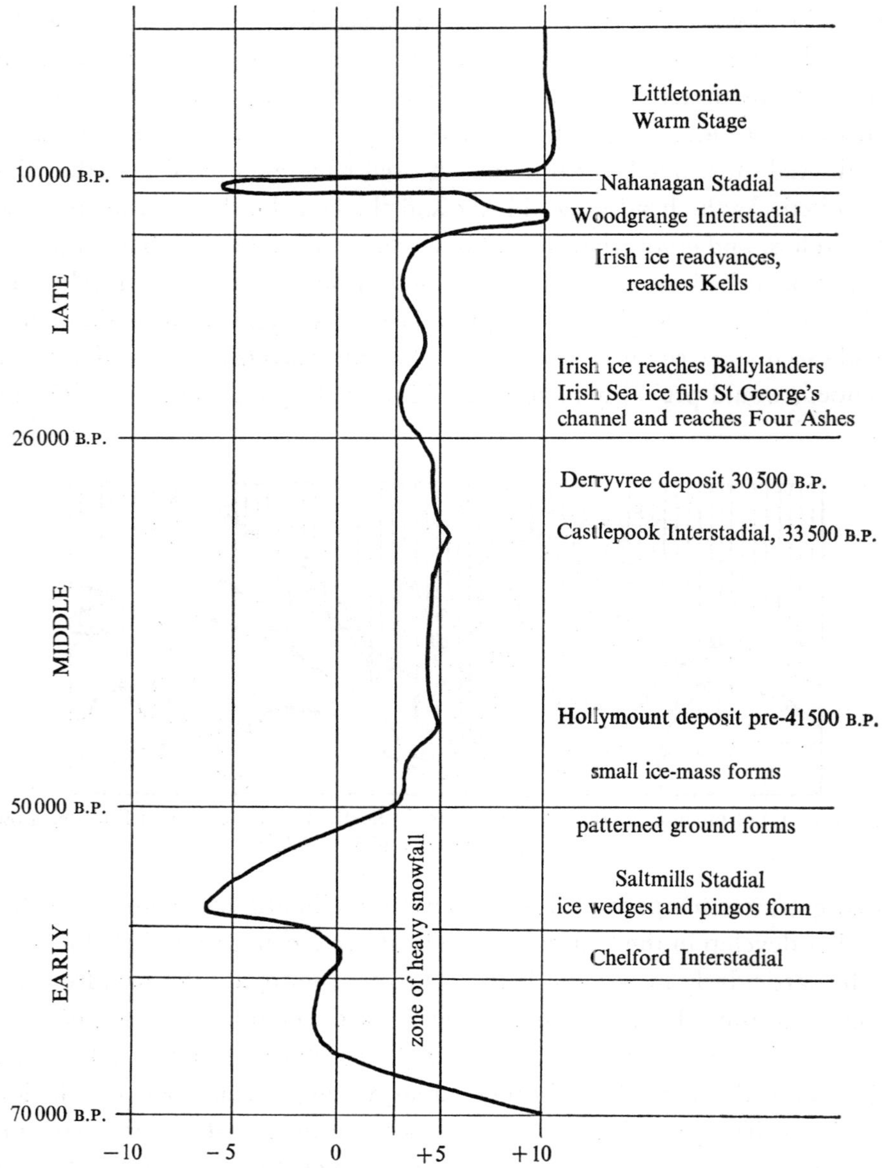

FIGURE 3. Hypothetical curve to suggest the movement of mean annual air temperatures from one level to another during the Devensian Cold Stage.

In Ireland, about one-fifth of the country, chiefly in the south, was not covered by ice at any stage of the Devensian (see figure 4), and it is in this ice-free area that we have our best development of periglacial structures, on a scale unmatched anywhere on deposits of Devensian age. If we accept the position set out above, that the surface of the Wolstonian deposits was free of any periglacial structures at the opening of the Devensian, then these structures will have developed over the whole area of Ireland, and have been subsequently destroyed or buried by the action of ice in those parts of the country where Devensian ice masses did later form. Outside those ice limits we have:

(i) fossil ice-wedge polygons, and numerous fossil pingos, both indicating temperatures below -7 °C,

(ii) well-developed patterned ground, with disturbance extending to 2 m below the surface, indicating easy cryoturbation in a deep active layer, and

(iii) where the ground was sloping, a landscape with concave slopes, where solifluxion on a massive scale has moved large quantities of material down into the valley bottoms.

In Britain we get our first view of Devensian climatic conditions at Chelford in Cheshire (Simpson & West 1958), where an organic mud whose fossils suggest a mean annual air temperature between $+2$ and -3 °C is contained in current-bedded sand, perhaps deposited in a braided river channel. The sand above the mud is penetrated by ice-wedge casts (Worsley 1966), which show that severe frost conditions developed after the mud had been formed.

It may well have been this cycle of cold that produced the massive frost structures of the south of Ireland, and so figure 3 places the cycle at the end of the Early Devensian. The deeply developed patterned ground, and the large quantities of solifluxion-earth or 'head' suggest that a deep active layer must have persisted for some time. As we have seen, we do not know how long our frost cycle took to complete a revolution, nor if it rotated at a constant rate. The massive features just referred to suggest that on this occasion the process of re-warming, and the deep active layer it produced, may have lasted for a considerable length of time, and to emphasize this possibility the curve has been drawn asymmetrically at this level. To achieve the survival of ice-wedge casts the converse would have to hold; re-warming would have to have had been rapid, and no deep active layer could have formed. At Saltmills in Co. Wexford erected stones can be traced down to a considerable depth, and patterned ground is also suggested; there are fossil pingos and ice-wedge polygons in the immediate vicinity, and so this cycle of cold and frost could be called the Saltmills Stadial (for location, see figure 4*a*).

We now arrive at a major problem. As we have seen, at both Hollymount (pre-41 500 B.P.) and Derryvree (30 500 B.P.) in Co. Fermanagh there is no indication in the materials that underlie the organic deposits that they had been either deeply weathered during the Ipswichian Warm Stage, or severely disturbed by frost action during the Saltmills Stadial. In each case the organic material seemed to rest on an unweathered undisturbed till, which should be older than 41 500 B.P. This position is accepted here, and it is envisaged that at the beginning of the Middle Devensian a small ice mass formed in the northwest part of the Irish midlands (see figure 4*a*), and deposited a layer of till. The deposits of this small early ice mass were later completely buried by a larger Devensian ice mass, which either removed the earlier till, or concealed it beneath its own deposits. In England there is no record of ice being formed at this time.

The relatively mild Middle Devensian continued, and just as in England we have interstadial conditions recorded in the Upton Warren Complex, so in Ireland we have the Castlepook Interstadial at 33500 B.P. We have also the organic deposits at Hollymount and Derryvree, but these may do no more than indicate a landscape with arctic plants and invertebrates, of which a record has happened to survive because the deposits were later buried by till.

The Late Devensian opens at 26000 B.P., and ice masses gradually built up in Ireland, Wales and Scotland. A powerful flow down the basin of the Irish Sea developed, and this carried material of marine origin as far as St David's Head in Wales, and Carnsore Point in Ireland; on the east the ice pushed into the English midlands as far as Four Ashes, Staffs, where its

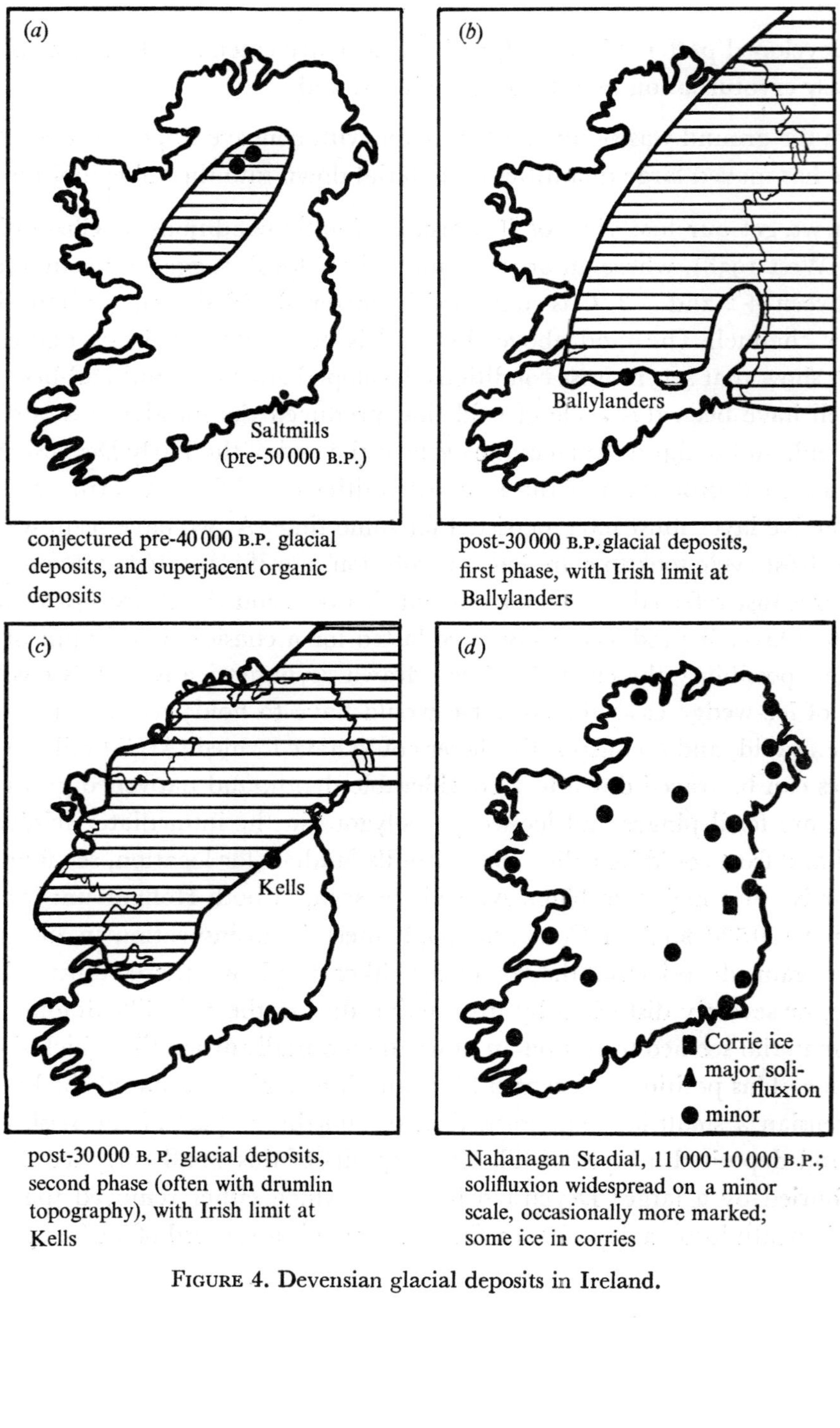

(a) conjectured pre-40000 B.P. glacial deposits, and superjacent organic deposits

(b) post-30000 B.P. glacial deposits, first phase, with Irish limit at Ballylanders

(c) post-30000 B.P. glacial deposits, second phase (often with drumlin topography), with Irish limit at Kells

(d) Nahanagan Stadial, 11000–10000 B.P.; solifluxion widespread on a minor scale, occasionally more marked; some ice in corries

FIGURE 4. Devensian glacial deposits in Ireland.

deposits rest on organic deposits of the Upton Warren Complex; on the west it kept ice from the Irish midlands pressed against the western shore of the Irish Sea. On the south, the Irish ice reached as far as Ballylanders, Co. Limerick (figure 4*b*).

At first the axis of this Irish ice mass ran approximately from Belfast to Limerick, but it later migrated slightly northwest, to run from the lower Shannon Estuary to Fair Head (figure 4*c*). Parts of western Ireland, hitherto free of ice, now became buried, while in the southeast some areas emerged from beneath the ice. The ice then began to melt away, revealing numerous drumlins as it did so, and had probably completely disappeared by 14000 B.P.

We thus have in the central midlands two areas of glacial deposits of slightly different ages, separated along a line that runs northeast/southwest through Kells, Co. Meath. Both areas show only very slight signs of periglacial activity, in marked contrast to the south of the country; rare ice-wedge casts and ice-wedge polygons occur in both areas; pingos are unknown in the younger northwest area, but do occur just inside the morainic limit in Co. Wexford, in the southeast area: patterned ground is unknown in either area; cryoturbation, if it can be recognized, is generally shallow and feebly developed: solifluxion is generally on a minor scale, and has nowhere produced concave curves; primary depositional features are still preserved in essentially mint condition. It seems clear that the two phases of the ice mass cannot be separated by a cycle of frost, neither are any organic deposits known to occur between the glacial materials of the two phases.

About 14000 B.P. a very dramatic amelioration of climate began, and convincing beetle evidence (Coope in Colhoun & Mitchell 1971) shows that temperature climbed quickly to a level at least as warm as that of today. However, the burst of heat was short-lived, and temperatures began to fall again, at first slowly. But about 11000 B.P., or perhaps a little later, temperature crash-dived to a very low level. In Scotland an ice mass of relatively modest size formed, and glaciers flowed south to the basin of Loch Lomond, and also the Lake of Menteith.

In Ireland so far ice at this time has been proved in only one corrie basin – Lough Nahanagan at 500 m in the Wicklow Mountains, and here the basin was only partly reoccupied. The emphasis seems to have been on cold, rather than ice. In the north of Ireland the Late Devensian ice mass surrounded, and perhaps buried, the Sperrin Mountains, and the dissolution of that ice can only have come late in the Devensian. Conditions mild enough to melt away an ice mass could not simultaneously have seen the growth of ice wedges. But the outwash gravels laid down by the decaying ice sheet are penetrated by ice-wedge casts (Colhoun 1971), which must be younger than the gravels, and probably formed in the Nahanagan Stadial.

When a pingo collapses, a central hollow is formed, and this is often occupied by a pond. Investigations of the muds in such ponds in Wales have not been able to produce mud older than 10000 B.P. (Watson 1975), and less detailed investigations in Ireland (Mitchell 1973) point in the same direction. Some pingos may thus have formed in the Nahanagan Stadial. In case it is thought that the duration of the stadial, probably less than 1000 a, was too short for the extensive development of pingos, recent studies (Ryckborst 1975) have shown that pingos up to 15 m in height – and there is no reason to think that many of the pingos in Ireland were on a much larger scale – can form in no more than 75 a.

When solifluxion begins to affect a slope whose soil is bound by the roots of grasses and herbs, lobate masses contained by plant roots begin to move down the slope. In the process mats of vegetation become buried, and may be preserved beneath the lobes of soil. At sea level in a cliff-section at Old Head, Co. Mayo, mats of vegetable debris lie below 2 m of 'till'. The plant material was first thought to be of interglacial age (Synge 1968), but a radiocarbon determination gave it an age of 10000 years, and therefore the solifluxion that buried it must have taken place in the Nahanagan Stadial. At Howth Demesne, Co. Dublin, also, a 'till' 2 m thick rested on a mud of Woodgrange age, with a ^{14}C date of 12000 B.P.; while in this case it is not impossible that the 'till' reached its present position in the course of landscape gardening, it may have been emplaced by solifluxion.

In the Nahanagan Stadial solifluxion on a minor scale was universal in Ireland, even out to the Atlantic coast, as is shown by figure 4*a*. At one site recently reinvestigated, Drumurcher, Co. Monaghan, the solifluxion-earth, which had a ^{14}C age of 10500 B.P., contained remains of about 100 species of beetle (Coope & Mitchell, in prep.). Many of the beetles indicated an arctic or sub-arctic regime, such as occurs today in the lower alpine regions of the mountains of Scandinavia, or the tundra regions of the far north. The site also produced the first seed so far found in Ireland of the arctic poppy (*Papaver radicatum* s.l.); today the distribution of this plant is strictly arctic-circumpolar.

The temperate muds that rest on the Nahanagan deposits have a ^{14}C age of 10000 years, and therefore this stadial lasted no more than 1000 years. Nevertheless in that short time, a frost cycle ran its full course from temperate conditions to conditions that allowed ice wedges and pingos to develop, and back to temperate conditions once more.

We can now give at least a tentative answer to the question we have posed, and we can say that the current limited amount of evidence that we have suggests that periglacial conditions occurred in Ireland twice during the Devensian Cold Stage. The first frost cycle, the Saltmills Stadial, came relatively early in the stage, after the Chelford Interstadial, and seems to have been prolonged and severe. The second cycle, the Nahanagan Stadial, came at the very end of the stage, between 11000 and 10000 B.P.; though short, its conditions were sufficiently severe for ice wedges to form. No Devensian periglacial features are known in Ireland whose stratigraphical position would suggest that they had formed in a frost period, other than one of the two described.

References (Mitchell)

Colhoun, E. A. 1971 Late Weichselian periglacial phenomena of the Sperrin Mountains, Northern Ireland. *Proc. R. Ir. Acad.* B **71**, 53–71.

Embleton, C. & King, C. A. M. 1968 *Glacial and periglacial geomorphology*. London: E. Arnold.

Flint, R. F. 1971 *Glacial and Quaternary geology*. New York: John Wiley.

Mitchell, G. F. 1973 Fossil pingos in Camaross Townland, Co. Wexford. *Proc. R. Ir. Acad.* B **73**, 269–282.

Mitchell, G. F. 1976 *The Irish landscape*. London: Collins.

Mitchell, G. F., Penny, L. F., Shotton, F. W. & West, R. G. 1973 A correlation of Quaternary deposits in the British Isles. *Geol. Soc. Lond., Spec. Rep.* No. 4.

Péwé, T. L. (ed.) 1969 *The periglacial environment*. Montreal: McGill-Queen's University Press.

Ryckborst, H. 1975 On the origin of pingos. *J. Hydrol.* **26**, 303–314.

Sparks, B. W. & West, R. G. 1972 *The Ice Age in Britain*. London: Methuen.

Tricart, J. (transl. Watson, E.) 1969 *Geomorphology of cold environments*. London: Macmillan.

Washburn, A. L. 1973 *Periglacial processes and environments*. London: E. Arnold.

Watson, E. 1975 Guide to Field Excursions. *Periglacial Symposium IGU, Aberystwyth.*

Discussion

R. J. O. HAMBLIN (*Institute of Geological Sciences, 5 Princes Gate, London SW7*). I was interested to hear Professor Mitchell suggest a lower as well as an upper Devensian till, as we have found the same in the I.G.S. survey of Telford, Salop. Here the Upper Devensian till as at Four Ashes is underlain by lacustrine and fluvioglacial sands, clays and gravels and then by a lower Irish Sea till. The 'Middle Sands' run out into the Main Terrace of the Worfe and hence of the Severn, and as Upton Warren Interstadial dates are known from tributaries of the Severn Main Terrace at Upton Warren and Fladbury we believe that our Middle Sands and Lower Boulder Clay indicate an ice advance immediately preceding the interstadial and probably younger than the Chelford deposit.

Phil. Trans. R. Soc. Lond. B. **280**, 211–228 (1977) [211]
Printed in Great Britain

Records of sea levels during the Late Devensian

By F. M. Synge

Geological Survey of Ireland, 14 Hume Street, Dublin 2, Ireland

Shorelines of both eustatic and isostatic type record sea-level changes over the last hundred thousand years. Marine cut rock platforms at roughly 25–40 m, 6–13 m, and just below mean sea level, appear to represent successive interglacials. Since the culmination of the Last Glaciation, the steady rise of sea level from below −100 m has been punctuated by rapid eustatic surges, notably at 16000–15000, 13000–12000, shortly after 10000, and round about 9000 a B.P.

Retardation of crustal recovery due to ice-loading about 15000–14000 a B.P. in Scotland parallels that which occurred in Scandinavia about 11000–10000 a B.P. In each case a localized transgression of considerable magnitude took place. The ensuing deglaciation resulted in the very rapid phases of uplift that led to the low marine levels recorded at *ca.* 10500 in Scotland, and *ca.* 8000 a B.P. in Scandinavia.

Introduction

Evidence for sea-level change is widespread throughout the coasts of the world. Marine deposits and the traces of former coastlines occur both above and below present sea level. Such changes of level can be explained in three different ways: *eustatic*, in response to the growth or melting of ice sheets; *isostatic*, caused by crustal deformation due to loading or unloading by ice or water; and *tectonic*, as a result of rapid earth movements that displaced crustal blocks.

Eustatic movements, by their very nature, are worldwide, so that theoretically they should provide a reliable basis for correlation. Isostatically and tectonically affected areas are particularly important, as they provide visible evidence of earlier sea-level changes which elsewhere may be drowned beneath the waters of the sea. These movements indicate a gradual general fall of sea level throughout the Quaternary era, but interrupted by temporary and rapid regressions of some 70–100 m during the main glacial episodes. At the outset of the Last Cold Period the level of the oceans started to fall well below its former stand a few metres above their present level. At the time of the greatest expansion of the ice sheets of this period a level of about 100 m below present datum was reached.

With ensuing deglaciation sea level again rose, but because the marginal areas of the great ice sheets did not immediately respond to isostatic rebound, they were submerged. Once crustal rebound gained momentum the formerly drowned peripheral areas started to emerge from the sea. But as sea level continued to rise, some of these marginal areas were again submerged, while new land was emerging rapidly in territory closer to the former ice centres. It has been suggested that this simple pattern has been interrupted at certain times by a very rapid downwarp close to the shrinking margins of decaying ice sheets (Wright 1937). The difficulty of explaining such a phenomenon and, indeed, finding good evidence of it, has led to the general abandonment of this idea. Yet, considerably more evidence has now come to light suggesting that marine transgressions did occur in such areas, quite unrelated to eustatic movements.

Once isostatic recovery was complete or considerably slowed down, the very rapid rise of sea level caused by the final melting of ice sheets all over the world led to a major transgression, which is represented in most places by a very clear shoreline. This event, which coincided with a marked climatic amelioration – the climatic optimum – culminated shortly after 7000 a B.P. and can be used as a convenient termination to the story of sea-level changes outlined in this paper.

Dating methods

Numerous methods have been evolved in order to date the various phases of sea-level change, and thus to establish a reliable chronology. Of these methods that of the radiocarbon (^{14}C) dating of organic material, such as peat or wood, bracketing or embedded within a marine sequence, has proved most effective. Thus a transgression can be dated by establishing the age of the upper layers of the underlying peat and that of the basal layers of the overlying one. But care must be taken to establish the altimetric height of the transgression by measuring the level of the actual shore notch, which could be considerably above the level of the tidal flats on which the upper peat later developed. This method has been used successfully in Britain, Ireland and Scandinavia to date land/sea-level changes during the past 9000 years.

Early archaeological sites can prove useful when they are closely associated with the coastline. But care is necessary to establish the exact level of the sea while the site was used. One ingenious method involves the determination of the phosphatic contamination of the soil caused by effluent from kitchen middens. A sharp lower limit to the contaminated soil fixes precisely the position of high water mark while the midden was being used (Simonsen 1968).

For earlier periods, the date of deglaciation has been established from ^{14}C determinations of marine shells derived from the basal layers of marine sequences deposited in front of a retreating icefront. In a falling sea-level sequence such dates refer to the marine limit. Correlation between age and sea level is particularly exact when the shells occur in clay layers interbedded in fluvio-glacial deltas. Further, the faunal characteristics of the shells may indicate the nature of the climate at the time. Generally the shells are absent from the actual shore deposits, probably because of subsequent weathering. Moraines representing oscillations of the icefront commonly contain lenses and masses of shelly marine clay. In Norway the dating of such shells have been found to date glacial advances and the associated sea level (Andersen 1968; Mangerud 1970). When such datings are incorporated in a shoreline diagram – i.e. a diagram of shoreline heights projected on to a base line running normal to the isobases – a dated uplift curve can be constructed to show the relation between sea level and time at a particular locality.

When it is possible to tie a dated glacial event to a 'varve chronology', the exact duration of ice recession and rate of fall of sea level can be measured in years. In this context it must be noted that a varve represents the annual load of sediment deposited on the sea floor each summer near the edge of the ice sheet. As the ice margin recedes, each annual varve is deposited overlapping that of the year before in the wake of the shrinking ice sheet.

In Greenland, Svalbard and Arctic Canada ^{14}C datings of driftwood and whalebone have proved very effective. It should be noted, however, that certain anomalous dates can be explained through the process of redeposition, either by man or a subsequent transgression of the sea. In Britain and Ireland driftwood does not seem to have survived the ravages of man or the weather from glacial times.

The oldest shorelines

Such is the effectiveness of glacial erosion that the survival of pre-existing marine features cut in bedrock, such as cliffs and shore platforms, within the areas covered by the Last Glaciation, is not usually expected. Yet in Scotland, in areas exhibiting signs of considerable glacial erosion, a marine cliff and platform with a notch at 25–40 m has survived in the Inner Hebrides (Wright 1937; McCann 1968). These features occur on the west side of areas of high relief, protected from the full force of glaciation from the east. In some cases till overlies the platform.

Much lower cliffs with a platform, or platforms extending from present sea level up to 14 m have a much wider distribution. Such features are specially noticeable outside the limits of the Last Glaciation along the south coasts of England and Ireland. Recently the very well developed platform associated with a cliff notch at 10 m at Oban in Scotland has been considered to belong to a late interglacial phase; the former correlation of this shoreline with the Main Postglacial Beach (Wright 1937) cannot be upheld, as the former tilts westward at a much greater gradient (0.15 m/km) than the latter (0.07 m/km), according to detailed levelling (Gray 1974*a*). The freshness of this feature, and its accordance in level with a buried gravel beach in the Forth dated about 10500 B.P., suggest that it may have formed during cold conditions prevailing during the last advance of the Scottish glaciers between 10800 and 10300 a B.P. (Sissons 1974). On the other hand the short time available for its formation, and the presence of ice moulding and striae that predated the latest advance of the glaciers, on the platform, argue for a much earlier age (Synge 1966; Gray 1974*a*).

The glacial shorelines

Many of the shorelines associated with deglaciation tend to be poorly developed compared to those of the present day, because of the short time available for their formation. Certain ones, however, stand out more clearly; either because they mark significant stillstands of sea level, or because they represent active transgressional phases. Such shore marks are most clearly developed in soft materials, and only very rarely occur in bedrock. The highest effects of marine wave action – the marine limit – can be represented by a very minor feature that may lie many metres above the highest obvious shoreline. Indeed the presence of marine clay above what was formerly regarded as the marine limit has been documented (Armstrong, Paterson & Browne 1975). The recognition and mapping of such washing limits is common practice in Scandinavia. Such minor features are associated with the upper limit of 'washed drift', represented by a metre thick layer of coarse sand and gravel which blankets the drift. The limit can also be located as the lower limit of channelled ablation moraine, braided channels on deltas, and perched blocks (figures 1 and 2).

These shorelines commonly occur above sharp fresh kettle holes, showing conclusively that the lowest kettle rims do not necessarily denote the marine limit, although they commonly accord with the level of a distinctive shoreline. Also perfectly sharp esker crests may have originally formed beneath sea level, and survive uplift without being bevelled by wave action. In one particular example observed in south Finland uplift was so rapid that the crest of the esker passed through the wave zone of the Baltic in the space of three years (figure 1). This might explain why the drift slopes which we may observe round many reservoirs in Britain today are notched by a clear cliff notch, while no similar feature can be found on the

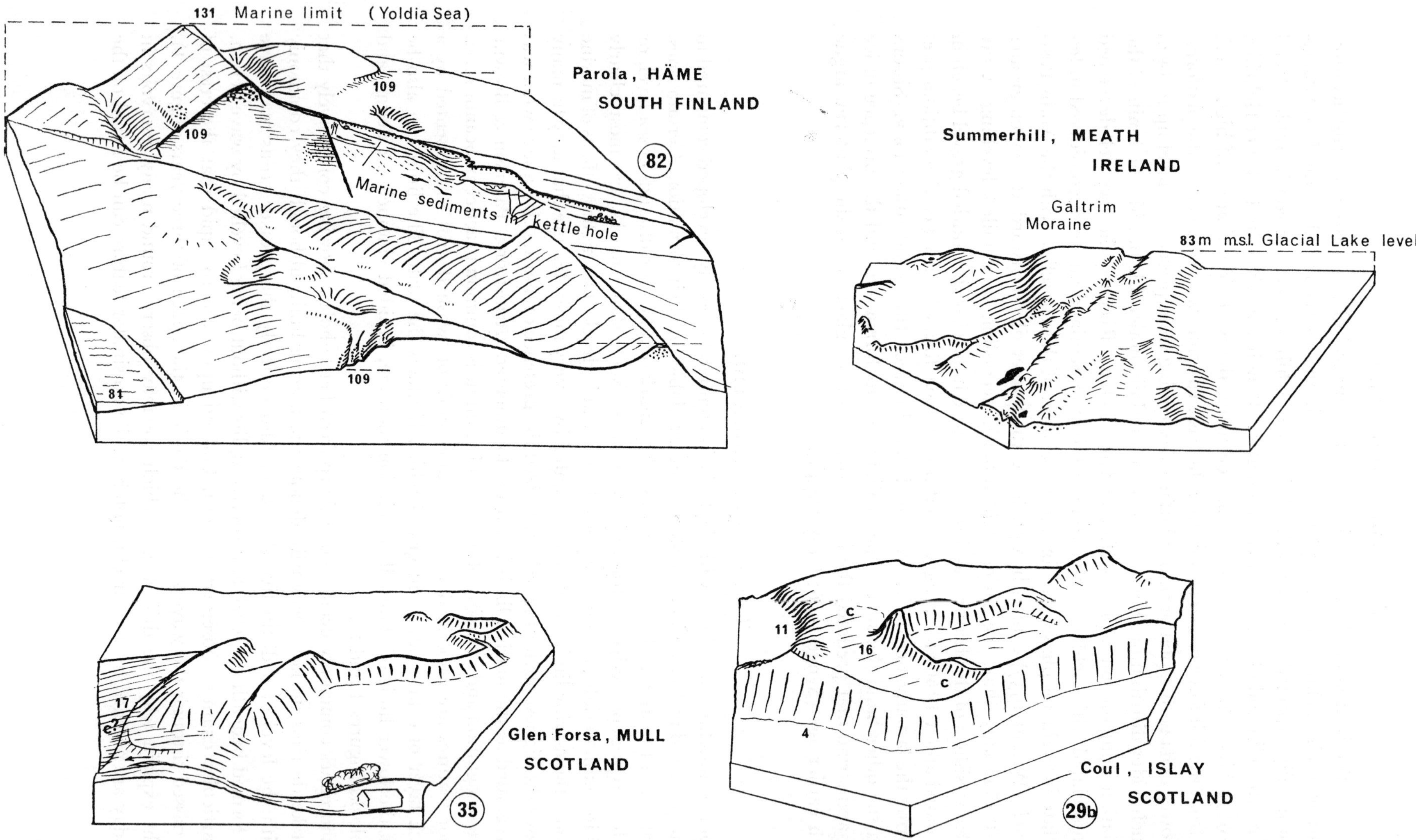

FIGURE 1. Deformation of eskers by wave erosion. Top left (82): parts of the crest of the Hattula esker, in south Finland, are bevelled by wave erosion at the washing limit (131 m). Open kettle holes and sharp drift forms occur below this level because of the continued presence of buried blocks of ice and the rapidity of uplift – namely, about 20 m per century. During the later stand of sea level at 109 m kettle holes were filled with littoral sediments and the drift hillocks were cliffed or removed by wave action (based on field observations by the writer). Top right: typical drift landforms associated with esker terminating in an ice-marginal delta moraine. Bottom left (35): esker truncated by later glacifluvial or marine action in Glen Forsa, Isle of Mull. Bottom right (29b): esker cliffed by wave erosion in an exposed location at the marine limit (16 m) on Islay.

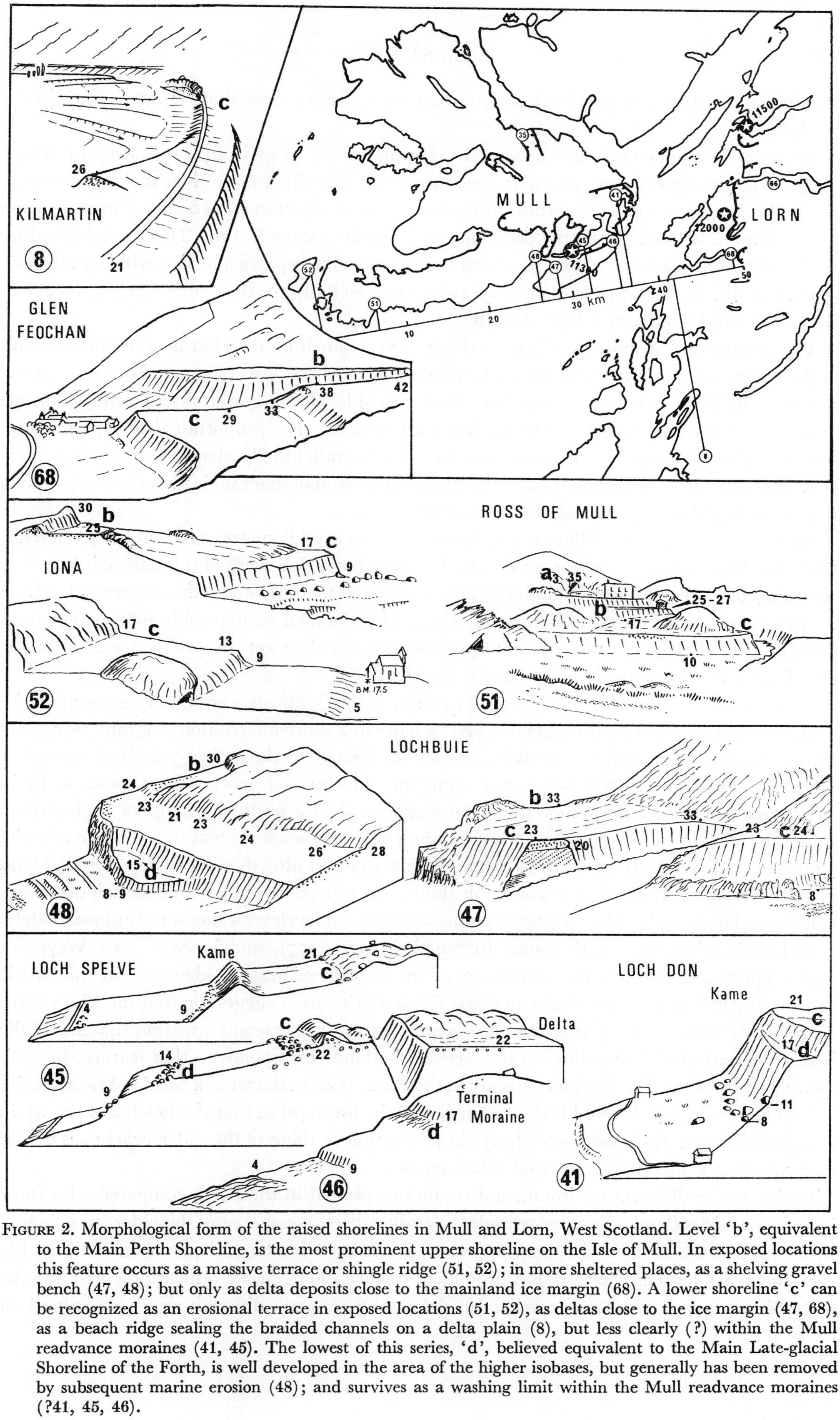

FIGURE 2. Morphological form of the raised shorelines in Mull and Lorn, West Scotland. Level 'b', equivalent to the Main Perth Shoreline, is the most prominent upper shoreline on the Isle of Mull. In exposed locations this feature occurs as a massive terrace or shingle ridge (51, 52); in more sheltered places, as a shelving gravel bench (47, 48); but only as delta deposits close to the mainland ice margin (68). A lower shoreline 'c' can be recognized as an erosional terrace in exposed locations (51, 52), as deltas close to the ice margin (47, 68), as a beach ridge sealing the braided channels on a delta plain (8), but less clearly (?) within the Mull readvance moraines (41, 45). The lowest of this series, 'd', believed equivalent to the Main Late-glacial Shoreline of the Forth, is well developed in the area of the higher isobases, but generally has been removed by subsequent marine erosion (48); and survives as a washing limit within the Mull readvance moraines (?41, 45, 46).

sea coast; clearly the reservoirs have endured for a longer time than many of the glacial sea levels!

Isostatic shorelines may be analysed by plotting them graphically on a height/distance diagram across the isobases. The older ones are invariably tilted more than the younger ones. As the broken nature of most terrain disrupts the continuity of any particular shoreline, some method has to be used to identify different portions of the same feature. This is established by the longitudinal tilt of the shoreline across the isobases. Each particular line will have its own particular tilt. In this way a system of shorelines can be built up. If some are dated, the age of the others can be determined graphically.

The accuracy of a shoreline diagram depends very much on the classification and identity of the features that are levelled. Very often such diagrams are also constructed on the assumption that uplift was regular, without the occurrence of local tilting and block faulting. For this reason some investigators prefer to analyse their data by the 'proportional height' method. According to this principle the main shorelines can be individually identified by their proportional height, one to the other. Thus a shoreline in this kind of diagram is not necessarily drawn to pass through the greatest number of measured points.

Shorelines that can be followed continuously for many kilometres give the best reference levels for determining the true isostatic tilt. The very important work of this nature initiated in the valley of the Forth in Scotland by Sissons underlined the need and value of precise levelling in an estuarine environment. This work, along with detailed stratigraphic information from numerous boreholes, still continues, and provides data for the most comprehensive and reliable dated uplift curve for Scotland (Sissons & Smith 1965).

Ten years ago, following on the work of Donner (1963), efforts were made to assemble the Scottish and Irish data on the glacial raised beaches in a shoreline relation diagram, using new measurements from widely scattered localities in the west of Scotland (Synge & Stephens 1966). Using the Main Postglacial Beach as datum, the older glacial beaches were found to lie in clearly marked zones. Each shoreline zone was related to a particular stage of deglaciation, from the oldest or 'a' series, dating from the time when the ice sheet was very extensive, to the youngest or 'f' series, from the time when the glaciers had already shrunk into the Highland glens of Scotland. At the time considerable doubt was expressed on the value of such diagrams, constructed from such widely scattered points where beach levels were measured approximately.

Although subsequent work round Inverness (Synge 1977), and in Skye (von Weymarn 1970), appears to confirm the correctness of the pattern of beaches identified in that work, certain conclusions appear to be erroneous. The writer's earlier suggestion that the Main Perth Shoreline of Sissons is not a true shoreline is not correct; it now seems apparent that one of the best developed glacial shorelines in the west of Scotland corresponds to that feature. But evidence that this shoreline tilts more steeply in the Forth than in the west of Scotland seems valid.

Another error, which should be corrected, was the assumption that the heights of the Main Postglacial Beach throughout are always in proportion to those of the older beaches. There is now sufficient evidence to show that this is not so.

Owing to the different environmental conditions present in the Forth compared with those in the west of Scotland, there are considerable difficulties in correlating the older shorelines of these two regions. Many of the 'beaches' in the Forth have developed as estuarine infill, so that each successive level only occurs as a prominent feature where glacial outwash merged into the firth (Sissons 1974). The present writer has defined beaches in a different way – i.e. as levels at which wave action was sufficiently prolonged to produce beach bars and cliffs in drift deposits. In this case, therefore, the prominence of any particular beach does not depend on the volume of glacial outwash, but rather depends on the duration of the stand of sea at a particular level.

Along the rocky coasts of the west of Scotland the raised beaches commonly occur as shelving deposits of shingle and gravel between irregular bosses of rock; in such instances the main beach levels may not be represented by either terrace flats or cliff notches, but only by the upper level of infill. In many places the sloping banks of beach material lie against steep rock faces simulating cliff notches (figure 2). Thus levelling such features to determine former stands of sea level requires special techniques.

Some of the above considerations may explain why recent studies in Mull and Lorn have failed to establish any clear pattern of glacial strandlines by a programme of detailed levelling of terrace flats and notches, in spite of the fact that clear evidence of consistent marine levels at '100 feet' and '75 feet' occur in the immediate neighbourhood, according to the observations of Bailey *et al.* (1924). This detailed investigation did, however, establish very accurately the heights and distribution of postglacial shorelines (Gray 1974*b*). The problem of relating sea-level changes to the glacial history of this fascinating area will be discussed later.

In this paper it is hoped to demonstrate the importance of mapping in some detail the morphology of strandline systems on a large scale, in order to establish the levels at which significant stands of the sea occurred. This kind of work was first undertaken in 1967 in the Varanger district of Norway, in collaboration with J. Rose and J. Smith with considerable success. Suitable areas for this kind of work occur also in Scotland, for example beside the Moray Firth from Elgin to Dornoch (Ogilvie 1923) and on the Ayrshire coast. But as yet this type of investigation has not been undertaken.

Sea level during the maximum of the last glaciation

Hitherto sea level during the maximum advance of the ice of the last glaciation has always been assumed to lie at levels well below that of the present time. But there is now some suggestion that this level did not lie far below present sea level. This is suggested by the disposition of the lateral moraine of the Irish Sea glacier along the Irish coast. These morainic deposits descend from 280 m on the Dublin Hills to 150 m some 20 km further south, but thereafter these maintain an upper limit at about 100 m for the next 90 km until the terminal moraine is reached at Kilmorequay on the south coast of Wexford. Such a horizontal ice margin suggests that this part of the glacier was afloat as shelf ice in the southern part of the Irish Sea basin. A similar distribution of glacial till of the same age in east Lincoln suggests an analogous situation in the southern part of the North Sea basin (figure 3).

The youngest marine shell dates of 30000–29000 a B.P. from the younger Irish Sea till on the west coast of Wales (John 1967), and the presence of organic beds dated 30655 and 36340 a B.P. in fluvial sediments beneath the terminal zone of the same till sequence in Staffordshire (Shotton 1967), give a maximal date for the glacial advance. This may suggest that the maximum ice advance in Britain and Ireland occurred some 9000 a before an important second re-advance. The latter is identified by a much more prominent end-moraine with no evidence of a high sea level. At that time the Irish Sea glacier was much less dominant so that the ice failed to extend much beyond the Lleyn Peninsula (Synge 1963).

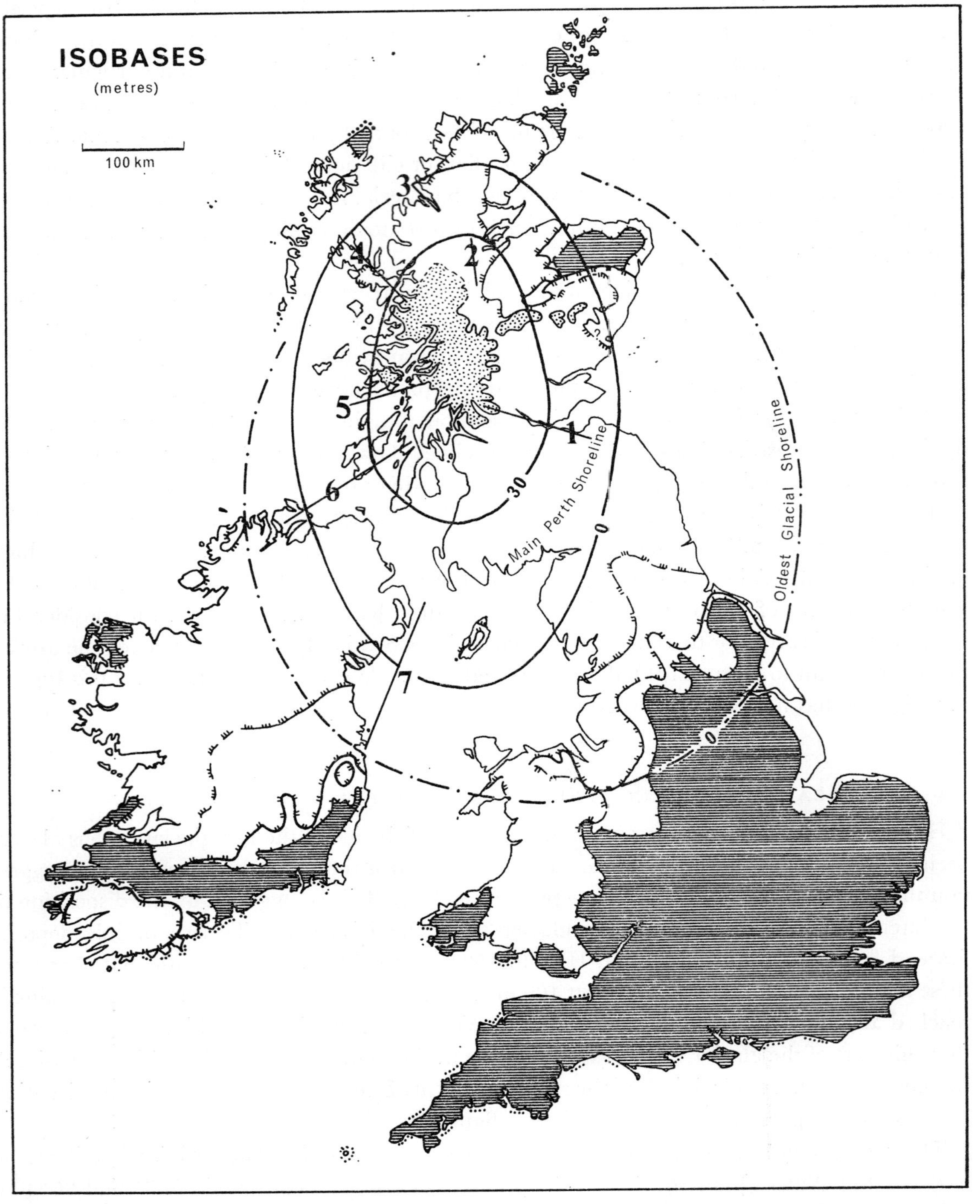

FIGURE 3. Relation between isobases and glacial limits. Unglaciated areas are heavily shaded, while that covered by the ice of the last readvance is stippled. Three isobases are shown: 0 m line of the oldest glacial, or 'a_1' shoreline; 0 m line and 30 m line of the '*b*' or Main Perth Shoreline. Base lines used in the construction of the isobases are numbered 1–7.

Oldest Late-glacial shorelines

The oldest Late-glacial shorelines occur well within the limits of the former ice sheet. The highest and oldest shore marks occur outside the limits of a very clear and prominent end-moraine on the east coast of Ireland – the Drumlin or Dunleer Readvance (McCabe 1973; Stephens, Creighton & Hannon 1975). This shoreline is very clearly displayed on Clogher Head, as a zone of bare rock from which the drift has been removed by wave action up to 19 m m.s.l. To the south the waves scoured a broad platform in drift about 1 km wide. South of the Boyne this shoreline is associated with a gravel beach ridge and cliff notch, that had been traced for 25 km along a gradient of 0.24 m/km.

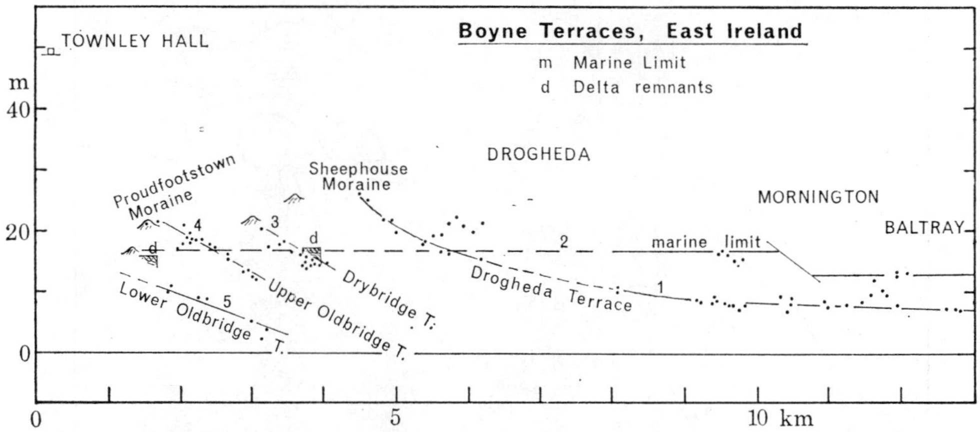

FIGURE 4. Relation between the oldest glacial shoreline ('a_1') and glacial outwash terraces in the Lower Boyne, Ireland. (1) *The Drogheda terrace*, outwash from the Sheephouse Moraine, deposited when sea level stood at about 7 m m.s.l. (2) *Marine limit* at about 16–17 m m.s.l. represented by the dissected fragments of a large delta between Proudfootstown and Drybridge (d). (3) *Drybridge* outwash terrace, deposited during a time of low sea level. (4) *Upper Oldbridge* outwash terrace. (5) *Lower Oldbridge* terrace, deposited after the area became ice-free.

In the Boyne valley the sea transgressed deltaic outwash by 10 m. This rise of level is represented by fragments of a large delta at 16–18 m m.s.l. constructed as the ice margin shifted some 4 km further upstream. As sea level fell a series of outwash terraces cut down through the former delta surface (figure 4).

Before this transgression occurred, or rather reached its maximum, Irish Sea ice readvanced up the Boyne valley. Glacio-marine deposits containing shells near the limit of this readvance, and originally regarded as a marine sequence *in situ* from an earlier glacial period (Colhoun & McCabe 1973), may in fact have been transported as an erratic at this time. If this is so the age of this transgression could be determined by a radiocarbon date from these shells.

The relation between this transgressive shoreline 'a_1' and those further north cannot be determined exactly. In the diagram the writer suggests that the high Boyne beach is older than the highest one near Carlingford. Both shorelines are truncated by readvance moraines. The older moraines immediately postdate shoreline 'a_1' and correspond with the moraine that encircles the Carlingford Mountains and the Mournes. The same moraine in the Isle of Man is dated as older than 18900 a B.P. This date was obtained from organic material from a kettle hole in the above morainic belt (Shotton & Williams 1971). The probable age of this first Late-glacial transgression is therefore older; probably by a few centuries. The apparent

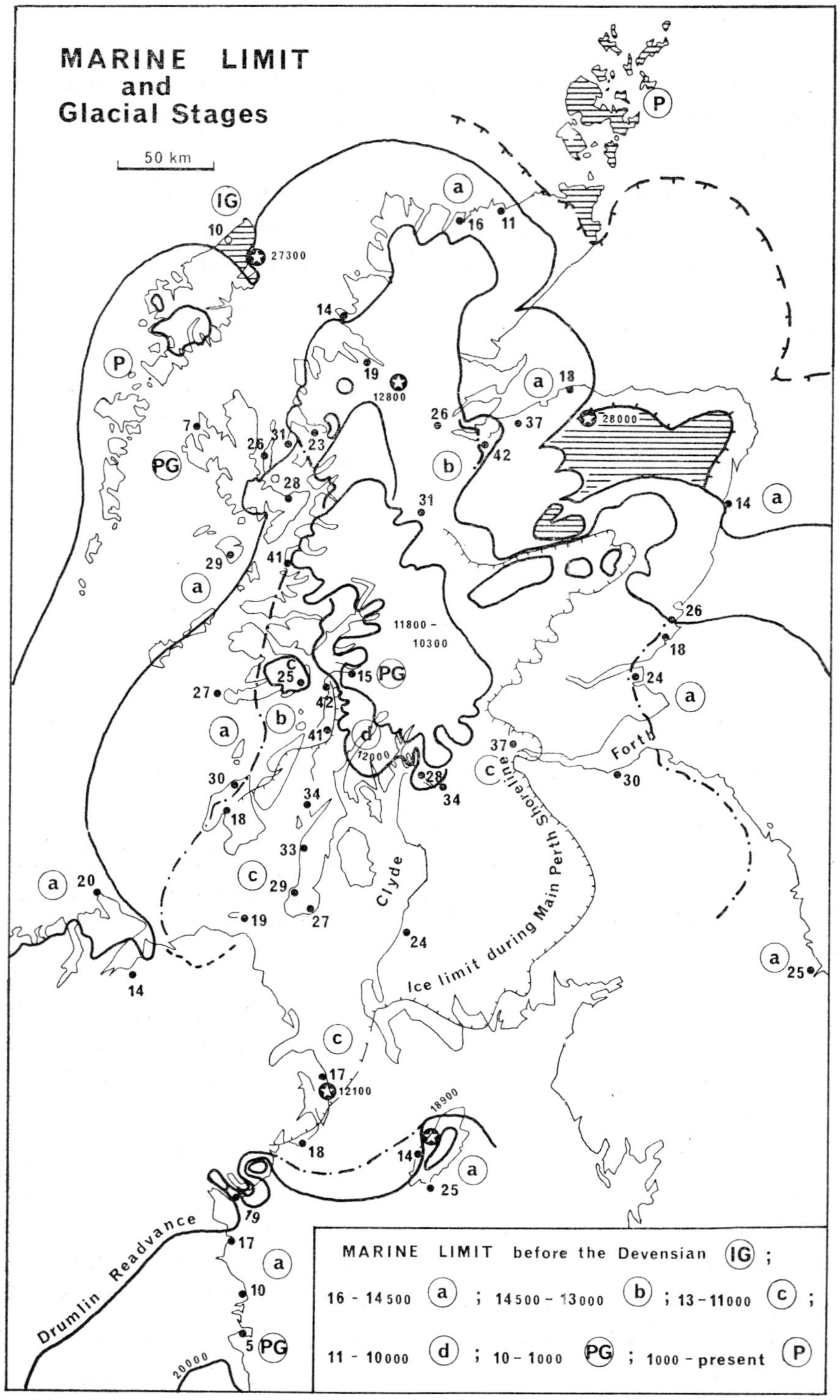

FIGURE 5. Height in metres and approximate age of the marine limit in Scotland and Ireland. The shaded areas (North Lewis, most of Orkney, Buchan and the Cairngorms) are regarded as being ice-free during the Last Glaciation. Star symbols refer to the oldest ^{14}C dated organic horizon at or near the drift surface; those in Scotland are based on Donner (1970) and Sissons (1976). Corrigendum: the number 28000 in the upper right of the figure should read 28100.

continuation of a marine limit up the east coast of the Isle of Man and into the highest outwash terraces of the Bride moraine would support this view. At Peel the marine limit is 10 m lower, suggesting that ice still lay against the west coast while the east coast was free of ice (figure 5).

The relation between the raised beaches and end-moraines in the northern part of the Irish Sea basin is shown in table 1.

TABLE 1

shorelines	South Louth and Boyne valley, Ireland	Carlingford, Ireland	Isle of Man (terminology after Mitchell 1965)
	Gilltown formation Sea level *low*	Windy Gap Moraine	Orrisdale series (kettle hole open *ca.* 18900)
	Sheephouse moraine; 7–8 m	?	?
a_1	Proudsfootstown delta; 17 m		
		19 m beach	21–5 m washing limit, on east coast
a_2	Dunany Readvance Moraine	Rathcor I/ Cooley Pt. Moraine	Jurby Ridge and Upper Ballateare till
	10 m beach	Rathcor II/ Cooley Pt. Moraine; 13 m beach	Crawyn sand plain and 15 m beach
a_3	Port beach 8 m	Greenore Moraine; 10–11 m beach	14 m marine limit at Peel

The highest beach, 'a_1', also seems to occur in northern England, where it has been noted at 25 m round the Tees estuary (Agar 1954). Near Liverpool a marine limit has been recorded at *ca.* 5 m as the 'Hillhouse shoreline' (Gresswell 1957) and originally regarded as postglacial; but now proved much older (Tooley 1974). In Inishowen in north Donegal the marine limit is very prominent, typified by large ridges of shingle up to 22 m m.s.l.; the bedding in these ridges has been heavily disturbed by frost-heaving (Stephens & Synge 1965).

Closely spaced ice-marginal stages, each associated with a particular outwash terrace and beach terrace, typify the withdrawal of the ice from the important Drumlin or Dunany Re-advance limit (figure 3). Analogous ice-marginal sequences were noted at Carlingford and Inishowen in Ireland; at the north end of the Isle of Man; and in East Fife and Inverness in Scotland.

THE PERTH SHORELINES – 'B' SERIES

Before 14000 a B.P. deglaciation appears to have occurred with great rapidity, particularly in the east of Scotland. This is apparent because of the presence of high level marine clays at 40–46 m near Crieff some 5–6 m above the very prominent Main Perth Shoreline, believed to have been formed about 13500–13000 a B.P. This suggests that isostatic response was at first slow, but then increased until it was again checked, possibly by an ice mass lingering in the Western Highlands and the Lower Clyde. During the ensuing transgression the decaying ice sheet still completely filled the North Channel and Firth of Clyde.

Such a prominent feature as the Main Perth Shoreline, which is tilted at 0.43 m/km in the Forth, should be recognizable throughout Scotland. The very manner in which it terminates

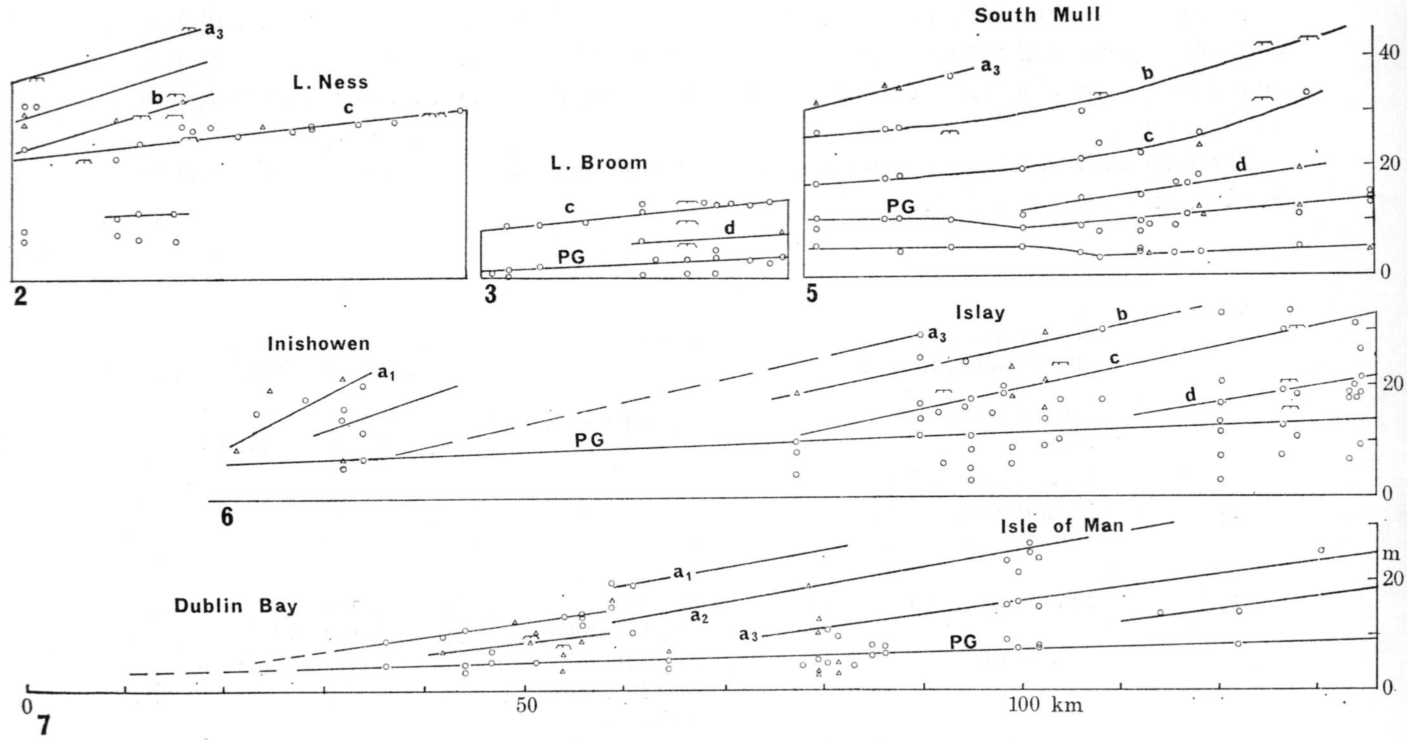

FIGURE 6. Shoreline diagrams from base lines 2, 3, 5, 6 and 7 in figure 3. The main reference levels are 'a_1' (Highest Glacial Shoreline), 'b' (Main Perth Shoreline) and 'PG' (highest Postglacial shorelines). The diagrams 2 and 7 are based on instrumental levellings; those numbered 3, 5, 6 are based on the hand-levelled data in Synge & Stephens 1966. In the diagrams, circles indicate shore notches; and triangles indicate beach ridges.

TABLE 2

	Ullapool	Coul, Islay	Inverness	Iona	Oban	(Sissons 1974) Plean, Forth
'b'	—	29 m, 0.23 m/km	—	25 m, 0.22 m/km	43 m, 0.75 m/km	35 m, 0.43 m/km Main Perth Shoreline
'c'	15 m, 0.15 m/km	—	24 m, 0.13 m/km	—	—	—
PG	5 m, 0.06 m/km	8 m, 0.03 m/km	10 m, ?	10 m, 0.07 m/km	13 m, 0.07 m/km	13 m, 0.08 m/ km Main Postglacial Shoreline

at the head of the Forth in a large fluvioglacial delta, that precedes a sudden drop of the marine limit by about 20 m at Stirling, should make identification easy. This supposition would seem to be correct; in many parts of the west of Scotland the only well developed shoreline above that of the Main Postglacial level would appear to be this very feature.

Previously the writer failed to make such a correlation because no allowance was made for variations in tilt from area to area. This is because isobases are not everywhere the same distance apart from one another. Once these clear upper shorelines are regarded as one and the same feature, a convincing isobase map could be constructed, which clearly shows that the focus of the ice cap was now shifting to the Western Highlands from a position further south (figures 3 and 6) (table 2).

Sea level between 13000 and 11800 a b.p.

After the sea receded from the Main Perth Shoreline lower shorelines were formed. The remnants of the former ice sheet melted rapidly in the Firth of Clyde, as the centre of dispersal shifted to the Western Highlands. At the same time waters with a boreal fauna were displacing the ice in the firth, as the marine limit was steadily falling owing to isostatic rebound. By 12600 b.p. the entire firth was ice-free, as shown by the presence of a shelly fauna in the marine clays of the Glasgow basin (Peacock 1971), associated with the marine limit at 34 m (Rose 1975). Generally it has been assumed that the sea connection was through the Lochwinnoch Gap, with a clearance of only a few metres. This seems most unlikely, for the obvious access route would have been up the Clyde. Therefore it is suggested that the upper red till further downstream represents a readvance of the L. Lomond glacier that later cut off the sea connection. This event could have occurred shortly before 11800 a b.p., the age of the shelly marine clay associated with a marine limit of 28 m along this part of the Clyde.

The transgressive nature of shoreline 'c' is clearly demonstrated in the Ness valley near Inverness, where the sea flooded open kettle holes. These are readily distinguished from those at higher levels by their less abrupt contours. The same shoreline also cut across the lower end of a fluvial terrace that was formed by the outlet of the youngest Glacial L. Ness; in this case the Ness river was graded to a level below that of Shoreline 'c'.

Although no obvious moraines are associated with the readvance postulated above, the sudden drop of the marine limit observed further west, in Loch Long, Glendaruel and near Otter Ferry in Loch Fyne suggests a significant halt of the ice margin (Sissons 1974). Barnacles *in situ* at 20.5 m at Glen Cruitten, dated *ca.* 12000 b.p. (Shotton & Williams 1973) pre-date a similar drop in level at Oban.

Sea level between 11800 and 10800 a b.p.

At this time, roughly corresponding to the milder climatic conditions represented by Pollen Zone II, or Allerød period, the significant *Shoreline 'd'* was formed. Near Oban it was observed as a clear notch on the distal side of a morainic deposit at 19–21 m. Previously the writer claimed that a higher shore – the equivalent of Bailey's '75 foot' level on Mull – occurred within the limits of the glacial readvance so clearly visible on that island (Synge 1966), as washing limits. The subsequent dating of marine shells collected from the terminal moraine at Kinlochspelve to *ca.* 11300 a b.p. was believed to prove that the readvance occurred during the Late-glacial climatic deterioration, *ca.* 10800–10300 a b.p. (Gray & Brooks 1972). As sea level during this latter period was below 15 m the presence of higher raised beaches of this date is impossible.

Critical examination of the evidence suggests alternative explanations. For example, the terminal moraines on Mull are quite unlike their supposed counterparts on the mainland. Instead of producing large outwash fans graded to a low sea level, they straddle their respective valleys – L. Don, L. Spelve and L. Bà – as narrow steep-sided ridges, rather typical of subaqueous moraines. Also, it should be noted, that the L. Don moraine is composed of sand, not gravel or till and blocks. Evidence of higher sea levels – 'b' level at 30–35 m and 'c' level at 21–25 m – lie immediately adjacent to these moraines, unlike those of the mainland bordering Loch Linnhe and the Firth of Lorn.

Bailey considered that the L. Don moraine was formed while the sea still stood at the 23 m ('75 foot') level, because the outer face of this feature is notched at that level. Gray (1975) confirmed the presence of a bench or notch beneath thick peat, but did not offer any explanation as to its origin. The writer also visited Wright's classical site in Glen Forsa, where an esker was interpreted as terminating in a fan that extends down to the level of the Main Postglacial Shoreline. This evidence shows that sea level was already low while ice still stood in the glen. Gray (1975) confirmed the fan form of the feature below the esker and pointed out the channels developed on its surface.

Although the writer would accept the idea that dead ice lingered in the glen as sea level fell, he points out that the notch cutting across the end of the esker at *ca.* 17 m (not 23 m as previously stated (Synge 1966)) is quite atypical of esker fans. Furthermore, the coincidence of this level with that of the 'c' shoreline is striking. This evidence, coupled with identification of an upper washing limit at a comparable level in L. Don and L. Spelve, suggests that after the readvance the ice cap went dead, and the glaciers occupying the sea lochs melted away very rapidly before sea level fell. As uplift continued dead ice remained in the valleys for a long time.

Additional evidence of the presence of the sea at the 'c' level during the advance of the Mull glaciers is furnished by the occurrence of fluvioglacial delta bedding adjoining the moraine at Kinlochspelve. Bailey attributed this to the presence of an ice-dammed lake on the site of the present L. Uisg, although no outlet channel was named. The level of this feature appears to be so close to that of the L. Spelve washing limits that the sea, rather than a lake, most probably occupied the valley at the time.

The ^{14}C dates obtained from marine shells caught up in the readvance moraines of the mainland valleys are all grouped around 11800 a B.P. (Peacock 1971). Therefore the readvance cannot have occurred before that date, but was it much later? Although the palynological and radiocarbon record points to a major readvance between 10800 and 10300 a B.P., the occurrence of other earlier ones remain a possibility. The total absence of dates between 11800 or 11300 and 10800 a B.P. is striking. This suggests that, between those dates, a marine fauna was unable to enter the inner sea lochs. This can best be explained by the continued presence of glacier ice.

The above situation exactly parallels that experienced in North Norway. Here, within the terminal zone of the Late-glacial readvance moraines, earlier ones beneath the moraines have been dated 11800 and 11300 a B.P., the very same dates that were obtained in connection with the moraines mentioned above. This indeed suggests that the milder conditions of the Allerød period was reflected by an increase in precipitation of snow in the West Highlands. This caused a rapid build up of snowfields which caused an advance of the mainland glaciers *ca.* 11800 a B.P. and a maximum advance of the Mull glaciers *ca.* 11300 a B.P. During the later colder conditions between 10800 and 10300 a B.P., the mainland glaciers advanced further than before, and built their outwash into a much lower sea level. Further east the drier and warmer summers of the Allerød was unfavourable to the survival of glaciers (Sissons 1974).

The suggested correlation between sea-level change and glacial advance is suggested in table 3.

TABLE 3

	Lorn	SE Mull	L. Lomond	
'b'	41–45 m outwash deltas			13000 B.P.
	?		34 m readvance max.	12000
'c'	first advance		28 m second advance	11800
	28–33 m Oban Moraine	20–24 m readvance		11300
'd'	18–20 m	15–16 m		10500
	12–14 m L. Etive Moraines		main readvance	10800/ 10300

SEA LEVEL DURING THE LOCH LOMOND READVANCE, BETWEEN 10800 AND 10300 a B.P.

In the Forth, deposits of a gravel beach beneath postglacial estuarine muds and peats denote a stable phase of marine activity at *ca.* 10500 a B.P. (Sissons 1974). This beach, now named the Main Late-glacial Shoreline, has been identified with the rock-cut cliff and main rock platform standing at 10 m at Oban. Reasons have already been given to suggest that the low level rock-cut shore features were originally fashioned much earlier (Gray 1974*a*).

At a few places near Oban, a very well developed shoreline cut in drift was observed some 5–6 m above the Main Postglacial Shoreline. In most localities this beach had been destroyed by later marine erosion. The clarity of this feature – shoreline 'd' according to the writer's classification – as the only obvious raised shore above the level of the postglacial series, suggests that it may be the correlative of the buried gravel layer of the Forth. This beach appears to correspond with Gray's LS2 line, inclined at 0.3 m/km, the only obvious zone of shore marks observed above the postglacial series near Oban (Gray 1974*b*).

SEA LEVEL AFTER THE LOCH LOMOND READVANCE

Because of the general low sea levels recorded after the last significant advance of the glaciers, little information is available for the thousand or so years after 10300 a B.P. In the Forth a transgression has been recorded, for outwash from the Menteith Moraine is overlain by marine silt at 12 m, some 7 m above the estimated level of the buried gravel layer. Afterwards the level of the sea fell.

CONCLUSION

This investigation of sea-level change throughout the Last Cold Period has shown at least four phases of stillstand. If the data from Great Britain and Ireland are plotted on time/elevation graphs, and compared with graphs from Finland, Norway, Svalbard and Arctic Canada against a standard eustatic curve, the shorelines caused by significant rises of the oceans can be isolated (figure 7). Three of these appear to be eustatic, 'a_3' at *ca.* 16000–15000 a B.P.; 'c' at *ca.* 11800–11300 a B.P. (Allerød interval); and 'd' at *ca.* 10500 a B.P. (Late-glacial).

Very significantly the 'b' shoreline (Main Perth) is not eustatic, but represents an ice marginal downwarp that postdated the Drumlin Readvance of Ireland, in much the same way that Portlandia transgressions in Scandinavia followed the Late-glacial Readvance. According

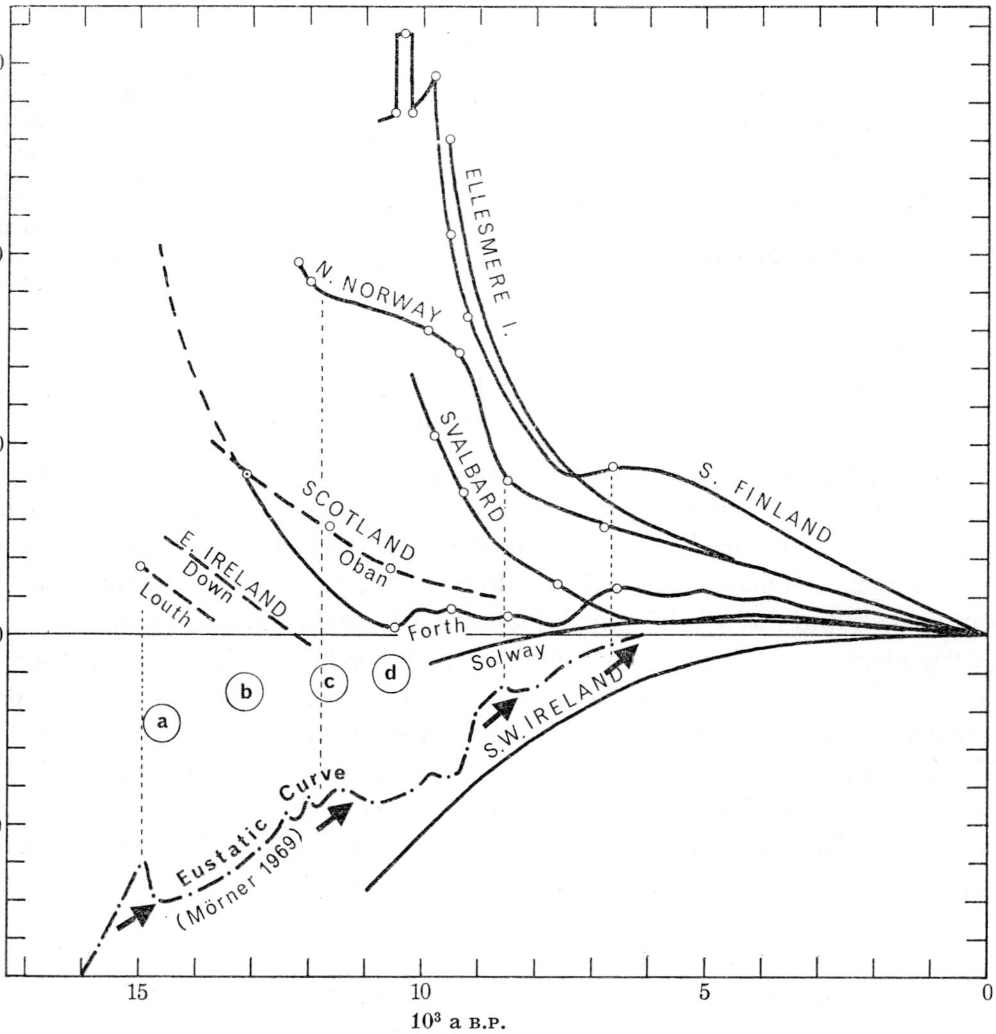

FIGURE 7. Time-elevation diagrams for stations round the North Atlantic. For comparison, Mörner's 1969 eustatic curve has been added. Data for *Ellesmere Island*, Arctic Canada, from Blake 1975; for *south Finland*, from the writer's field observations in association with Donner 1964; for *North Norway*, from the writer's field observations in association with Andersen 1968; for *Svalbard*, from Feyling-Hanssen & Olsson 1959; for *Scotland*, the *Forth* from Sissons 1974; *Oban* and *Lorn*, from the writer's field observations; the *Solway*, from Jardine 1975; and for *Ireland*, *Down*, from Synge & Stephens 1966; *Louth* from the writer's observations; and *south west Ireland* from Stillman 1968. The main eustatic surges coincide with shorelines 'a_1' (Highest Glacial Shoreline) at *ca.* 15000 a B.P. and 'c' at *ca.* 12000 a B.P. Shorelines 'b' (Main Perth Shoreline), 13500–13000 a B.P., and 'd' (Main Late-glacial) *ca.* 10500 a B.P. in Scotland seem to be isostatically controlled. Likewise retardation of uplift, and even downwarping, has produced significant shorelines at *ca.* 10300 and *ca.* 9800 in Scandinavia, that seem to coincide with a eustatic rise.

to some authorities this sinking occurred up to five hundred or so years after the readvance had taken place.

There are some indications that the Late-glacial raised beach that is so prominent in North Norway, indicates a combination of eustatic rise and crustal sinking. The sinking being a response to a lengthy halt in glacial recession, namely from *ca.* 12000 a B.P. to 10300 a B.P.

References (Synge)

Agar, R. 1954 Glacial and Post-Glacial geology of Middlesbrough and the Tees Estuary. *Proc. Yorks. geol. Soc.* **29**, 237–253.

Andersen, B. G. 1968 Glacial geology of Western Troms, North Norway. *Norges geol. Unders.* **256**, 155 pp.

Armstrong, M., Paterson, I. B. & Browne, A. E. 1975 Late-glacial ice limits and raised shorelines in East-Central Scotland. *Quaternary Studies in North East Scotland*, (QRA Excursion Guide) (ed. A. M. D. Gemmell, Geography Department, University of Aberdeen).

Bailey, E. B., Clough, C. T., Wright, W. B., Richey, J. E. & Wilson, G. V. 1924 The Tertiary and Post-Tertiary geology of Mull, Loch Aline and Oban. *Mem. geol. Surv.*

Blake, W. Jr. 1975 Radiocarbon age determinations and Postglacial Emergence at Cape Storm, Southern Ellesmere Island, Arctic Canada. *Geogr. Ann.* **57** A, 1–71.

Colhoun, E. A. & McCabe, A. M. 1973 Pleistocene glacial, glaciomarine and associated deposits of Mell and Tullyallen townlands, near Drogheda, Eastern Ireland. *Proc. R. Ir. Acad.* B **73**, 165–206.

Donner, J. J. 1963 The Late- and Post-glacial raised beaches in Scotland II. *Ann. Acad. Scient. Fenn.* A 3, **68**, 12 pp.

Donner, J. J. 1964 The Late-Glacial and Post-Glacial emergence of South-Western Finland. *Soc. Sci. Fennica, Commentationes Phys.- Math.*, **30**, No. 5, 47 pp.

Donner, J. J. 1970 Land/sea level changes in Scotland. *Studies in the vegetational history of the British Isles* (eds D. Walker & R. G. West). Cambridge University Press.

Feyling-Hanssen, R. W. & Olsson, I. 1959 Five radiocarbon datings of post glacial shorelines in central Spitsbergen. *Norsk. Geogr. Tidsskr.* **17**, 121–131.

Gray, J. M. 1974*a* The main rock platform of the Firth of Lorn, western Scotland. *Trans. Inst. Brit. Geogr.* **61**, 81–99.

Gray, J. M. 1974*b* Lateglacial and postglacial shorelines in western Scotland. *Boreas* **3**, 129–138.

Gray, J. M. 1975 The Loch Lomond Readvance and contemporaneous sea-levels in Loch Etive and neighbouring areas of western Scotland. *Proc. geol. Ass. Lond.* **86**, 227–238.

Gray, J. M. & Brooks, C. L. 1972 The Loch Lomond Readvance moraines of Mull and Menteith. *Scott. J. Geol.* **8**, 95–103.

Gresswell, R. K. 1957 Hillhouse coastal deposits in south Lancashire. *Lpool Manchr geol. J.* **2**, 60–78.

Jardine, W. G. 1975 Chronology of Holocene marine transgression and regression in south-western Scotland. *Boreas* **4**, 173–196.

John, B. S. 1967 Further evidence for a Middle Würm interstadial and a Main Würm glaciation of south-west Wales. *Geol. Mag.* **104**, 630–633.

Mangerud, J. 1970 Late Weichselian vegetation and ice-front oscillations in the Bergen district, Western Norway. *Norsk Geogr. Tidsskr.* **24**, 121–148.

McCabe, A. M. 1973 The glacial stratigraphy of eastern counties Meath and Louth. *Proc. R. Ir. Acad.* B **73**, 355–382.

McCann, S. B. 1968 Raised rock platforms in the Western Isles of Scotland. *Geography at Aberystwyth* (eds E. G. Bowen, H. Carter & J. A. Taylor), pp. 22–34. University of Wales Press.

Mitchell, G. F. 1965 The Quaternary deposits of the Ballaugh and Kirkmichael districts, Isle of Man. *Q. Jl geol. Soc. Lond.* **121**, 359–381.

Mörner, N.-A. 1969 The Late Quaternary history of the Kattegatt Sea and the Swedish West Coast. *Sveriges Geol. Undersökn.* C **640**, 487 pp.

Ogilvie, A. G. 1923 The physiography of the Moray Firth coast. *Trans. R. Soc. Edinb.* **53**, 377–404.

Peacock, J. D. 1971 Marine shell radiocarbon dates and the chronology of deglaciation in western Scotland. *Nature Phys. Sci.* **230**, 43–45.

Rose, J. 1975 Raised beach gravels and ice wedge casts at Old Kilpatrick, near Glasgow. *Scott. J. Geol.* **11**, 15–21.

Shotton, F. W. 1967 Age of the Irish Sea glaciation in the Midlands. *Nature Phys. Sci.* **215**, 1366.

Shotton, F. W. & Williams, R. E. G. 1971 Birmingham University radiocarbon dates: V. *Radiocarbon* **13**, 141–156.

Shotton, F. W. & Williams, R. E. G. 1973 Birmingham University radiocarbon dates: VII. *Radiocarbon* **15**, 451–468.

Simonsen, P. 1968 Fosfatanalyser fra Varangerbopladserne. *Tromsø Museums Skrifter* **7**, (6), 23–36.

Sissons, J. B. 1974 The Quaternary in Scotland: a review. *Scott. J. Geol.* **10**, 311–337.

Sissons, J. B. 1976 The geomorphology of the British Isles – Scotland (eds E. H. Brown & K. Clayton). London: Methuen.

Sissons, J. B. & Smith, D. E. 1965 Raised shorelines associated with the Perth Readvance in the Forth valley and their relation to glacial isostasy. *Trans. R. Soc. Edinb.* **66**, 143–168.

Stephens, N., Creighton, J. R. & Hannon, M. A. 1975 The Late-Pleistocene period in noth-eastern Ireland: an assessment 1975. *Ir. Geogr.* **8**, 1–23.

Stephens, N. & Synge, F. M. 1965 Late-Pleistocene shorelines and drift limits in North Donegal. *Proc. R. Ir. Acad.* B **64**, 131–153.

Stillman, C. J. 1968 The Post Glacial change in sea level in southwestern Ireland: new evidence from freshwater deposits in the floor of Bantry Bay. *Sci. Proc. R. Dublin Soc.* A **3**, 125–127.

Synge, F. M. 1963 A correlation between the drifts of south-east Ireland and those of west Wales. *Ir. Geogr.* **4**, 360–366.

Synge, F. M. 1966 The relationship of the raised strandlines and main end-moraines on the Isle of Mull, and in the District of Lorn, Scotland. *Proc. Geol. Ass., Lond.* **77**, 315–328.

Synge, F. M. 1977 Land and sea level changes during the waning of the last regional ice sheet in the vicinity of Inverness. *Transactions of the Inverness Scientific Society and Field Club.* (In the press.)

Synge, F. M. & Stephens, N. 1966 Late- and Post-glacial shorelines and ice limits in Argyll and north-east Ulster. *Trans. Inst. Brit. Geogr.* **39**, 101–125.

Tooley, M. J. 1974 Sea-level changes during the last 9000 years in North-West England. *Geogrl J.* **140**, 18–42.

von Weymarn, J. A. 1970 Spät- und Postglaziale Strandterrassen auf der Insel Skye und auf dem Angrenzenden Nordwest-Schottland. Unpublished thesis, Ruprecht-Karl Universität, Heidelberg.

Wright, W. B. 1937 *The Quaternary Ice Age.* (2nd edn), 478 pp. London: Macmillan.

Phil. Trans. R. Soc. Lond. B. **280**, 229–246 (1977) [229]
Printed in Great Britain

FLORA AND FAUNA

Early and Middle Devensian flora and vegetation

BY R. G. WEST, F.R.S.
Subdepartment of Quaternary Research, University of Cambridge

WITH AN APPENDIX BY
R. ANDREW AND R. G. WEST, F.R.S.

The record of palaeobotanical data from the Early and Middle Devensian (to 18000 B.P.) is reviewed, with additional pollen data from the Devensian type site at Four Ashes presented. Pollen and macrofossil assemblages derive principally from herb vegetation, but woodland episodes are known from the Early Devensian. Correlations of herb and woodland biozones are made with events in the Weichselian sequences of the Netherlands and north central Europe, and comparisons are made with the north American Wisconsin interstadial sequence and events in the North Atlantic cores. The environment of the herb and woodland biozones is discussed. The effect of a cool Atlantic as a modifying factor affecting the longitudinal zonation of Middle Weichselian vegetation across north central Europe is considered. The relation between environmental evidence based on flora/vegetation and that on fauna is discussed. Apparent discrepancies result from inadequate solutions of the problems associated with interpreting palaeoclimates from fossil assemblages. An interpretation of the data in the context of variation of assemblages of the same age across north Europe may offer better solutions for these problems.

1. INTRODUCTION

We will be concerned here with flora and vegetation of the period from the beginning of the Devensian to 18000 B.P. This period may cover *ca.* 60000–80000 years, depending on the placing of the Ipswichian/Devensian boundary. It precedes the period of major Devensian ice expansion, which took place in the east of England a few thousand years after 18000 B.P. The stratigraphical relations of the many sites which show Early and Middle Devensian fossil floras are difficult to determine. The sites are isolated from one another and are usually associated with low terraces of rivers, which often have a complex stratigraphy themselves. In the latter half of the period we are considering (to *ca.* 45000 a B.P.) the prime method of determining the sequence of fossil floras has been by radiocarbon dating, applied with due regard to the difficulties of dating material with very low radiocarbon content. Beyond that, biostratigraphy plays a certain role, particularly the alteration of woodland and herb biozones in the pollen record. Sufficient is known of the outline sequence of such biozones in northwest Europe to use them as a basis for biostratigraphy. For example, Amersfoort, Brørup and Odderade (pollen-based) interstadials have been described in the Early Weichselian, all with woodland, and each separated by herb biozones. No woodland biozones have yet been identified from the radiocarbon dated part of the sequence (later than *ca.* 45000 a B.P.) in northwest Europe, and Early Devensian woodland biozones in Britain have been correlated with these Early Weichselian woodland biozones of the continent. The absolute age of these woodland biozones has not yet been satisfactorily determined.

With this background of dating we can now consider the fossil floras in more detail.

2. The fossil record

Table 1 lists sites with Early and Middle Devensian floras, indicating analyses of pollen and macrofossil floras, the presence of herb or woodland biozones, and the relation of these to insect assemblages. Figure 1 shows the location of these sites. It will be noted that most sites are in the Midlands of England, and only one is in Ireland. They thus cover a rather limited area of the British Isles.

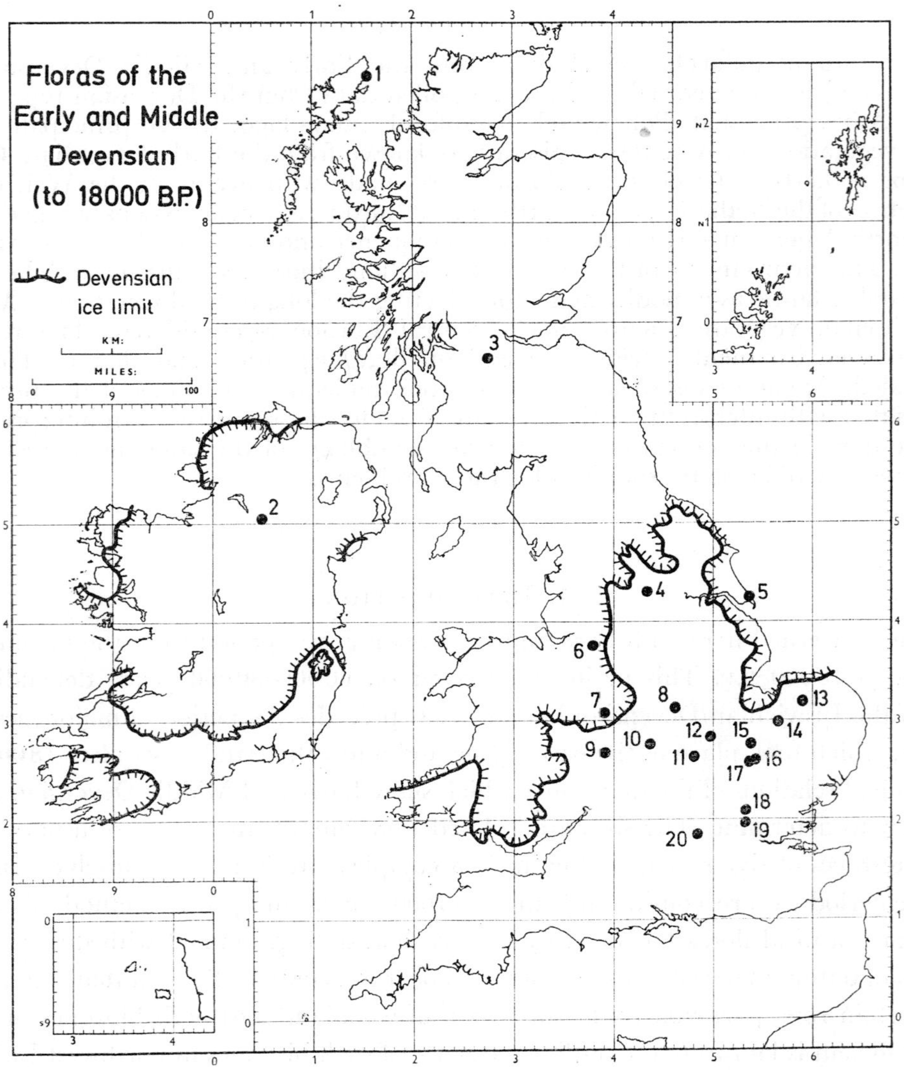

FIGURE 1. Location of sites with Early and Middle Devensian floras. Site numbers refer to the list in table 1.

TABLE 1. FOSSIL FLORAS OF THE EARLY AND MIDDLE DEVENSIAN

site no. in figure 1	site	age (^{14}C B.P.)		author	flora pollen	flora macro	biozone	fauna (insect)
				1. ^{14}C-dated sites				
5	Dimlington, Yorks.	18500 ± 400 18240 ± 250	I 3372 Birm 108	Penny, Coope & Catt 1969; Bell 1968	.	+	h	post-UW, cold
16	Barnwell Station, Cambs	19500 ± 600	Q 590	Bell & Dickson 1971; Coope 1968	.	+	h	post-UW, cold
1	Tolsta Head, Lewis	27333 ± 240	SSR 87	von Weymarn & Edwards 1973	+	.	h	
18	Broxbourne/Nazeing, Herts	28000 ± 1500	Q 25	Godwin & Willis 1960	.	+	h	
12	Great Billing, Northants	28225 ± 330	Birm 75	A. Morgan 1969	.	+	h	post-UW, cold
2	Derryvree, Co. Fermanagh	30500 ± 1170	Birm 166	Colhoun, Dickson, McCabe & Shotton 1972	(+)	+	h	post-UW, cold
7	Four Ashes, Staffs (45)	30500 ± 440	Birm 195	A. Morgan 1973	+†	.	h	post-UW, cold
7	Four Ashes, Staffs (2)	30655 + 765 − 700	Birm 25	A. Morgan 1973	+†	.	h	post-UW, cold
10	Brandon, Warwicks	30766 + 537 − 520 32270 + 1029 − 971	Birm 27 Birm 10	Shotton 1968; Kelly 1968	.	+	h	post-UW, cold
7	Four Ashes, Staffs (3)	36340 + 770 − 900	Birm 24	A. Morgan 1973	+†	.	h	post-UW, cold
8	Syston, Leics	37420 + 1670 − 1390	Birm 78	Bell, Coope, Rice & Riley 1972	.	+	h	post-UW, cold
4	Oxbow, Yorks	38600 + 1720 − 1420	NPL 163B	Gaunt, Coope & Franks 1970	+	+	j	post-UW, cold
7	Four Ashes, Staffs (34)	40000 + 1400 − 1200	Birm 196	A. Morgan 1973	+†	.	h	UW
9	Upton Warren, Worcs (Band 2)	42520 ± 1300 42100 ± 800	GRN 595 GRN 1245	Coope, Shotton & Strachan 1961	+	+	h	UW
15	Earith, Hunts (E7)	42140 + 1890 − 1530	Birm 88	Bell 1970	.	+	h	UW
7	Four Ashes, Staffs (4)	42530 + 1345 − 1115	Birm 56	A. Morgan 1973	+†	.	h	UW
7	Four Ashes, Staffs (20)	> 43500	Birm 74	A. Morgan 1973	+†	.	h	pre-UW, cold
15	Earith, Hunts (E9)	> 45000	Birm 86	Bell 1970	+	+	h	pre-UW, cold
6	Chelford, Cheshire	60800 ± 1500	GRN 1475 (enriched)	Simpson & West 1958	+	+	w	Ch
				2. Sites relatively dated				
7	Four Ashes, Staffs (19)	post-UW		A. Morgan 1973	+†	.	h	post-UW, cold
20	Marlow, Bucks	> 31000 UW		Bell 1969; Shotton & Williams 1971	+	+	h	UW
7	Four Ashes, Staffs (10)	Chelford ws		A. Morgan 1973	+†	.	w	Ch

TABLE 1 (*cont.*)

site no. in figure 1	site	age (^{14}C B.P.)	author	flora: pollen	flora: macro	biozone	fauna (insect)
14	Wretton, Norfolk	herb s III	West *et al.* 1974	+	+	h	cool
		Chelford ws		+	+	w	tundra
		herb s II		+	+	h	
		Wretton ws		+	+	w	
		herb s I		+	+	h	cool
13	Beetley, Norfolk	herb s II	Phillips 1976	+	+	h	
		Beetley ws (= Chelford ws?)		+	+	w	
		herb s I		+	+	h	
	Sidgwick Avenue, Cambridge	Early Devensian	Lambert, Pearson & Sparks 1963	.	+	h	+
			3. Other sites				
11	Thrapston, Northants	Middle Devensian in part	Bell 1968	+	+	h	
19	Barrowell Green, Middx	Middle Devensian	Reid & Chandler 1923	.	+	h	
3	Airdrie, Lanark	Devensian	Reid 1899; Coope 1962	.	+	.	+

Abbreviations: h, herb; w, woodland; s, substage; ws, woodland substage; UW, acme biozone of Upton Warren beetle-based interstadial; Ch, Chelford pollen or beetle interstadial. †, see appendix for pollen analyses.

(*a*) *Sediments and environments represented*

Early and Middle Devensian floras have been found mainly in organic infills of channels in fluviatile environments depositing sands and gravels. The channels are of varying size and contain sediment of varying organic content. Cut and fill processes may result in the redeposition of fragments of such channel fillings as 'erratics'. Floras have also been found in infills of hollows formed by the melting of ground ice, as at Beetley, Norfolk.

Observation of the sedimentary regime is important for the interpretation of these fossil floras. Most Devensian macrofossil floras derive from communities growing in situations close to braided river systems, while the pollen floras are derived from local and regional vegetation. A comparison of macroflora, pollen flora and sediment is therefore needed to analyse the provenance of the fossils, and to determine local and regional components (West *et al.* 1974, p. 391). The studies reported here do not all give the same degree of consideration to provenance. Nevertheless, the floras of the herb biozones have a considerable uniformity over a wide area, indicating local diversity within a regional vegetation of tundra physiognomy, probably predominantly grassland.

(*b*) *Pollen floras*

Pollen floras from the Early and Middle Devensian are broadly divisible into two categories, those dominated by tree pollen (a.p.) and those dominated by non-tree pollen (n.a.p.). The former give pollen assemblage biozones denoting regional woodland (woodland biozones), the latter give biozones denoting regional herb communities (herb biozones). The existence of a woodland biozone gives a basis for describing an interval as a pollen-based interstadial.‡

‡ The use of the term interstadial based on differing lines of evidence leads to confusion (see West *et al.* 1974, p. 398).

(i) *Woodland biozones*

High a.p. pollen spectra have been recorded at Chelford (site 6), Four Ashes (site 7), Wretton (site 14) and Beetley (site 13). All show high frequencies of *Pinus*, lesser frequencies of *Betula*, and low frequencies of *Picea*. The Chelford assemblage was compared with *Pinus–Betula–Picea* forest of northern Fennoscandia by Simpson & West (1958). The macroscopic remains recorded at Chelford and Beetley indicate the species concerned were *Pinus sylvestris* L., *Betula pubescens* Ehrh. and/or *Betula pendula* Roth., and *Picea abies* (L.) Karst. cf. ssp. *obovata* (Ledeb.) Hulten (Phillips 1976). An interesting taxon represented in the pollen flora is pollen referable to that of the dwarf shrub *Bruckenthalia spiculifolia* (Salisb.) Reichb., a calcifuge species of woods and subalpine pastures in the Balkan mountains (Phillips 1976). This has also been recorded from Early Weichselian woodland biozones in northwest Europe (Behre 1974).

High a.p. biozones from Four Ashes, Wretton and Beetley have been correlated with that from the earlier-described site at Chelford, and all have been grouped to a substage which may be called the Chelford pollen-based interstadial. The Beetley pollen diagram shows the most complete expression of this substage. At Wretton earlier biozones with *Betula–Pinus* and *Betula–Pinus–Alnus*–Gramineae–*Calluna* assemblages occur. These appear to represent an older woodland substage, named the Wretton pollen-based interstadial, separated from the younger interstadial by a herb biozone. Both are placed in the Early Devensian. In the light of the difficulties of stratigraphy and correlation of the Early Devensian, this grouping of sites into interstadials can obviously be only provisional, and further biostratigraphical studies will certainly modify the present arrangement.

(ii) *Herb biozones*

High n.a.p. biozones have been recorded (table 1) at Wretton from before the Wretton pollen-based interstadial, between this and the Chelford Interstadial, and after this interstadial. At other sites high n.a.p. spectra have been recorded at the following radiocarbon ages B.P. (site numbers refer to figure 1): > 45000 (site 15, E9), > 43500 (site 7), > 31000 (site 20), 42530 (site 7), 42140 (site 15, E7), 42100/42520 (site 9), 40000 (site 7), 38600 (site 4), 36340 (site 7), 30655 (site 7), 30500 (site 7), and 27333 (site 1). Other sites, undated, with high n.a.p. spectra are at Four Ashes (Middle Devensian, site 7) and Earith (Middle Devensian, site 15, E1, E2, E4, E5).

All these pollen spectra show above 80 % n.a.p., many above 95 % (% total land pollen). A discussion of the interpretation of such pollen spectra in terms of regional vegetation was given by West *et al.* (1974). Much information is now available on pollen rain in areas of herb vegetation (arctic, alpine (Birks 1973); steppe (Levkovskaya (1973)) and it seems clear that these high n.a.p. spectra derive from herb vegetation, dominated probably by Gramineae and with a tundra physiognomy. There is no evidence for widespread vegetation with shrubs, though at Tolsta Head *Juniperus* and *Salix* pollen frequencies indicate shrub vegetation. In most spectra Gramineae and Cyperaceae pollen make up a high percentage of total pollen, but the provenance of Cyperaceae varies. At many sites high frequencies of Cyperaceae pollen are accompanied by Cyperaceae macroscopic remains and thus are probably of local origin. The spectra may show considerable variety in kind and number of herb taxa. Occasional high frequencies of *Armeria*, Compositae, Cruciferae and *Plantago* occur. These variations reflect the variety of plant communities associated with braided river plains and of the neighbouring uplands, and the changes of provenance associated with changes of sedimentary régime.

As will be discussed later, pollen spectra from sites with insect faunas indicating the acme of the beetle-based Upton Warren interstadial (sites 7, 9, 20) show no increase of tree pollen frequencies. The pollen spectra so far obtained from this period appear to show no variation in regional vegetation that is not seen in the other herb-rich parts of the Middle and Early Devensian.

(*c*) *Macrofossil floras*

There is a long history of the study of plant macrofossils in last glaciation deposits. Clement Reid (1899) and E. M. Reid (1949) made substantial contributions to the 'full-glacial' flora. More recently, C. A. Dickson (1970) and F. G. Bell (1969, 1970) have, through careful taxonomic work, made great improvements in standards of identification. As a result, there is a considerable body of evidence of species of the Early and Middle Devensian flora, adding substance to the evidence given by the pollen flora.

Macroscopic floras (table 1) are known from both the woodland biozones (e.g. Chelford (Simpson & West 1958); Beetley (Phillips 1976)) and from the herb biozones. It is the assemblages from the herb biozones which have been more frequently found and which are of particular interest. These floras have been found in varying sediment types, ranging from 'drift-muds' or leaf beds, to laminated silts and clays and detritus muds. The taphonomy of the floras may thus differ considerably (West *et al.* 1974). Nevertheless, there is a considerable uniformity of the flora, as there is with the pollen spectra of the herb biozones. The absence of woodland is borne out by the macrofossil flora, but there is evidence for local dwarf shrub communities at certain sites (e.g. Earith), recorded by the sediment (leaf-beds) and the species lists, which include *Betula nana* L., *Salix herbacea* L., *S. polaris* Wahlenb., *S. phylicifolia* L. and *S. viminalis* L. However, these communities do not seem to have been very extensive since they have not been recorded substantially in the pollen spectra. Various habitat and vegetation categories have been identified in the macrofossil flora, including aquatic and marsh communities, tall herb communities, ruderal and weed species associated with disturbed and open ground (related to freeze-thaw phenomena and floodplain processes) and grassland (Beetley, Wretton). These components of the Devensian flora have been discussed in detail by Bell (1970) and Godwin (1975).

Three phytogeographical or ecological elements have merited particular attention. These are the southern, steppe and halophyte elements in the macrofossil floras. Their occurrence has been fully discussed by Bell (1969). Bell concluded the southern element, a heterogeneous group (including true thermophiles, and some species southern because of habitat availability), indicated summer (July) temperatures near to 16 °C, with a continental type of climate, since many of the southern species have a continental distribution in Europe as a whole. A number of taxa are characteristic of the steppe region of the southwest of the U.S.S.R. They may not be members of steppe communities as such, but their presence may indicate a xeromorphic situation with grasses, other herbs and dry mineral soils. Species considered both as obligate halophytes (e.g. *Glaux maritima* L.) and facultative halophytes are of widespread occurrence in these macrofossil floras. Their presence may indicate saline soils, which are likely to have been induced by evaporation (which may be related to wind and insulation) and low precipitation in the summer, associated with impeded drainage. Such soils can be associated with permafrost in northern Eurasia. It is notable that floras with halophytes are found at Earith and Wretton in association with ice-wedge casts, though the stratigraphical evidence cannot prove contemporaneity. Thus at Wretton (site WUB) a high frequency of halophyte macrofossils occurred in a sediment very rich in terrestrial flora penetrated by an ice-wedge cast.

We have seen that it is not possible yet to separate different categories of herb biozone on the basis of pollen assemblages. It seems that it is also difficult to show that different macrofossil floras indicate differing vegetational conditions of substance (Bell 1970, p. 374). Variations in taphonomy and habitat, together with phytosociological variation of open plant communities, represent a multivariate situation which has not yet been analysed in detail, and which requires, for analysis, the combined evidence of stratigraphical context, sediment, macrofossils and pollen.

(*d*) *Continuity of the record*

The record of Early and Middle Devensian fossil floras, as shown in table 1, covers in outline the period of time involved, but there are considerable gaps of time in which there is no knowledge of the flora. We may expect these gaps to be filled as new records accumulate. The present record indicates the probability of survival of a large flora during the Devensian in the area south of the ice limit, and such a conclusion is supported by records of floras closely related in time to glacial advance (Dimlington) and to structures indicating permafrost (Wretton). We may note that the pollen record of the Weichselian in the Netherlands (Zagwijn 1961, 1974) is more complete, but on the basis of lack of organic deposits in certain periods, especially 13000–27000 B.P., a polar desert environment has been postulated. This point is discussed in § 4.

3. Correlations

(*a*) *Netherlands*

The most detailed and best dated sequences from the last glaciation come from the Weichselian of the Netherlands (van der Hammen & Wijmstra 1971; Zagwijn 1961, 1974). The sequence from this area is summarized in figure 2. A correlation of the Chelford woodland substage with the Odderade or Brørup interstadials of northwest Europe was suggested by West *et al.* (1974), with the latter correlation being the more likely. A correlation of the Wretton woodland substage with the Amersfoort Interstadial was made by the same authors. These correlations are indicated in figure 2. There is no clear evidence from the flora or vegetation in the herb biozones of pollen-based interstadials which might be correlated with the Moershoofd, Hengelo and Denekamp interstadials of the Netherlands. This raises the important problem of the interpretation of high n.a.p. spectra in terms of vegetation and climate, especially since it has now been amply demonstrated that climatic regimes with a parameter of warmth indicated by beetle faunas are not evident so far from changes within high n.a.p. spectra. From the palaeobotanical point of view, the best interpretation will be obtained from studies of pollen, macroscopic remains, and sedimentary conditions and environment.

(*b*) *North central Europe*

A number of Middle Weichselian floras are known from southern Poland (Srodon 1968). These are of particular interest in showing forest-tundra with coniferous forest. For example at Dobra (Srodon 1968, ^{14}C age 32500 a B.P. (GrN 5111), *Pinus cembra*, *P. sylvestris*, *Picea abies* and

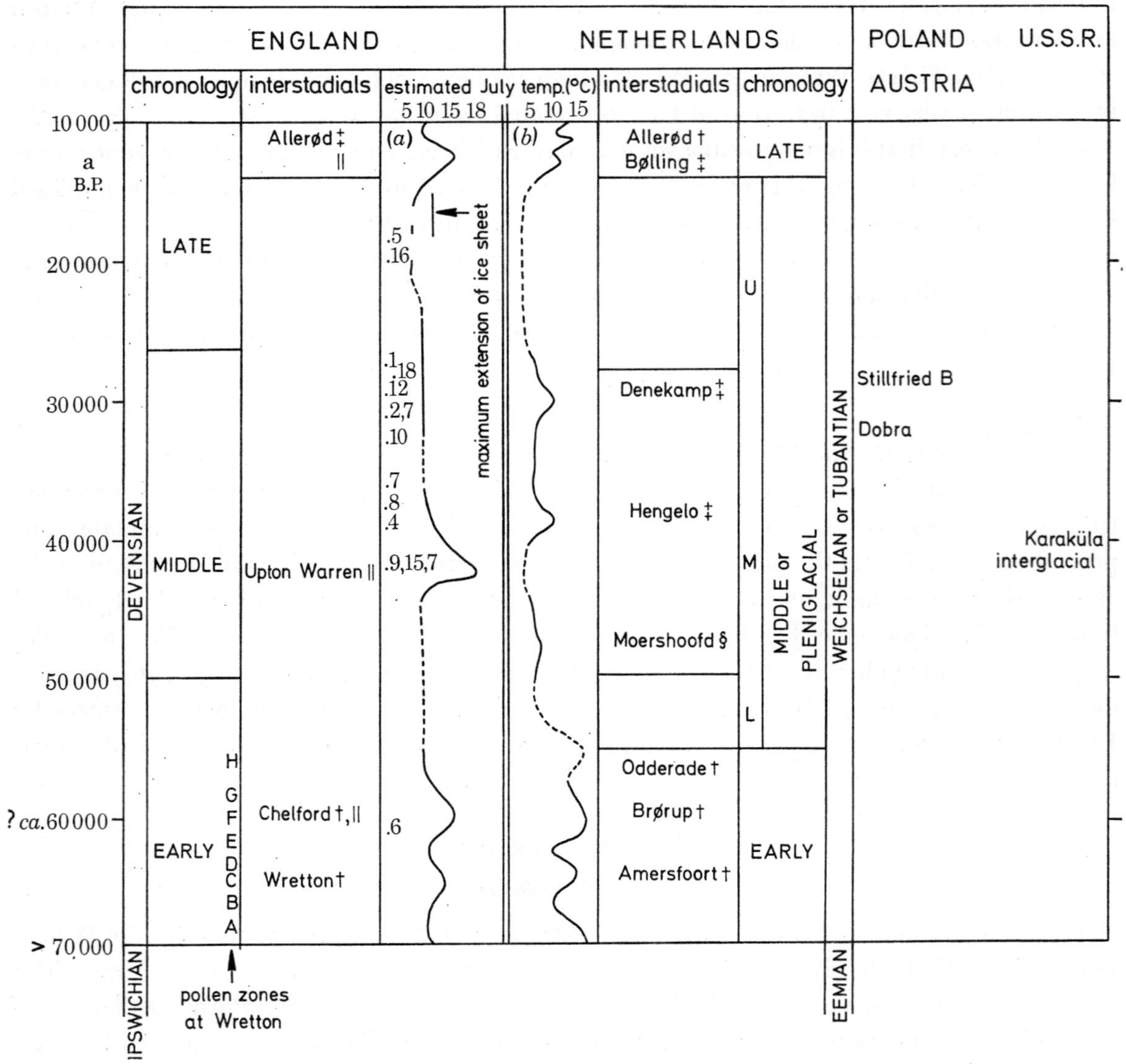

FIGURE 2. A classification of the Devensian, showing the age of dated (absolute and relative) Early and Middle Devensian floras. Correlation with the Netherlands and with some central and north European sites are shown. The figure shows the position of interstadials and the basis for their definition: †, pollen-based interstadial, woodland; ‡, pollen-based interstadial, shrub-tundra; §, pollen-based interstadial, tundra; ||, beetle-based interstadial. *a*, curve based on beetles back to Chelford Interstadial (Coope, Shotton & Strachan 1971; Coope 1975*a*) and below this on palaeobotany; *b*, curve based on palaeobotany.

Larix occur, together with a rich herbaceous flora recalling that of the alpine meadows in the Tatra Mountains. Srodon (1968) points out that this difference from the flora and vegetation of northwest Europe is paralleled in the Late-glacial and in the Early Weichselian interstadials, in both of which periods the flora and vegetation in south Poland showed a warmer climate than that indicated by floras and vegetation of northwest Europe. The Polish Middle Weichselian interstadials are correlated with the 'Paudorf' of central Europe and Stillfried 'B' soil complex of Austria (location in figure 3).

When we go further east we find evidence reported of forest interglacial conditions in the mid-Weichselian. This interglacial, the Karaküla interglacial of the Russian Plain

(Serebryanny, Raukas & Punning 1970), has ^{14}C dates from its sediments, indicating an age *ca.* 50000–25000 a B.P., and shows a vegetational history with high *Picea* values and low frequencies of *Tilia*, *Ulmus*, *Quercus* and *Corylus*.

This apparent longitudinal zonation of vegetation in north central Europe deserves further study from botanical and climatic viewpoints, especially since it appears that the Middle Weichselian vegetational history of north central and eastern Europe shows a complexity of change not so far found in the British Middle Devensian.

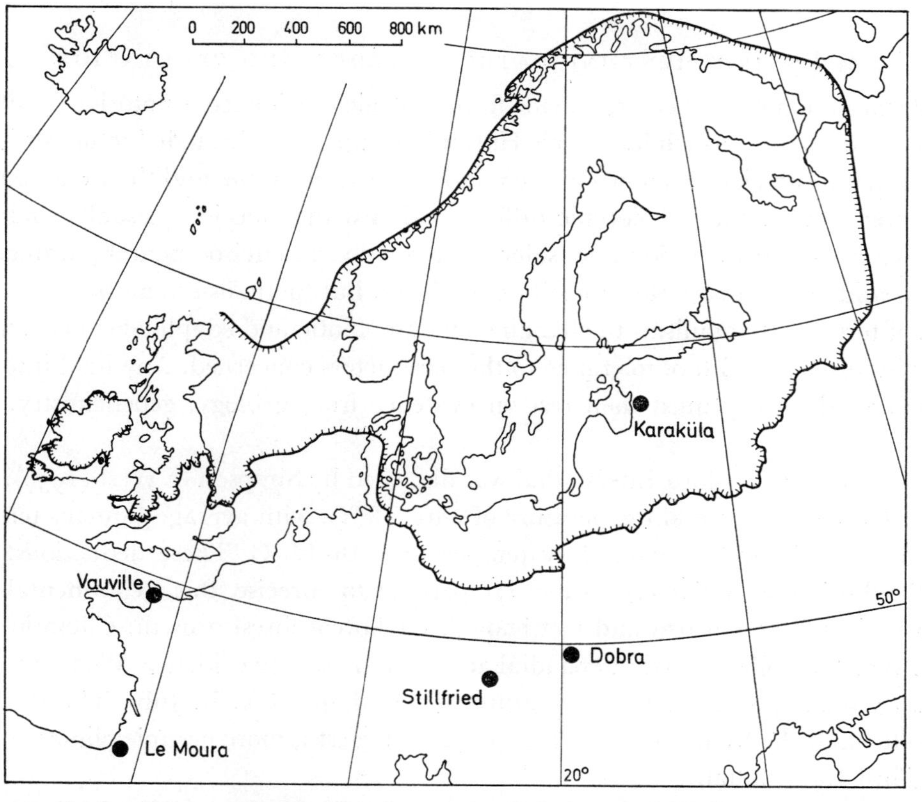

FIGURE 3. Map of western Europe showing areas and sites considered in text, with margin of Devensian/Weichselian ice.

(*c*) *North Atlantic and North America*

A further consideration of this last point leads to the conclusion that the cold water masses present in the northeast Atlantic in Middle and Early Devensian time (McIntyre & Ruddiman 1972, depression of annual temperature of NE Atlantic surface water by 5 °C; see also Manley 1951) had a cooling effect in northwest Europe and gave a climate which affected the vegetation zonation. This effect may be paralleled by mid-continent and (east) coast differences in north America, where Late-glacial mid-continent coniferous forest and east coast herb vegetation present a contrast (Davis 1967), mirrored in northwest Europe by Late-glacial herb vegetation in the west and coniferous forest in European U.S.S.R. (West 1961). In the Middle Wisconsin, in the eastern Great Lakes region (Berti 1975), the Port Talbot and Plum Point interstadials (Port Talbot I, 55000–50000 a B.P.; Port Talbot II and Plum Point interstadials 48000–23000 a B.P.) show coniferous forest with *Quercus* and forest-tundra with *Pinus* and *Picea* respectively.

The situation is comparable to that in southern Poland, and again offers a great contrast to the situation in northwest Europe.

A further point in relation to the ocean record is the discrepancy between the age of Early Devensian (Weichselian) interstadials as determined by radiocarbon dating (*ca.* 60000 a B.P.) and the ages of possible correlative fluctuations in the palaeotemperature curve (Emiliani stages 5 a–d) dated at *ca.* 75000–115000 a B.P. (Shackleton & Opdyke 1973). Difficulties of dating both organic terrestrial material and marine cores prevent a resolution of this problem at present.

4. Environment indicated by flora and vegetation

The interpretation of past environments from biological evidence is notoriously difficult. We have to start from the principles of the Huttonian approach through 'actualism', with the limitations that insufficient is known of the relation between species and their environment and that the extent and nature of biotypic differentiation within species is usually unknown. We have to take present distributions, consider their relation to environmental parameters, draw conclusions and apply the results to fossil assemblages. But 'actualism' can be a trap, since the responses of terrestrial organisms to changing climate, biotic and edaphic factors can rarely be evaluated in terms of width of tolerance to the parameters concerned. Any final interpretation of environmental history must then rest on evidence from geology, geochemistry, flora and fauna.

The climate of the Chelford Interstadial was discussed by Simpson & West (1958). The fossil flora indicated a mean annual temperature of 2 to −3 °C, with average January temperatures of −10 to −15 °C, and average July temperature 16–12 °C. These deductions, based on present distribution of similar vegetation, are perhaps too precise. But a continental climate is suggested with summers warm and wet enough for boreal forest growth. Climatic inferences about the Wretton pollen-based interstadial are based on scanty evidence. Woodland spread at this time, suggesting a summer temperature higher than 10 °C in July. The strong rise of *Calluna* pollen and the frequencies of *Alnus* perhaps suggest a more oceanic climate than in the later Chelford Interstadial.

The interpretation of the climate of the herb biozones is extremely difficult. Bell (1970) considered that the presence of thermophile species at Earith indicated July temperatures approaching 16 °C, with winters very cold, as indicated by permafrost indicators. However, we have to explain the lack of trees (i.e. the lowering of tree-line below today's sea level), as well as the presence of a wide variety of species of present day distribution from the Arctic to southern parts of Europe. The combination of species in the flora of the herb biozones is not one found today. The explanation will lie in the nature of the climate.

As discussed above, we have a situation in the Middle Devensian with a cold Atlantic Ocean, with the climate of the British Isles governed partly by this and partly by variation in the continental weather systems. This situation parallels the present, but must offer a completely different climatic regime to our area. We must view the climate of the Middle Devensian against this background, and against the evidence for the longitudinal zonation across northern Europe which we have already discussed. According to the evidence from flora, the climate must have had warm summers, but on flora alone it is difficult to estimate winter cold, since present distributions of northern plants are determined not only by temperature but also competition.

The position of the boundary between the tundra and boreal woodland is determined by a complex of factors, with the structure of the vegetation (e.g. its relation to albedo) and the climate playing an interlocking role (Hare & Ritchie 1972). The explanation of the treelessness is difficult to determine. The long period covered by herb biozones indicates a 'climax' of treelessness, which could be caused by a number of factors – low winter temperatures, low summer temperatures, soil instability, drying winds, waterlogged soils in the summer and so on. Perhaps persistent cool winds, with high evaporation, were an important cause. It is possible that the high albedo associated with snow cover and herb vegetation caused a shorter growth season in spite of the summer temperature, with the persistence of snow cover associated with the cool Atlantic. Such a combination would explain the presence of thermophiles in herb biozones, such as is found in the Middle Devensian. In the coldest parts of the sequence, when the polar front was far south in the Atlantic, the treelessness has a more obvious explanation in the increased severity of the climate.

An analogous situation may exist now where a cold ocean borders a continental mass. For example, in parts of Siberia, where the coastal vegetation is tundra, with taiga further south in more extreme climatic conditions, probably related to the median position of the Arctic front in the summer (Krebs & Barry 1970); or the treeless maritime tundra of the Aleutians and the Alaska peninsula (Hare & Ritchie 1972); or the barren lands of northern Labrador (57° N) where the cold Labrador current plays an important part in determining the climatic regime (Wenner 1947). We may perhaps envisage a climate with cold westerlies in the summer, possibly with low air humidity, and very cold anticyclonic conditions in the winter.

Further east, in south Poland, forest-tundra existed in the Middle Weichselian (Stillfried 'B' or 'Paudorf' Interstadial). The exclusion of trees from northwest Europe is probably associated with a climate dependent on the cold Atlantic, a hypothesis supported by the extension of 'tundra' in the Devensian far south on the French coast (figure 3; Vauville, Le Moura, van Campo 1969). Clearly a further study of the cold maritime climate of western Europe in the Devensian is required. At present we merely note that the occurrence of the mixture of species in the Devensian must relate to such a climate, vastly different from the climate of today.

Taking the Middle and Early Devensian as a whole, we see an early period with pollen-based interstadials giving way to a long period with herb biozones. The sequence gives the impression of an unbalanced situation achieving stability at the beginning of the Middle Devensian. In central and eastern northern Europe the fluctuations of the Early Weichselian are known from southern Poland, but in the Middle Weichselian a climate favourable to boreal forest growth was present in that area at times and further east there appears evidence at times for more thermophilous vegetation (Karaküla interglacial). It is perhaps these oscillations of Middle Weichselian climate, damped to the west by the stability induced by the Atlantic, which are expressed, not by the herb biozones, but by the thermophilous beetle faunas occurring in the Upton Warren beetle-based interstadial in Britain.

In the Late Devensian, after 30000 a B.P., with the expansion of the Devensian (and Scandinavian†) ice sheets herb biozones extend south in Poland and 'polar desert' occurred in the Netherlands. The evidence from well-dated floras in this part of the Devensian record is poor, but what there is suggests the presence of the typical herb biozone flora.

† The later Weichselian ice advances of the Scandinavian ice sheets certainly include the Frankfurt and later stages, but the number and extent of early Weichsel ice advances is a matter for discussion (see Serebryanny *et al.* 1970).

5. Comparison with faunas

Here we will briefly note some comparisons of palaeobotanical data of environments with that from non-marine molluscs and insects. The molluscs associated with Devensian floras have been discussed by Sparks (West *et al.* 1974). Species of southern molluscs persist into the Early Devensian, with some species thought to indicate cold conditions. At Wretton, the faunas are associated with herb biozones alongside evidence of waste mantle structures requiring considerable cold. Even so some of the southern species are freshwater (e.g. *Planorbarius corneus*), but they range east into Russia and Siberia, and in such a distribution may lie the key to their presence in the treeless phases of the Devensian. They will apparently resist cold winters provided summers are warm, a conclusion compatible with that from the plants. The high percentages of *Pupilla muscorum* in the herb biozones at Wretton tally with these climatic indications, since the species is a characteristic dominant of European loess faunas. The molluscs indicate a continental climate. In the continental parts of the U.S.S.R. where Wretton species occur winter temperatures range from −10 to −20 °C, with summer temperatures of 10–21 °C. A general coldness and aridity is suggested by the Devensian (and Weichselian) mollusc faunas (Ant 1969).

The climate from the insect faunas of the Chelford Interstadial (Coope 1959) correspond with that inferred from the plant remains from Chelford. At Wretton, however, the insect faunas of the sediment with the Chelford woodland substage deposits was of a tundra type. The discrepancy has not been explained, but is probably the result of differing provenance of the pollen flora and the insect fauna, perhaps in terms of local variation in habitats of the sand plains of the R. Wissey floodplain at the time. A second type of variation between the floras and the insect remains is found in the assemblages from the Upton Warren Interstadial (acme *ca.* 42000 a B.P.) (Coope 1975*b*). Table 1 shows that several sites show the characteristic fauna of this time with thermophilous beetles, but the herb assemblage is similar, as far as we now know, to that elsewhere in the Middle Devensian. There is no sign of the development of forest. A possible explanation is that time was not sufficient to allow the development of forest-tundra or forest. However, there is no sign of increased representation of shrub or herb taxa which respond to the late-glacial amelioration, and in the absence of measurements of pollen deposition rate, which would indicate increased pollen productivity as a result of amelioration of climate, it is impossible to prove an amelioration of this type. It seems more likely that the climatic régime of the interstadial favoured the representation of thermophilous beetles, but did not affect the physiognomy of the vegetation. Climatic causes for this situation were discussed in § 4.

6. Conclusions

Difficulties in the interpretation of Devensian fossil floras in terms of vegetation and environment have been mentioned in the course of this discussion. When we consider that the vegetation we have described existed in Britain for 70000 of the last 80000 years, we realize the importance of resolving these difficulties for a proper appreciation and interpretation of environmental history. The desiderata for the solution of the problems are an increased knowledge of the ecology of the present day representatives of the species concerned, improved pollen morphological studies to gain a better knowledge of the Devensian floras and to reach a standard of identification similar to that reached for macroscopic fossils, and a better appreciation of the combined pollen, macro and sediment evidence from particular sites. Only with such studies can the taphonomy of the floras be better understood and so the vegetation and environment be better understood. The work must necessarily be complemented by studies of the origin of pollen and macro assemblages from vegetation at the present time. It is clear that sites must be investigated in this way, and must be reliably dated, across the longitude of northern Europe. We can then expect much further revelation of the remarkable vegetational and environmental history in the Devensian cold stage.

References (West)

Ant, H. 1969 Zur Würm-glazialen Überdauerung europäischer Landgastropoden in Eisrandnähe. *Malacologia* **9**, 249–250.

Behre, K.-E. 1974 Die vegetation im Spätpleistozän von Osterwanna/Niedersachsen. *Geol. Jb.* A**18**, 3–48.

Bell, F. G. 1968 Weichselian glacial floras in Britain. Ph.D. dissertation, University of Cambridge.

Bell, F. G. 1969 The occurrence of southern, steppe and halophyte elements in Weichselian (last-glacial) floras from southern Britain. *New Phytol.* **68**, 913–922.

Bell, F. G. 1970 Late Pleistocene floras from Earith, Huntingdonshire. *Phil. Trans. R. Soc. Lond.* B **258**, 347–378.

Bell, F. G., Coope, G. R., Rice, R. J. & Riley, T. H. 1972 Mid-Weichselian fossil-bearing deposits at Syston, Leicestershire. *Proc. Geol. Ass. Lond.* **83**, 197–211.

Bell, F. G. & Dickson, C. A. 1971 The Barnwell Station Arctic Flora: a reappraisal of some plant identifications. *New Phytol.* **70**, 627–638.

Berti, A. 1975 Palaeobotany of Wisconsinan interstadials, eastern Great Lakes region, North America. *Quat. Res.* **5**, 591–619.

Birks, H. J. B. 1973 Modern pollen rain studies in some arctic and alpine environments. In *Quaternary plant ecology* (eds H. J. B. Birks & R. G. West), pp. 143–170. Oxford: Blackwell Scientific Publications.

Colhoun, E. A., Dickson, J. H., McCabe, A. M. & Shotton, F. W. 1972 A Middle Midlandian freshwater series at Derryvree, Maguiresbridge, County Fermanagh, Northern Ireland. *Proc. R. Soc. Lond.* B **180**, 273–292.

Coope, G. R. 1959 A late Pleistocene insect fauna from Chelford, Cheshire. *Proc. R. Soc. Lond.* B **157**, 70–86.

Coope, G. R. 1962 Coleoptera from a peat interbedded between two boulder clays at Burnhead, near Airdrie. *Trans. Geol. Soc. Glasgow* **24**, 279–286.

Coope, G. R. 1968 Coleoptera from the 'Arctic Bed' at Barnwell Station, Cambridge. *Geol. Mag.* **105**, 482–486.

Coope, G. R. 1975*a* Climatic fluctuations in northwest Europe since the last interglacial, indicated by fossil assemblages of Coleoptera. In *Ice Ages: ancient and modern* (eds A. E. Wright & F. Moseley), pp. 153–168. Liverpool: Seel House Press.

Coope, G. R. 1975*b* An ecological study of a temperate interlude in the middle of the last glaciation, based on fossil Coleoptera from Isleworth, Middlesex. *J. anim. Ecol.* **44**, 365–391.

Coope, G. R., Morgan, A. & Osborne, P. J. 1971 Fossil Coleoptera as indicators of climatic fluctuations during the last glaciation in Britain. *Palaeogeogr., Palaeoclimatol., Palaeoecol.* **10**, 87–101.

Coope, G. R., Shotton, F. W. & Strachan, I. 1961 A late Pleistocene fauna and flora from Upton Warren, Worcestershire. *Phil. Trans. R. Soc. Lond.* B **244**, 379–421.

Davis, M. B. 1967 Late-glacial climate in northern United States: a comparison of New England and the Great Lakes region. In *Quaternary palaeoecology* (eds E. J. Cushing & H. E. Wright), pp. 15–43. Yale University Press.

Dickson, C. A. 1970 The study of plant macrofossils in British Quaternary deposits. In *Studies in the vegetational history of the British Isles* (eds D. Walker & R. G. West), pp. 233–254. Cambridge University Press.

Gaunt, G. D., Coope, G. R. & Franks, J. W. 1970 Quaternary deposits at Oxbow opencast coal site in the Aire valley, Yorkshire. *Proc. Yorks. Geol. Soc.* **38**, 176–200.

Godwin, H. 1975 *History of the British flora.* 2nd edn. Cambridge University Press.

Godwin, H. & Willis, E. H. 1960 Cambridge University natural radiocarbon measurements: II. *Radiocarbon* **2**, 62–72.

Hammen, T. van der, Maarleveld, G. C., Vogel, J. C. & Zagwijn, W. H. 1967 Stratigraphy, climatic succession and radiocarbon dating of the Last Glacial in the Netherlands. *Geol. en. Mijnb.* **46**, 79–95.

Hammen, T. van der & Wijmstra, T. A. 1971 The Upper Quaternary of the Dinkel valley (Twente, Eastern Overijssel, the Netherlands). *Meded. Rijks Geol. Dienst, N.S.* **22**, 55–213.

Hare, F. K. & Ritchie, J. C. 1972 The Boreal bioclimates. *Geogr. Rev.* **62**, 333–365.

Kelly, M. R. 1968 Floras of Middle and Upper Pleistocene age from Brandon, Warwickshire. *Phil. Trans. R. Soc. Lond.* B **254**, 401–415.

Krebs, J. S. & Barry, R. G. 1970 The arctic Front and the tundra-taiga boundary in Eurasia. *Geogr. Rev.* **60**, 548–554.

Lambert, C. A., Pearson, R. G. & Sparks, B. W. 1963 A flora and fauna from Late Pleistocene deposits at Sidgwick Avenue, Cambridge. *Proc. Linn. Soc. Lond.* **174**, 13–29.

Levkovskaya, G. M. 1973 The zone peculiarity of the modern vegetation and the subfossil pollen spectra of the western Siberia. (In Russian, English summary.) In *Methodical problems of palynology*, pp. 116–120. Moscow: Nauka.

McIntyre, A. & Ruddiman, W. F. 1972 Northeast Atlantic post-Eemian paleooceanography: a predictive analog of the future. *Quat. Res.* **2**, 350–354.

Manley, G. 1951 The range of variation of the British climate. *Geogr. J.* **117**, 43–68.

Morgan, A. 1969 A Pleistocene fauna and flora from Great Billing, Northamptonshire, England. *Opuscule Entomologica* **34**, 109–129.

Morgan, A. 1973 Late Pleistocene environmental changes indicated by fossil insect faunas of the English Midlands. *Boreas* **2**, 173–212.

Penny, L. F., Coope, G. R. & Catt, J. A. 1969 Age and insect fauna of the Dimlington silts, east Yorkshire. *Nature, Lond.* **224**, 65–67.

Phillips, L. M. 1976 Pleistocene vegetational history and geology in Norfolk. *Phil. Trans. R. Soc. Lond.* B **275**, 215–286.

Reid, C. 1899 *The origin of the British flora*. London: Dulau & Co.

Reid, E. M. 1949 The late-glacial flora of the Lea valley. *New Phytol.* **48**, 245–252.

Reid, E. M. & Chandler, M. E. J. 1923 Barrowell Green (Lea Valley) arctic flora. *Q. Jl geol. Soc. Lond.* **79**, 604–605.

Serebryanny, L., Raukas, A. & Punning, J.-M. 1970 Fragments of the natural history of the Russian Plain during the Late Pleistocene with special reference to radiocarbon datings of fossil organic matter from the Baltic region. *Baltica* **4**, 351–366.

Shackleton, N. J. & Opdyke, N. D. 1973 Oxygen isotope and palaeomagnetic stratigraphy of equatorial Pacific core V28-238: oxygen isotope temperatures and ice volumes on a 10^5 year and 10^6 year scale. *J. Quat. Res.* **3**, 39–55.

Simpson, I. M. & West, R. G. 1958 On the stratigraphy and palaeobotany of a Late Pleistocene organic deposit at Chelford, Cheshire. *New Phytol.* **57**, 239–250.

Shotton, F. W. 1968 The Pleistocene succession around Brandon, Warwickshire. *Phil. Trans. R. Soc. Lond.* B **254**, 387–400.

Shotton, F. W. & Williams, R. E. G. 1971 Birmingham University radiocarbon dates: V. *Radiocarbon* **13**, 141–156.

Srodon, A. 1968 On the vegetation of the Paudorf Interstadial in the western Carpathians. *Acta Palaeobotanica* **9**, 1–27.

van Campo, M. 1969 Végétation wurmienne en France. Données bibliographiques. Hypotheses. In *Etudes Francaises sur le Quaternaire*, pp. 104–111. Paris: INQUA.

Wenner, C. G. 1947 Pollen diagrams from Labrador. *Geogr. Annlr.* **29**, 137–374.

West, R. G. 1961 Late and postglacial vegetational history in Wisconsin, particularly changes associated with the Valders Readvance. *Am. J. Sci.* **259**, 766–783.

West, R. G., Dickson, C. A., Catt, J. A., Weir, A. H. & Sparks, B. W. 1974 Late Pleistocene deposits at Wretton, Norfolk. II. Devensian deposits. *Phil. Trans. R. Soc. Lond.* B, **267**, 337–420.

Weymarn, J. von & Edwards, K. J. 1973 Interstadial site on the Island of Lewis, Scotland. *Nature, Lond.* **246**, 473–474.

Zagwijn, W. H. 1961 Vegetation, climate and radiocarbon datings in the Late Pleistocene of the Netherlands. Part I. Eemian and Early Weichselian. *Meded. Geol. Sticht.*, *N.S.* **14**, 15–45.

Zagwijn, W. H. 1974 Vegetation, climate and radiocarbon datings in the Late Pleistocene of the Netherlands. Part II. Middle Weichselian. *Meded. Rijks Geol. Dienst*, *N.S.* **25**, 101–110.

Appendix. Pollen analyses from Four Ashes, Worcs.

By R. Andrew and R. G. West

Subdepartment of Quaternary Research, University of Cambridge

The stratigraphy of the gravel pits at Four Ashes, Worcs, has been described by A. V. Morgan (1973) and the insect faunas which occur in organic lenses within the sands and gravels at Four Ashes have been described by A. Morgan (1973). The sections showed up to 4.6 m of sand and gravel overlying Triassic sandstone and forming a terrace-like feature. The site provided a sequence of faunas from Ipswichian to Middle Devensian in age. The site has become the type site for the Devensian cold stage (Mitchell *et al.* 1973). Nine samples for pollen analysis were provided by Dr A. Morgan, five of which had finite dates. The locality numbers of the samples are those of Dr A. Morgan. The samples were prepared according to the schedule used in the Subdepartment of Quaternary Research, with final mounting in glycerine jelly.

The results of the analyses are shown in figures A 1 and A 2. Notes on the pollen morphology of some taxa are given below. The assemblages form three groups.

(i) Site 44. The assemblage shows high tree pollen (a.p.) percentages, with *Quercus*, *Alnus* and *Corylus* well represented. This spectrum is correlated with Ipswichian zone Ip IIb.

(ii) Site 10. The assemblage shows high a.p. percentages, constituted by pollen of *Betula*, *Pinus* and *Picea*. This assemblage is very similar to that from Chelford (Simpson & West 1958), and is therefore correlated with the Chelford Interstadial.

(iii) All the other assemblages show high non-tree pollen (n.a.p.) with up to 15 % *Betula* and below 7 % *Pinus*. The dominant n.a.p. taxa are Gramineae and Cyperaceae, with a wide variety of sparsely represented n.a.p. taxa.

Of the arboreal taxa, the frequency of *Betula* indicates the local presence of birch. In the absence of macroscopic fossils it is not possible to say whether this is dwarf or tree birch. The frequencies of *Pinus* are low and may result from long-distance transport rather than local growth. *Salix* pollen also occurs in low frequency. Its presence probably indicates local willow scrub. *Calluna*, and to a lesser extent *Empetrum* are represented by low pollen frequencies. Their local presence is uncertain.

Gramineae and Cyperaceae pollen form the bulk of the n.a.p. total. Their frequencies bear some relation to the sediment composition (see below). Pollen of aquatic plants is scarce but consistently present. *Caltha*, *Hippuris*, *Littorella*, *Myriophyllum alterniflorum* and *Sparganium* are represented. Spores of *Botrychium*, *Equisetum*, *Selaginella* and *Sphagnum* are present in low frequencies. *Selaginella* is the most abundant of these.

Figure A 2 shows the variety of n.a.p. taxa other than Gramineae, Cyperaceae and aquatics. They must represent plants from a variety of habitats in the region, both drier, open ground (*Armeria*, *Artemisia*, *Campanula*, Compositae, Liguliflorae, *Helianthemum*, *Plantago maritima*, *P. media/major*, *Rumex acetosella*) and local wetter habitats probably associated with the surface of the sand and gravel plain (perhaps *Ranunculus*, *Thalictrum*).

The left-hand columns in figures A 1 and A 2 show the composition of the sediment containing the pollen, using the major divisions of sediment type proposed by Troels-Smith (1955). The highest frequencies of Cyperaceae occur in samples 34 and 4, which show a Dh (detritus herbosus) element probably originating from local sedge growth. Sample 4 contained *Carex* nutlets. Even with this local representation, the analysis shows an accompanying high variety of other n.a.p. taxa. Sample 19, a highly inorganic sediment, showed low frequencies of *Picea*, *Ulmus*, *Quercus* and *Alnus* pollen, which may be derived with the inorganic sediment from an older horizon. In spite of the variation of sediment composition and so taphonomy of the assemblages, the Gramineae pollen percentages remain between 20 and 36 %, indicating regional vegetation with abundant Gramineae.

Notes on pollen morphology. The following notes are based on observations made on pollen of British taxa in the pollen reference collection of the Subdepartment of Quaternary Research.

Acer. Tricolpate, oval, 39 µm, striate, with long furrows. Five grains of this type were found in sample 44. In their larger size, thinner wall and whorled striae they resemble *A. monspessulanum* L. rather than *A. campestre* L.

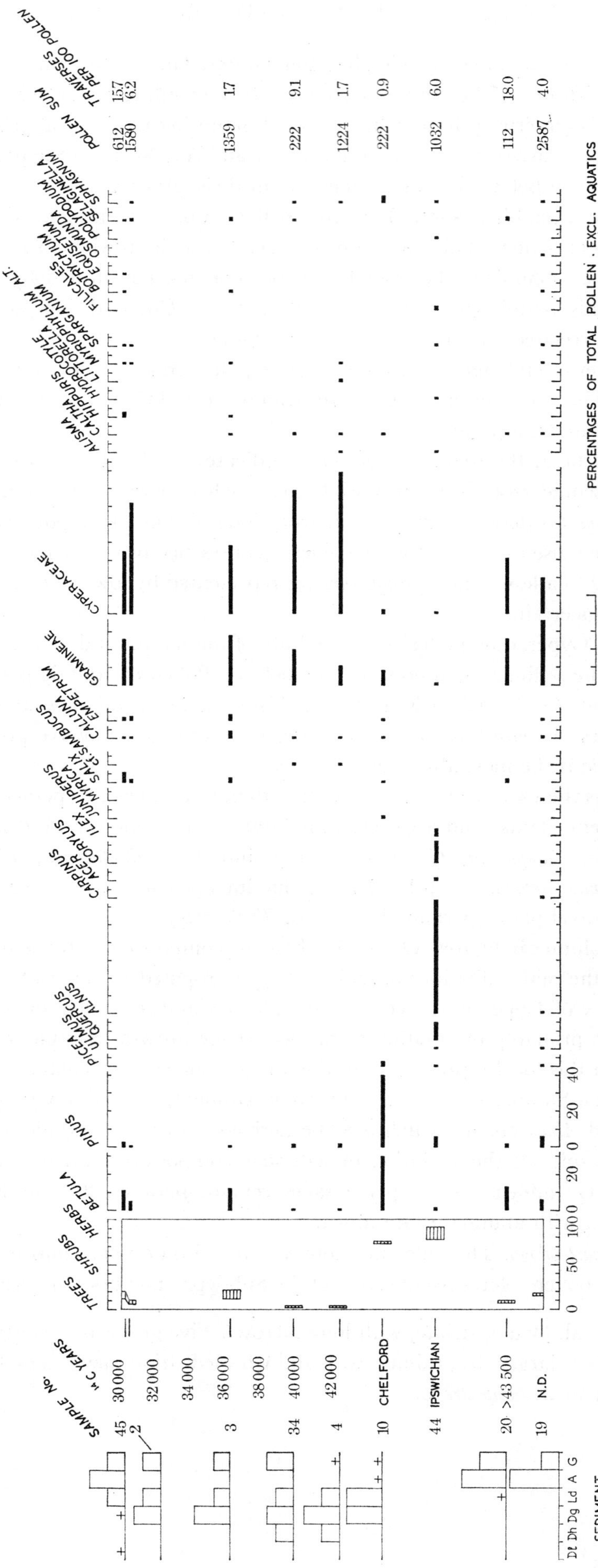

FIGURE A1. Four Ashes 1970: pollen diagram for arboreal, shrub, Gramineae, Cyperaceae and aquatic pollen, and spores.

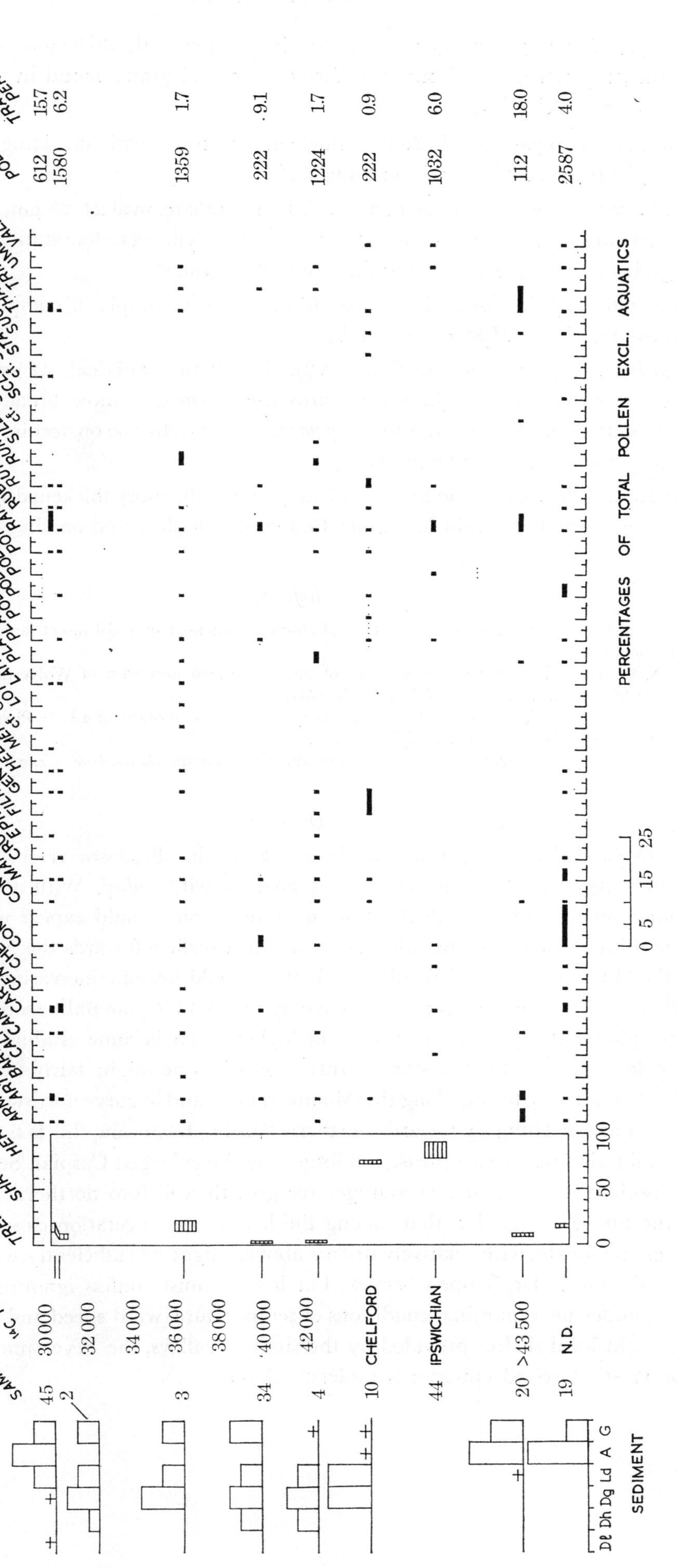

FIGURE A2. Four Ashes 1970: pollen diagram for non-arboreal pollen except Gramineae, Cyperaceae and aquatic pollen.

Bartsia/Euphrasia (*Bartsia* in figure A2). Tricolpate, spherical, 28–30 µm, very thin wall, very fine surface pattern, with long open furrows. Several grains found in sample 3 similar to *Odontites verna* (Bell.) Dum.

Bruckenthalia. Tricolpate, oval, 20 µm, thickened furrows, with crackling peculiar to many species of Ericaceae. One grain in sample 10.

Gentiana/Centaurium (Gentianaceae in figure A2). Tricolpate, oval 24–28 µm, thick wall, 'ladder-type' network, striae indistinct to nil. Grains of this type were found in sample 45 (1 grain), 2 (3 grains), 3 (15 grains), 4 (3 grains) and 19 (2 grains).

Hippuris. Sixteen grains found in sample 45 and two in sample 3, comparing well with the 5-furrowed grains of *Hippuris vulgaris* L.

Saussurea/Arctium (cf. *Saussurea* in figure A2). Tricolpate, spherical, 42 µm, thick wall, with coarse elements graduated in size towards the furrows, shallow broad-based spikes. One grain found in sample 20 similar to *S. alpina* rather than *Arctium* on account of smaller size and coarser elements in the exine patterns.

Lathyrus/Vicia. Tricolpate, oblong-oval, 27 µm, thin wall, short thickened furrows, large mesh shallow network. One grain in sample 45, two in sample 2 and one in sample 3.

References

Morgan, A. 1973 Late Pleistocene environmental changes indicated by fossil insect faunas of the English Midlands. *Boreas* **2**, 173–212.

Morgan, A. V. 1973 The Pleistocene geology of the area north and west of Wolverhampton, Staffordshire, England. *Phil. Trans. R. Soc. Lond.* B **265**, 233–297.

Simpson, I. M. & West, R. G. 1958 On the stratigraphy and palaeobotany of a Late Pleistocene organic deposit at Chelford, Cheshire. *New Phytol.* **57**, 239–250.

Troels-Smith, J. 1955 Characterisation of unconsolidated sediments. *Danm. Geol. Unders.* IV, **3**, no. 10.

Discussion

G. Manley (3 *Whitwell Way, Coton, Cambridge*). Following Professor West's demonstration one can readily suggest an operative climate by analogy with today. With a prevailing westerly flow from a much colder North Atlantic in summer, one would expect winds to be strong, with diminishing cloud; moving inland over a warm surface towards the North German Plain (where the 'Lüneburg Heath' is still found) they would become increasingly dry. It suggests the likelihood of summer air temperatures averaging 10–13 °C, not unlike some of our windswept northern uplands today upon which trees at higher levels become established with difficulty. With a polar front where Professor McIntyre puts it, one might fairly expect more summer depressions than today to run along the Mediterranean and to curve northward across the Black Sea, or even across Hungary towards northern Russia. Even now this is far from uncommon. Warm humid air from the southeast, not forgetting the enlarged Caspian Sea, would carry the summer isotherms north, and encourage tree growth well into northern Russia. Might one hazard the further suggestion that among the herbaceous vegetation close to the ground the summer microclimate, with relatively dry air above, might be sufficiently warm to explain the presence of some of Dr Coope's beetles; but here I must confess ignorance of their habits. Moreover, under such marginal conditions of temperature, wind speed and evaporation linked with the slight local shelter provided by the shallow valleys, the development and spread of Professor West's forested episodes is understandable.

Phil. Trans. R. Soc. Lond. B. **280**, 247–271 (1977) [247]
Printed in Great Britain

The Late Devensian flora and vegetation of Britain

By Winifred Pennington (Mrs T. G. Tutin)
Freshwater Biological Association, Ambleside, and *School of Biology, Leicester University*

[Pullout 1]

For the period *ca.* 15000–10000 years ago, which spans the interval between the latest presence of lowland ice and the final disappearance of mountain glaciers from Britain, so many botanical data are now available that it is possible to analyse plant distributions and vegetation composition in detail not yet possible for earlier periods. Species lists, provided by identification of pollen and of macroscopic remains, show a combination of diverse phytogeographic elements into an assemblage characteristic of the period – an admixture of apparently thermophilous plants with those now found in northern and arctic situations, and of oceanic and steppe elements. Ordination of data on floras has revealed much similarity between the regions of Britain, but pollen analysis continues to emphasize how strong must have been the regional differentiation of vegetation.

A comparison of pollen diagrams published since 1970, from sites lying on a broad north–south transect of western Britain, shows much variety in the pollen assemblage zones which have been distinguished, and in the vegetation interpreted from the pollen data by comparison with surface samples and modern vegetation. Sufficient ^{14}C dating is now available, however, to permit chronostratigraphic correlation of these pollen zones, and to show that there is consistent evidence for climatic amelioration at about 13000 B.P., and rapid deterioration at about 11000 B.P., leading to conditions of incomplete vegetation cover and universal soil disturbance which can be correlated with geomorphological evidence for the recrudescence of mountain glaciation in western Scotland, the Lake District, and North Wales, in the period called Younger Dryas on the continental mainland. Between 13000 and 11000 there is in western Britain evidence for a woodland biozone, or palaeobotanical interstadial, equivalent to Bølling plus Allerød of continental stratigraphers (Mangerud, Andersen, Berglund & Donner 1974), and divided by a very minor regression of vegetation during Older Dryas time (*ca.* 12000–11800). The now closely ^{14}C-dated site at Low Wray Bay, Windermere, is described in detail, and suggested as a reference site for this interstadial.

Introduction

A gap in the British plant record follows the deposition of an arctic moss flora at Dimlington *ca.* 18500 years ago. On lithostratigraphic evidence, this gap is assumed to correspond with the glacial stadial during which ice covered most of the area within the limits of Devensian till. Unfortunately there is as yet no evidence, from that part of Britain south of the ice margin, as to how much of the Middle Devensian flora survived within Britain the presumably severe conditions of this glacial maximum. The earliest post-Dimlington plant remains as yet discovered in Britain, with ^{14}C dates of *ca.* 14500, come from inside the limits of glaciation, from lake deposits accumulated within basins on the surface of Devensian gravels and till. From this time onwards there is an increasing volume of evidence on Late Devensian flora and vegetation, from both macroscopic remains and from pollen analysis, which demands more intensive comparison and analytical treatment than is possible for the sparser data from earlier periods.

From the time of the earliest investigations of the Late Quaternary deposits of northern Europe (summarized by Iversen (1973)) plant remains have been used to characterize the

Late-glacial environment. The threefold division of Late Weichselian deposits was first made in south Scandinavia on the evidence of plants identified from macroscopic remains, e.g. *Dryas octopetala*: this plant evidence agreed with lithostratigraphic divisions to indicate a threefold sequence of cold–temperate–cold environments. Pollen analysis of profiles from south Scandinavia subsequently identified the temperate substage, named for the type site at Allerød, with the zone of maximum percentages of arboreal pollen (mainly *Betula*) which became Pollen Zone II in a zone sequence I, II and III. Since then, the pattern of the percentage curve for *Betula* in pollen profiles from south Scandinavia has been widely adopted as a Late Weichselian temperature curve (see, for example, Van der Hammen, Maarleveld, Vogel & Zagwijn 1967), on the grounds, which were ably argued by Iversen (1954) that trees constitute the most reliable indicators of macroclimate and summer temperatures. Late Weichselian pollen diagrams have customarily been divided (zoned) into Pollen Zones I, II, III and their subdivisions on changes in proportions of *Betula* pollen, interpreted as showing the direct results of local responses of tree birches to the major climatic changes which independent evidence shows were included in this period.

In their Correlation of Quaternary Deposits in the British Isles, Mitchell, Penny, Shotton & West (1973) used Pollen Zones I, II and III for subdivision of the most recent Devensian deposits. Godwin (1975), in his recently published authoritative account of the history of the British flora, has continued to use these zones as the only practicable framework within which to order the great volume of Late-glacial plant records (from both macroscopic remains and pollen grains identified to species level) now lodged in the Data Bank of the Cambridge Botany School. But from 1963 onwards it had become clear that this could not be a *generally* adequate basis for division of the Late-glacial (Late Devensian) deposits of the British Isles without some further definition of Zone II. Watts (1963) showed that in some Irish profiles there is no expansion of percentages of *Betula* pollen, and Kirk & Godwin (1963) published analyses of a profile from Loch Droma in the NW Highlands of Scotland, in which there was no significant expansion of the curve for percentages of *Betula*, though macroscopic remains of tree birch indicated its presence within the Late-glacial period. Since then many published Late Devensian profiles from western Britain have shown an absence of any significant expansion of *Betula* (see, for example, Moar 1969; Pennington 1970, Fig. 6*c*, for high altitudes in the Lake District; Pennington, Haworth, Bonny & Lishman 1972, for northern mainland Scotland; Birks 1973, for parts of the Isle of Skye).

These discoveries point to an absence of any Late Devensian woodland from much of western Britain, and to pronounced differences between pollen zones from one part of Britain to another. In addition, radiocarbon dating has shown that at sites in western Britain where significant expansion of woodland (more than 20 % *Betula* pollen) has been found, the zone of maximum development of birch woodland must have been non-synchronous, falling within the period 12500–12000 ^{14}C a B.P. in the southern Lake District (Pennington & Bonny 1970; Pennington 1975, confirmed by hitherto unpublished dating) and within the period 11900–11000 ^{14}C a B.P. in North Wales (Simpkins 1974; Burrows 1974, 1975). It therefore seems well established that not only was there much regional differentiation within the Late Devensian vegetation of Britain, but that there is no generally applicable pollen chronology. This raises questions about interpretation of pollen data in terms of climate.

In the last five years it has become usual to describe Late Devensian pollen profiles in terms of pollen assemblage zones, and to refer these as far as possible to a ^{14}C chronology, as advocated

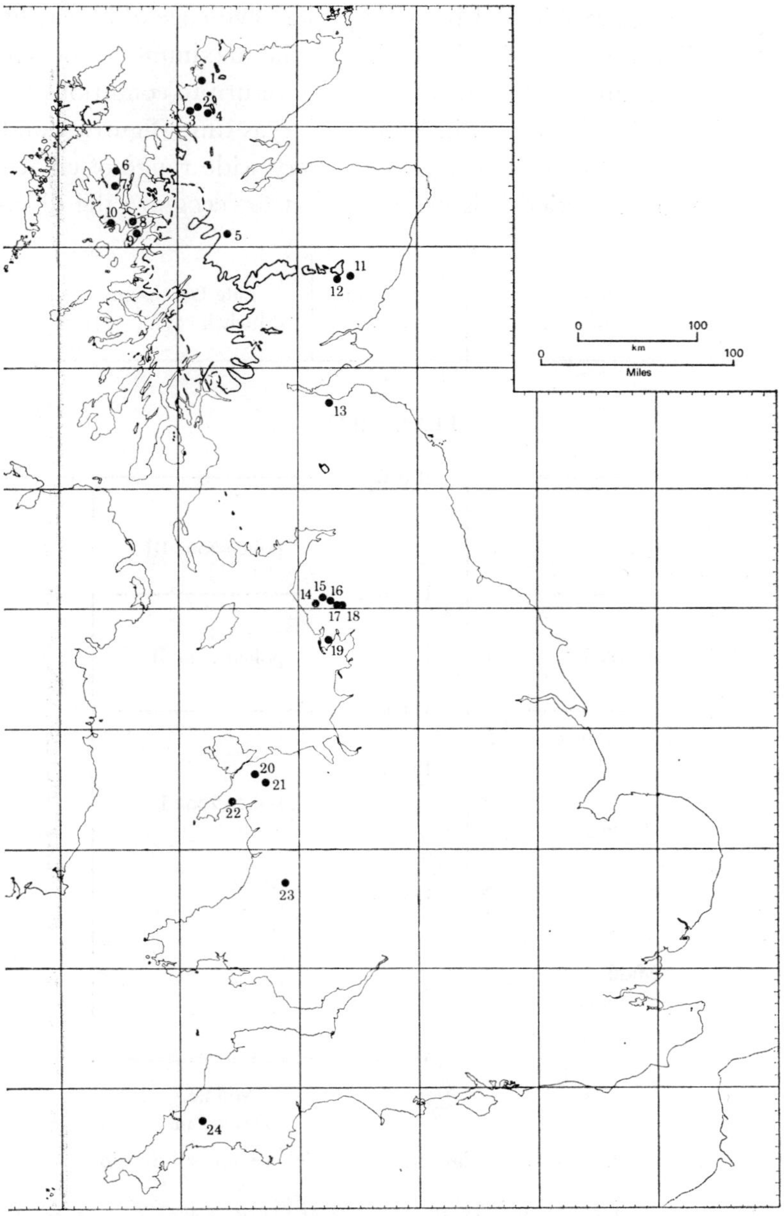

FIGURE 1. Map, showing (i) sites discussed in the text, and (ii) boundaries of the Loch Lomond Readvance (Sissons 1974) in Scotland. Sites of corrie glaciers of the same period in the mountains of Cumbria (Manley 1959) and North Wales (Seddon 1962) are not shown. Sites are: 1, Lochan an Smuraich (Pennington 1977); 2, Cam Loch (Pennington 1975); 3, Loch Sionascaig; 4, Loch Borralan; 5, Loch Tarff (all Pennington *et al.* 1972); 6, Loch Mealt; 7, Loch Fada; 8, Loch Cill Chriosd; 9, Loch Meodal; 10, Lochan Coir'a'Ghobhainn (all Birks 1973); 11, Blackness; 12, Roineach Mhor (both Walker 1975); 13, Corstorphine (Newey 1970); 14, Burnmoor Tarn (Pennington 1970); 15, Sty Head Tarn (Pennington, unpublished); 16, Blea Tarn, Langdale (Pennington 1970, 1973); 17, Blelham Bog (Pennington 1970; Pennington & Bonny 1970); 18, Low Wray Bay, Windermere (Pennington 1947, 1973; Godwin 1960, 1975); 19, Dalton-in-Furness (Johnson *et al.* 1972); 20, Nant Ffrancon (Seddon 1962; Burrows 1974, 1975); 21, Capel Curig (Crabtree 1972); 22, Glanllynnau (Coope & Brophy 1972; Simpkins 1974); 23, Elan valley (Moore 1970); 24, Hawks Tor, Bodmin Moor (Conolly, Godwin & Megaw 1950; Brown 1977).

by West (1970). Lithostratigraphic and biostratigraphic evidence alike has firmly established that the course of amelioration from the Devensian glacial maximum to the temperate environment of the Flandrian was interrupted by a temporary return to conditions sufficiently cold to re-establish mountain glaciers in Britain in Younger Dryas time (figures 1 and 2). The period of Late Devensian plant records therefore covers a very wide range of environments. In any discussion of the biota of the period it is necessary first to recognize the diversity of included

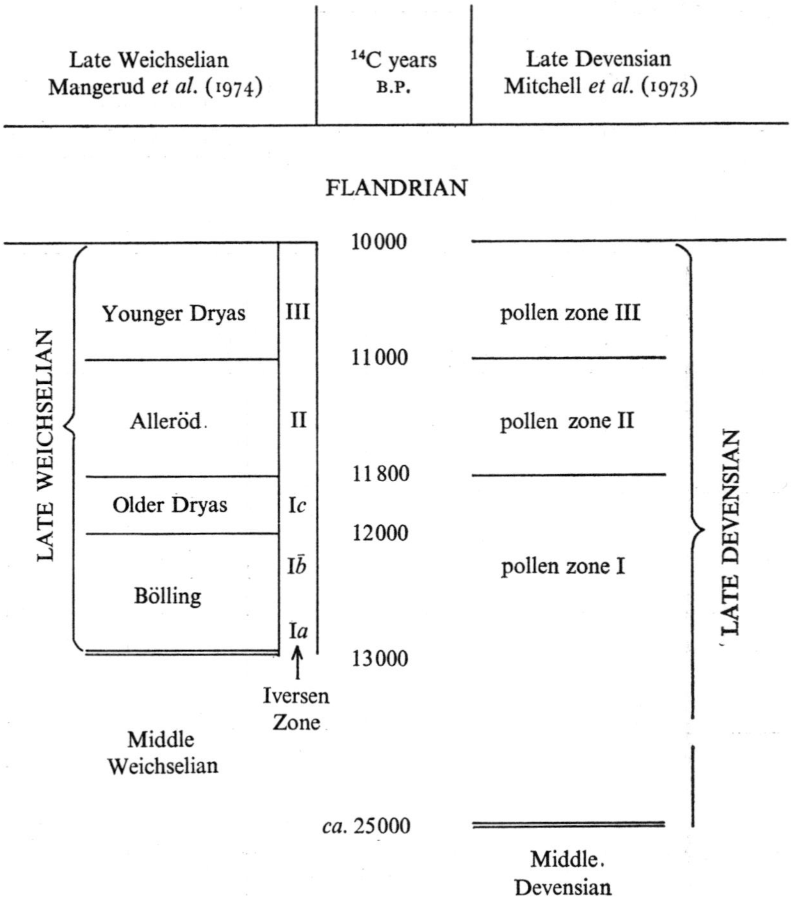

FIGURE 2. Late Devensian and Late Weichselian subdivisions.

environments (climates) and secondly, to find a clearly defined system for subdivision of Late Devensian profiles which can be generally applied in Britain. Scandinavian stratigraphers have now established a chronostratigraphic subdivision of the Late Weichselian period (Mangerud *et al.* 1974), and I have shown how this can be applied to two closely ^{14}C-dated profiles from western Britain (Pennington 1975). Figure 2 shows how use of this chronostratigraphy makes it possible to define the familiar Late Weichselian subdivisions (Allerød, etc.) as chronozones, and to correlate these with Late Devensian chronozones in dated British profiles.

FLORA

Little can be added to the comprehensive account of the Late Devensian flora which is presented in the second edition (1975) of Godwin's *History of the British flora*. This incorporates all the records for each taxon (from each of Pollen Zones I, II and III) which had been entered in the Data Bank of the Cambridge University Botany School by 1970. Godwin's discussion emphasizes and enlarges upon, with long species lists, the picture of the rich and varied Late-glacial flora of northwest Europe which has been consistently developed and expanded since the pioneer work of himself, Jessen and Iversen in the 1940s. The Cambridge records show that about half of the total number of species in the British flora has now been recorded from Late Quaternary deposits, and of these about 60 % were already in the country by the end of Late Devensian time.

Since it is unlikely that existing records of the Middle Devensian flora (see, for example, Bell 1969) represent more than chance-preserved samples, and since there are no plant records at all from the period of the Devensian ice maximum, it will probably remain impossible to decide how many of the Late Devensian records represent species which had survived within Britain from earlier periods. Taxa recorded from both Middle and Late Devensian deposits, and therefore among those most likely to have persisted in Britain, include, together with those of subarctic or montane distribution today, such as *Betula nana*, *Salix herbacea* and *Oxyria digyna*, a large number of weeds and ruderals whose primary requirement appears to have been for open habitats, aquatics such as *Groenlandia densa*, steppe elements such as *Helianthemum canum* and *Artemisia* spp., and species of predominantly southern distribution today such as *Lycopus europaeus* and *Onobrychis viciifolia* (cf. Godwin 1975, Tabs 37–41).

A large number of relatively warmth-demanding plants including the tree birches and *Populus tremula*, which are absent from all Middle Devensian records, must be supposed to have entered Britain in response to ameliorating conditions after the glacial maximum was past, across wide land bridges with the Continent which were then exposed by low Late-glacial sea levels. This low sea level would also have left exposed land areas to the west of Britain, and Godwin's suggestion that these could have provided glacial refuges for trees and other taxa receives support from the evidence, to be presented in the next section of this paper, for early immigration of tree birches into parts of western Britain after the Devensian ice maximum. The distribution of *Pinus sylvestris*, which appears confined to the Late Devensian woodland of eastern and southern England, indicates a very different immigration route for this tree – as Godwin comments, its Late-glacial distribution would appear to have been continuous, across the dry bed of the North Sea, with the mainland pine-birch region, in Allerød time.

The outstanding difference between the Late Devensian flora and the modern flora of Britain is the absence from the former of a large group of thermophilous trees and shrubs – including *Corylus*, *Ulmus*, *Quercus*, *Alnus*, *Tilia*, *Hedera*, *Ilex*, etc. The conventional interpretation of this absence in terms of low Late-glacial summer temperatures and slow dispersal from distant refuges may be questioned with respect to *Corylus* in western Britain. Birks (1973) shows that reconstruction of the Late Devensian environment of the Isle of Skye by detailed consideration of evidence from other plant data leads to the conclusion 'it seems inconceivable that conditions were not suitable for the germination, growth and flowering of *Corylus avellana*'. The rapid expansion of this species at the opening of Flandrian time in western Britain suggests that it must have been present at no great distance, yet some as yet unidentified factor inhibited its expansion during the Late Devensian period.

Late Devensian flora – phytogeography

Godwin (1975, pp. 446–450) presents a 'phytogeographic synopsis' which includes the results of numerical analysis of past and present British floras by Birks & Deacon (1973). Late Devensian sites from which data had been published by 1970 were allocated to 10 geographic regions of Britain (Birks & Deacon, Fig. 2), and for each region the percentage of plant records falling into each of five aggregate phytogeographic groups was calculated for (*a*) the Late Devensian flora and (*b*) the present flora. The phytogeographic groups include the following floristic elements:

(*a*) Arctic–alpine, Arctic–sub-arctic, Alpine;
(*b*) Northern montane, Continental northern;
(*c*) Continental northern, Continental southern, Continental widespread;
(*d*) Sub-atlantic northern, Sub-atlantic southern, Sub-atlantic widespread, Atlantic northern, Atlantic southern, Atlantic widespread, Mediterranean Atlantic;
(*e*) Continental southern, Sub-atlantic southern, Atlantic southern.

Table 5 in Birks & Deacon (1973) gives the percentages for each floristic element in each geographic region, and Figure 159 in Godwin (1975) presents the results in the form of histograms, in which the values for floristic elements are aggregated into the above five phytogeographic groups.

The conclusions were:

(i) Late Devensian plant records, considered as a whole for each of the ten regions, indicate a phytogeographical diversity as great as that found in the present British flora.

(ii) These ten regions of Britain differ less from each other with respect to the phytogeographical affinities of their Late Devensian floras than they do with respect to these affinities of their modern floras. This is shown by the histograms but was further demonstrated by numerical analysis by Birks & Deacon. By two-dimensional plots of the geographical regions, resulting from non-metric multi-dimensional scaling of the dissimilarity coefficients matrix, these authors were able to demonstrate a clear north–south gradient in the present British flora which cannot be found in the Late Devensian flora. Their conclusion for the Late Devensian was that 'there was little floristic differentiation within Britain at this time'.

(iii) The preponderant elements in the Late Devensian flora were those included in groups (*b*) and (*c*) – northern and continental – and these showed little geographic variation throughout Britain.

(iv) In so far as floristic differentiation can be recognized in the Late Devensian flora, it is found mostly with respect to the aggregate groups (*a*) and (*d*). Even at that time the effect of altitude is perceptible, in the higher values for group (*a*) in regions which include high mountains.

This elegant demonstration emphasizes the contraction of range of many species since Late Devensian time – a contraction which has differentiated those regions where they have survived from those in which they have been extinguished. The causes of extinction must include:

(i) The direct effects of temperature, as argued for arctic–alpine species by Conolly & Dahl (1970). Many species in this group now survive only in areas where summer temperatures do not exceed a certain maximum.

(ii) The reduction of habitats open to species of high light requirements or low competitive ability by the post-glacial spread of forest – cf. the example of the Upper Teesdale refugium for such species (Godwin 1956).

(iii) The decline in soil base-status from the generally high level characteristic of a Lateglacial stage, with subsequent limitation of many species to areas of calcareous bedrock (cf. Iversen 1954, 1973), e.g. the wider northward range in Late Devensian time of *Polemonium caeruleum* and *Scabiosa columbaria*.

Though it is convenient for some purposes to consider the Late Devensian period as a whole, aggregating floral records from all its subdivisions, it is salutory to remember that for many sites the aggregate species record covers a period of more than 3000 years, which included strong contrasts in climate. The apparent uniformity in geographical distribution within Britain of most of the components of the Late Devensian flora may reflect to some extent the consequences of combining data from cold periods (with expansion southwards of arctic–alpine and montane species) and more temperate periods (with northward expansion of more thermophilous species). As knowledge of Late Devensian chronology increases, we may look forward to more critical comparisons, restricted to a few centuries of Late Devensian time, which might well reveal stronger north–south gradients in the fossil floras.

Late Devensian flora – interpretation in terms of climate

The simultaneous presence in plant assemblages of species of such widely contrasted modern affinities that their present geographical ranges do not overlap, which is one result of the wider ranges of many plants in Late Devensian time than today, presents obstacles to interpretation of Late Devensian climate in terms of the present ranges of species. Iversen (1954) discussed one aspect of this problem, postulating that the primary factor which controls the present distribution of many light-demanding plants (heliophytes) is competition from other species which cast shade, so the presence of heliophytes in Late-glacial deposits gives no information about temperature, to which many of them appear indifferent. Bell (1969) discussed difficulties of reconstruction of the Middle Devensian environment, because of the presence in many plant assemblages of species of southern and steppe distributions today, together with northern elements. Godwin (1975, pp. 432–3) made the point that the absence of trees from a flora does not of itself indicate low summer temperatures, because there are features of the extreme continental type of climate which are adverse to trees irrespective of summer temperatures – very cold winters, severe wind exposure, severe spring thaw and floods, or highly unstable soils. Persistence of any one of these features may have limited the establishment and expansion of trees in any region of Britain in Late Devensian time, which would explain to some extent the accumulating evidence that the development of tree birches was not always in equilibrium with temperature, as deduced from the present ranges of other species present.

Bell (1969) discusses the question of the extent to which the present distribution limits of plant species reflect their physiological tolerances and are in equilibrium with climate. Perring & Walters (1962), discussing distribution maps of the present British flora, comment: 'The effect of climate upon different species varies widely, and it may take years of research to determine what is the critical factor which has limited a particular species to its present range.' Botanists have therefore accepted for some time that until this critical factor is determined by experiment, climatic interpretations based on the present range of a species represent hypothesis rather than fact.

Vegetation

Godwin (1975) described in detail the pattern of change in Late Devensian vegetation shown by results, published up to 1970, from sites where the profile had been divided by the author into Pollen Zones I, II and III on the basis of the *Betula* curve, or on the lithostratigraphic sequence in which organic sediment sandwiched between two minerogenic deposits was correlated with Zone II. Figure 156 in Godwin presents comparative percentage pollen spectra, in the form of sectorial diagrams, for each of these zones at a number of British sites; these demonstrate clearly the contrasted pollen spectra found (*a*) at the same site within successive periods of Late Devensian time, and (*b*) from one site to another within Britain during the periods of time represented by each zone. In his text Godwin discusses fully the composition of the plant communities represented by Late Devensian pollen spectra and reconstructed from assemblages of macroscopic remains – communities of woodland – which in southern and eastern England included *Pinus* as well as the more widespread tree birches and *Populus tremula*, aquatic, marsh, shrub and heath, grassland and other terrestrial habitats, together with the records of the many plants found today by waysides and in arable fields as weeds, and in open montane habitats.

In this paper I shall consider work on the Late Devensian vegetation of western Britain, published in 1970 or subsequently, in which definition and correlation of pollen zones have followed the principles advocated by West (1970) and some attempt has been made to correlate pollen zones to a chronostratigraphical framework. Figure 1 shows the position of these sites, which lie on a broad north–south transect from north Sutherland to Cornwall, and gives references in the legend. Table 1 sets out the pollen assemblage zones, correlated to the chronostratigraphic subdivisions shown in figure 2 by the available ^{14}C dating. In three areas it has been possible to prepare a synthesis of regional vegetation history by correlation of pollen assemblage zones from at least 5 sites: namely, (i) the English Lake District (Pennington 1970, 1973), table 1*a*(i); (ii) the mainland of Scotland from the Great Glen north to Sutherland (Pennington *et al.* 1972; Pennington 1975, 1977), table 1*a*(ii), and (iii) the Isle of Skye (Birks 1973), table 1*b*(ii). In (i) and (ii) the chronostratigraphy was established by detailed ^{14}C dating at one or two sites, and parallel changes in pollen stratigraphy and lithostratigraphy at *ca.* 13000, 12000 and 11000 B.P. were interpreted as evidence for climatic changes at these dates, justifying correlation with similar horizons in the undated profiles (table 1). In (iii) Birks adopted a different plan, assuming constant rates of sediment accumulation at each site (determined from early Flandrian ^{14}C dates) and using these to calculate dates for Late Devensian pollen zone boundaries and demonstrate their non-synchroneity (Birks 1973, Fig. 27). Pollen assemblage zones from these three regions have been compared firstly, within western Britain from the Lake District northwards, and secondly with sites from the Grampian Highlands, from North and Central Wales, and from Cornwall (table 1*b*).

In the synthesis of vegetation history which follows some attempt will be made to interpret pollen data objectively, avoiding the 'intuitive' European approach criticized by Livingstone (1968). Sufficient evidence is available to show that the pollen spectra of each zone are likely to include contributions from several different plant communities (Kirk & Godwin 1963; Birks 1973). In reconstruction of palaeoenvironments from pollen data, the assumed relationships between pollen spectra and vegetation and between vegetation and climate can be tested by comparison with spectra currently being produced by a variety of vegetation types. It will be shown how such methods have demonstrated that, whereas most of the Late Devensian pollen spectra found by Birks on Skye can be interpreted in terms of the present vegetation of that island, between sea level and the mountain summits, many Late Devensian pollen spectra from the British mainland can either be matched only by those now accumulating at higher latitudes or altitudes than are found in Britain, or cannot be matched in northern Europe today.

Synthesis of Late Devensian vegetation history

(a) *Before* 13000 B.P.

Lake deposits containing plant remains have been dated to well before 13000 at several sites near the coasts of the Irish Sea (see Pennington 1975), indicating that ice had withdrawn from that area before 14500. The basal sediments in a kettlehole basin sectioned by coastal erosion at Glanllynnau (figure 1) contained seeds of the family Caryophyllaceae and moss fragments, together with remains of an arctic beetle assemblage, and the seeds yielded a date of 14468 ± 300 which is unlikely to be erroneous (Coope & Brophy 1972). The earliest organic matter to be deposited in Lake District lake basins, which contains pollen spectra and annual pollen deposition rates indicative of treeless tundra, has been dated to 14330 ± 230 at Blelham Bog (Godwin & Willis 1964; Pennington & Bonny 1970) and to 14557 ± 280 and 14623 ± 360 (SRR-681 and 682 – figure 3) in the littoral sediments of Windermere. Though these lake sediments of the Lake District may contain ^{14}C-deficient carbon from aquatic photosynthesis, the close agreement between dates from these two contrasted environments (a small kettle hole and a large lake with an extensive mountain catchment) suggests that an unlikely combination of errors would have been required to produce this result. It may therefore be concluded that a sparse vegetation was developing by *ca.* 14500 on newly deglaciated land, at a time when faunal remains indicate low temperatures, as at Glanllynnau (Coope & Brophy 1972 – Coleoptera) and Blelham Bog (Harmsworth 1968 – Cladocera).

(i) *Salix herbacea*–Cyperaceae–*Lycopodium selago* pollen assemblage zone (p.a.z.): before *ca.* 14000

In Windermere this correlates broadly with the sample dated to 14557 ± 280 (figure 3) and with such low pollen concentration that annual deposition rates cannot have exceeded 100 grains/cm² per year, a figure low even for tundra (Davis, Brubaker & Webb 1973). In the correlated zone at Blea Tarn, higher in the same catchment, annual deposition of most taxa in this p.a.z. compares closely with that estimated for present deposition in a lake at 1400 m in Jotunheim, S. Norway, surrounded by *Salix herbacea*-dominated snowbeds of the middle alpine region (Pennington 1973). Faegri (1953) suggested that this assemblage, which he found at the base of two Late-glacial profiles from southwest Norway, represents chionophilous communities likely to have characterized the periglacial vegetation of oceanic Europe.

Birks (1973) compared with this p.a.z. at Blea Tarn his *Lycopodium*–Cyperaceae p.a.z. at the base of three profiles from Skye, but *Salix* percentages are lower on Skye than in the Lake District. He found within this zone combinations of taxa which indicated at one site chionophilous communities, and at other sites communities now found on exposed summit heath and fell-field. At Cam Loch in Sutherland the lowest p.a.z., dated to well before 13000 by the overlying ^{14}C date, contains *Oxyria–Rumex* type pollen together with *Salix herbacea* (Pennington 1975). This p.a.z. is not recorded from other sites; it may represent a type of environment during which only in certain lake basins was pollen recruited to the accumulating sediment.

(ii) *Rumex*–Gramineae p.a.z.: *ca.* 14000–13000

Figure 3 shows how a pollen zone dominated by 20 % *Rumex* (predominantly *R. acetosa*) and 25 % grass pollen is in Windermere dated to *ca.* 13000–13800. Pollen concentration is ten times that of the underlying zone; division by the mean Late-glacial rate of sediment accumulation

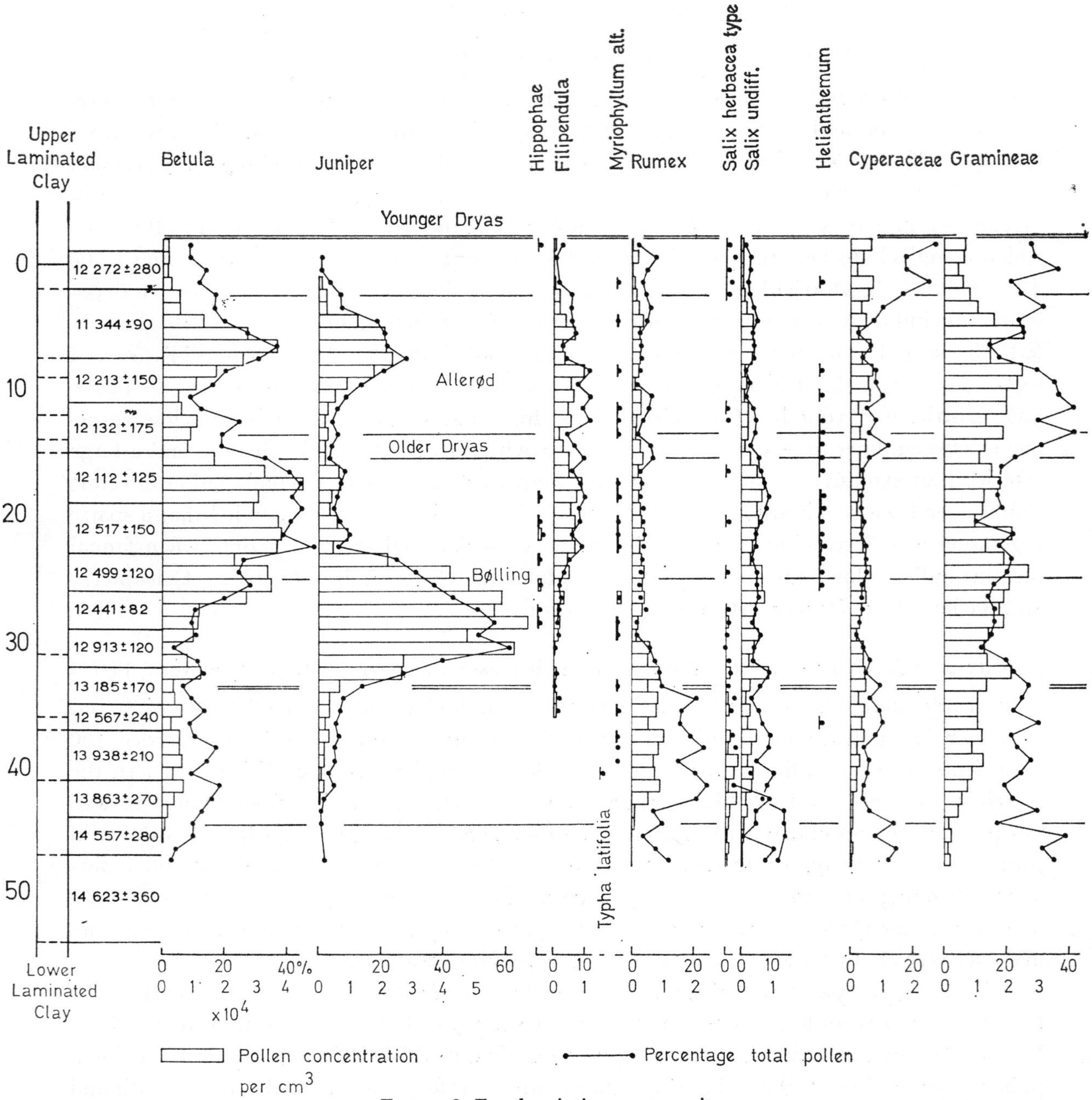

FIGURE 3. For description see opposite.

indicates an annual influx of *ca.* 1000 grains/cm². Within this zone appear many taxa suggestive of climatic amelioration, e.g. *Filipendula, Hippophae, Typha latifolia*, together with, at the base of the zone, a significant increase in percentage and absolute values for juniper and tree birch pollen. Through this zone *Salix herbacea* declines, Cyperaceae values remain low, and *Lycopodium selago* disappears. Some undoubted *Betula nana* pollen is present, together with that of *Empetrum, Thalictrum* and Rubiaceae. A rich assemblage of herbaceous taxa is represented by occasional grains (see figure 3); this includes *Helianthemum* and *Artemisia*.

A comparable zone is found at the base of many profiles from western Britain (see table 1 *b*).

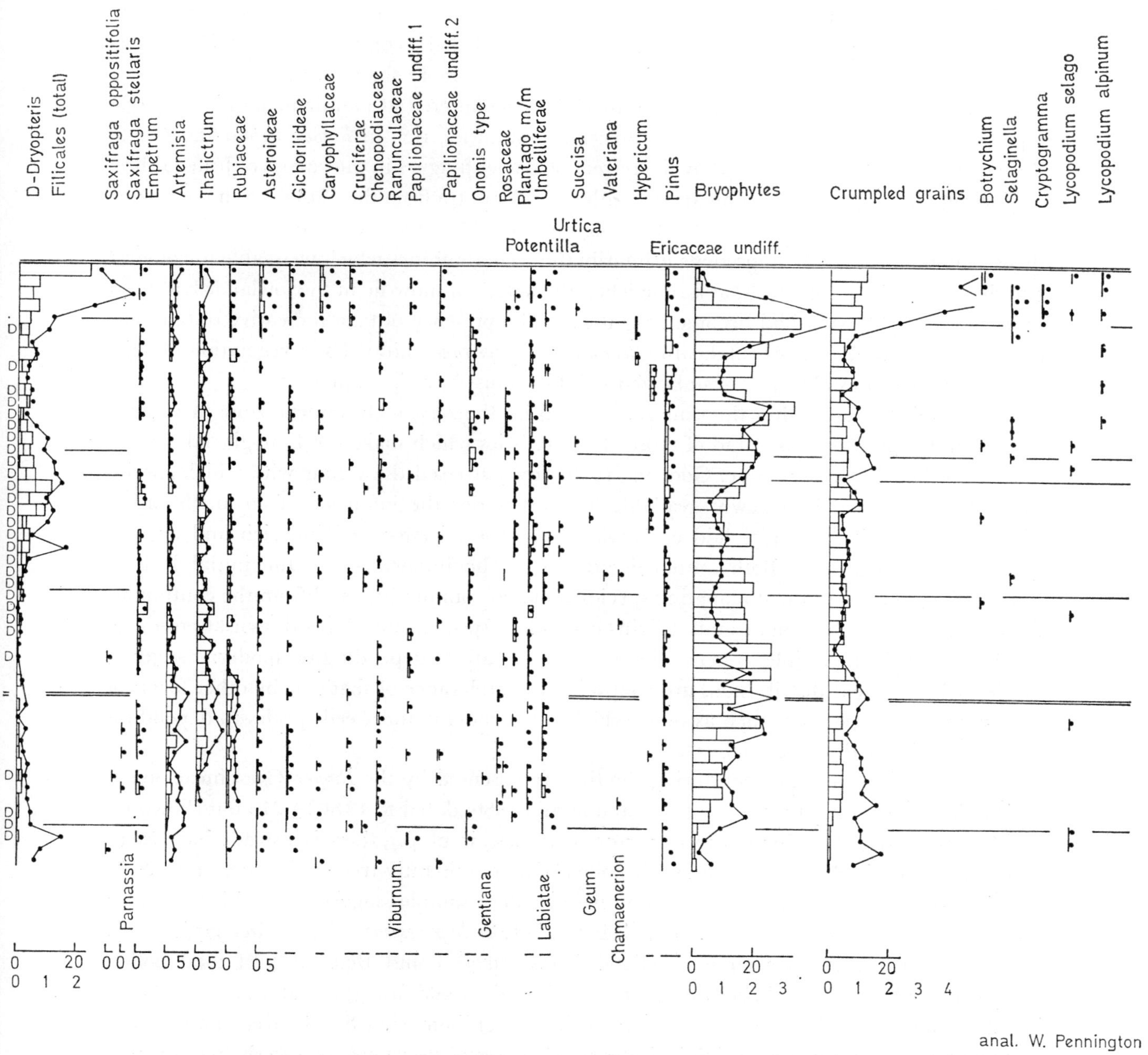

FIGURE 3. A percentage and absolute pollen diagram from Low Wray Bay, Windermere. This represents more detailed and absolute analysis of the profile previously investigated (Pennington 1943, 1947, Figs. 3 and 14, Godwin 1960, Fig. 1). This core was taken with a 12.5 cm diameter corer (Mackereth, Gilson & Findley, unpublished) which permitted ^{14}C dating of 15 samples. The line of best fit to these samples (figure 5) indicates a mean rate of sediment accumulation between 11000 and *ca.* 14000 of *ca.* 1 cm in 50 years. Division of the scale for pollen concentration (per cm^3) by 50 therefore converts concentration figures to grains deposited $cm^{-2}\ a^{-1}$. The pollen diagram covers the organic silts between lower and upper laminated (varved) clays (Pennington 1947).

It is interpreted by Birks, on the basis of surface samples from high-altitude vegetation, as the product of communities of species-rich grassland and tall herb communities, but *Rumex* percentages are lower on Skye than at mainland sites (never more than 10, often less than 5). I have found the high values for *Rumex* (up to 45 %) encountered at mainland sites in only one type of surface sample – that from the pioneer communities, rich in *Rumex acetosa* and *Oxyria digyna*, of ground recently exposed by the retreat of glacier tongues in the valleys of Jostedalsbre, S. Norway (cf. Faegri 1933): comparisons with typical spectra from the *Rumex*–Gramineae zone in NW Scotland are given in table 2 of Pennington (1977). I would therefore suggest that on the British mainland this zone represents primarily those communities which were able to colonize and develop on immature soils, at a time when the length of snow-lie in the lowlands had been reduced, and solifluxion had become less intense. The presence of taxa indicative of tall herb communities and of dwarf-shrub heath (*Betula nana*, *Empetrum*) shows that a vegetation mosaic, representative of unstable and of stabilized habitats, must have been present.

It seems likely that within the time represented by the *Rumex*–Gramineae zone in western Britain, full expression was found of those characteristics which make the Late-glacial vegetation of northern Europe unique. Godwin (1956, 1975) stressed the vigour with which species were able to expand into newly available habitats under the influence of an ameliorating climate; Iversen (1954, 1973) laid emphasis on the special effects of lime-rich and humus-deficient immature soils. Both authorities discussed the importance of the initial lack of competition, both from the dominating species of closed communities and from the continuous woodland whose shade would inhibit light-demanding species. Both have drawn attention to the fact that it is impossible to make climatic interpretations depending on modern ranges of species, when present distributions are controlled by intolerance of shade or by other forms of intolerance of competition, such as those which cause ruderal and weed species to depend on anthropogenically disturbed habitats.

What was happening to trees during the time represented by the *Rumex*–Gramineae zone? In Windermere, within the limits of ^{14}C sample SRR-680, dated to 13863 ± 270 but shown by its $\delta\,^{13}C$ value to incorporate hardwater error (Harkness, in prep.), there is a steady increase in percentage and absolute values for *Betula* pollen of predominantly tree birch type, up to 20 % and 200–400 grains $cm^{-2}\,a^{-1}$. Comparison with modern samples suggests that this represents the local arrival of tree birches – the 'establishment' of *Betula pubescens* (cf. Godwin 1975, p. 489). In moss polsters from open birch woodland, Birks (1973) found that 20 % of total pollen represents the lower limit of *Betula* percentages, while Fredskild (1973) found an annual *Betula* influx of 200–400 grains/cm^2 in the recent sediments of Comarum Sø, S. Greenland, within a vegetation of shrub tundra which includes small copses of *Betula pubescens* in sheltered places and appears to approximate to the 'park-tundra' postulated by Iversen (Degerbøl & Iversen 1945). This conclusion from pollen data is supported by the presence of fruits and catkin-scales of *Betula pubescens* at a horizon in Windermere now identified as *ca.* 4 cm above the base of the *Rumex*–Gramineae p.a.z. (compare figure 3 with Fig. 14 of Pennington 1947). It seems necessary to conclude that the local arrival of tree birches in the southern Lake District took place before 13000 B.P., and that some factor delayed the expansion of the species until *ca.* 12500. The factor could have been climatic or edaphic. For this site it is not, however, possible to explain the presence of a thermophilous insect fauna below the main *Betula* zone in terms of delayed immigration of tree birches (Coope & Brophy 1972; Coope 1976).

A sample of the vegetation of southeastern England, during the time represented by the

Rumex–Gramineae p.a.z. in the west, came from Colney Heath (Godwin 1964, 1975) where peat erratics in river deposits gave a date of 13560 ± 210. The long list of herbaceous taxa identified, regarded as highly characteristic of late-glacial plant assemblages, contains most of those found in the *Rumex*–Gramineae zone in Windermere, together with leaves of *Salix herbacea* and wood which was probably of juniper. Godwin's view, that the Colney Heath flora was most likely to represent that of the later part of the glacial period in southern Britain, is now confirmed, by ^{14}C correlation with the period of declining snow cover and new exposure of skeletal soils at sites within the limits of Devensian glaciation.

At Nant Ffrancon (figure 1), though critical ^{14}C dating is as yet unfortunately lacking, synthesis of the work of Seddon (1962) and Burrows (1974, 1975) suggests a situation there similar to that at Windermere, in that near the base of the organic deposits, a pollen zone containing a *Rumex*–Gramineae assemblage and some *Betula* and *Filipendula* pollen appears to correspond with the horizon where Burrows found the earliest macroscopic remains of *Betula pubescens*.

In northern mainland Scotland there is a contrast, in that the p.a.z. defined by maximum *Rumex* percentages is not a clearly defined *Rumex*–Gramineae zone, but overlaps with *Lycopodium selago* below and with *Juniperus*, or *Empetrum*, above (table 1*b*(i), (ii)). This can be interpreted as the effect of prolongation in higher latitudes, first of conditions favouring the chionophilous *Lycopodium selago*, and second of conditions favouring the preservation of disturbed habitats occupied by *Rumex acetosa* and other pioneer species. In the Grampian Highlands, the pollen zone in Walker's (1975) two sites which appears to correlate with the *Rumex*–Gramineae zone at other sites (table 1*b*(iii)) contains so much *Pinus* pollen that it is most readily interpreted as the result of secondary deposition of pollen from melting ice. A similar effect is seen at the base of this p.a.z. in profiles from higher altitudes in the Lake District (table 1*a*(i)).

Small percentage increases in *Betula* within this zone, which were found in several cores from Blelham Bog (Pennington 1970; Evans 1970) have been shown by absolute pollen analysis to represent statistical artefacts rather than any real increase in *Betula* (Pennington & Bonny 1970). The analysis of contiguous cm samples in the Windermere profile has confirmed this (see figure 3). Chance selection of certain samples on the percentage *Betula* curve could have produced an appearance of a fluctuation, but analysis of contiguous samples has shown no consistent trend in percentage or absolute values for *Betula* above the rise already described. Johnson, Tallis & Pearson (1972) incline towards the same explanation of the *Betula* curve within the lowest zone of their profile from Dalton-in-Furness, 30 km south of Windermere.

(*b*) 13000–11000 B.P.

(i) *Juniperus* p.a.z.: 13000–12500 in the Lake District; 13000–? elsewhere

Figure 3 shows the steep percentage and absolute rise in *Juniperus* which at Windermere coincides with the upper boundary of the *Rumex*–Gramineae p.a.z., and Fig. 2 in Pennington & Bonny (1970) shows an exact parallel at Blelham Bog. At both sites this zone boundary is dated to 13000, and in the Windermere ^{14}C samples the $\delta\,^{13}C$ values show no evidence of hardwater error. A p.a.z. of high values for *Juniperus* has been widely recognized below the main *Betula* zone at western sites (table 1*b*) and absolute pollen analysis has shown how a very significant increase in pollen deposition rates coincides with this first major expansion of juniper at 13000 (Pennington 1975, and figure 3). There is a date of 12556 ± 230 for the correlative zone at

Glanllynnau (Coope & Brophy 1972; Simpkins 1974) and inferred dates of just after 13000 for the base of the Gramineae–*Rumex–Juniperus* subzone at Loch Meodal on Skye (Birks 1973). At Cam Loch the base of the *Rumex–Juniperus* p.a.z. coincides with the mean position of the ^{14}C sample dated to 12956 ± 240 (SRR-253). There is therefore a body of evidence from western Britain to show that a widespread increase in pollen production by juniper coincided with the date of 13000, suggesting a climatic amelioration correlated with that used to define the base of the Bølling chronozone. Nevertheless a clear expansion of juniper has not been found at all sites in western Britain, and the distribution of this plant would appear to have been far from uniform; compare three neighbouring sites in northern Scotland, Lochs Sionascaig & Borralan (Pennington *et al.* 1972) and Cam Loch (Pennington 1975).

In these papers I have interpreted the sudden increase in pollen production by juniper where present, as a response to climatic amelioration by a species already present, as Iversen (1960) interpreted the later juniper maximum at the late-glacial/post-glacial boundary. At the Windermere site, where the evidence already quoted shows that tree birches must have been present before 13000, it remains to be explained what factor favoured expansion of juniper rather than birch at 13000. Godwin (1975) discussing the ecology of juniper, which is very imperfectly understood, gives the evidence for its present status as a dominant above the present tree-line in northern Britain, a habitat which suggests that in some respects it is less exacting than birch. The data presented by Birks (1973) on modern pollen rain within juniper communities shows, however, that the high percentages found within this *Juniperus* p.a.z. at many sites can only be matched in tall dense thickets of *Juniperus communis* ssp. *communis* in Britain today, and not by the more prostrate forms found at higher altitudes.

(ii) *Betula* p.a.z.: 12500–12000 in the Lake District, ? elsewhere

A clearly defined p.a.z. of maximum Late Devensian percentages of *Betula* is found immediately above the juniper zone at Loch Meodal on Skye (Birks 1973) and in Lake District lowland profiles (figure 3). Detailed regional studies by Dr Birks in Skye and by myself in the Lake District have shown first, that there is much variation between sites with respect to *Betula* percentages reached in this zone, these declining at the less edaphically and climatically favoured sites on Skye and with increasing altitude in the Lake District, and secondly that all the evidence points to 'an uninterrupted succession on stable soils, through a juniper phase ... to birch woodland' (Pennington 1970) or 'a progressive unidirectional vegetation succession ... presumably in response to climatic amelioration' (Birks 1973). Comparisons already made with modern samples suggest that 20 % of *Betula* pollen may approximate to a tree cover somewhere between the scattered birch copses of parts of southern Greenland (equivalent to 'park-tundra') and a canopy cover of 25–30 % (Birks 1973, Fig. 25), while the 30 % or more of *Betula*, found throughout this zone in the Lake District lowlands and corresponding with an annual influx of 800–1000 grains $cm^{-2}\,a^{-1}$ at two sites, exceeds by a factor of 4 the present deposition from park-tundra, and on a Birks regression line corresponds with a canopy cover of at least 50 %.

^{14}C dates from Blelham Bog and from Windermere (figure 3) agree to show that the lower boundary of the *Betula* p.a.z. in the Lake District falls at *ca.* 12500. It is postulated that this boundary, equivalent to that of the biostratigraphic zone which Iversen called Ib (Mangerud *et al.* 1974) does not represent a synchronous and climatically determined horizon such as those used as boundaries for Late Weichselian chronozones by Mangerud *et al.*, but the result of local vegetation development at each site.

The characteristic Late-glacial assemblage of *Helianthemum*, *Hippophae* and *Artemisia*, which Iversen noted as quite absent from arctic and sub-arctic regions today (in Degerbøl & Iversen 1945), reaches its maximum development in the Windermere profile within the zone of maximum *Betula* – i.e. the chronozone 12500–12000. This assemblage, which has not been found in Scotland, has not been recognized in modern vegetation but must represent a community of unshaded habitats. It seems probable that some local factor continued to inhibit the growth of tree birches on particular types of terrain, so that a mosaic of woodland and of open areas with unstable soils was present.

(iii) *Rumex–Empetrum* ± *Juniperus* p.a.z. Northern Scotland only: *ca.* 13000–12000

This pollen zone has been found at five sites in northern Scotland, is correlated at Cam Loch with the chronozone 13000–12000 (Pennington 1975) and is correlated with the biostratigraphic zone A2 of Pennington *et al.* (1972). A correlative zone can be recognized at Blackness in the Grampians (Walker 1975). Juniper values vary from site to site, reaching a maximum (20 %) at Loch Sionascaig, the westernmost site. Pollen deposition rates at Cam Loch during this chronozone resemble those found by Fredskild in two lakes, Isoetes Sø and Kløftsø, situated in dwarf-shrub tundra dominated by *Empetrum hermaphroditum* plus some juniper, near the coast of SW Greenland (Fredskild 1973). Percentages of those taxa present are also in fair agreement with Fredskild's data (Pennington 1977, Tab. 2). The absence of *Calluna* from the Late Devensian pollen spectra distinguishes them from all modern samples from dwarf-shrub heath in Britain or Norway today. I postulate that the *Empetrum*-rich deposits at Loch Droma, ^{14}C dated to 12810 ± 155 (Kirk & Godwin 1963) correlate with this zone. The evidence is that dwarf-shrub tundra, similar in pollen productivity though not in species composition to that found on the outer coasts of SW Greenland today, was the prevailing vegetation of the northern Scottish mainland during the Bølling chronozone.

(iv) *Betula–Rumex* p.a.z.: 12000–11800 in the Lake District

At many sites, including Loch Meodal (Birks 1973) the *Betula* zone is narrow and only a few samples were analysed. The longer profiles (indicative of more rapid sediment accumulation) from Blelham Bog and Windermere have shown that the *Betula* zone as a whole is divided by a narrow zone of reduced annual deposition of the pollen of the more thermophilous taxa, which at both sites is correlated with the chronozone *ca.* 12000–11800 and so with the Late Weichselian Older Dryas chronozone (Pennington 1975; and figure 3). Comparison of the pollen data with geochemical data (figure 4) shows that on both criteria this narrow zone represents partial regression towards the composition of pre-13000 deposits. A rather small percentage increase in *Rumex* defines the p.a.z. Confirmation of this interpretation of climatic deterioration at this horizon comes from the insect fauna of the Windermere profile (Coope 1977).

(v) *Rumex–Artemisia* p.a.z.: *ca.* 12000–*ca.* 11800 in northern Scotland

This p.a.z. is found at the five sites from NW Scotland, is correlated with a distinct paler lithostratigraphic unit, and with a distinct diatom zone which includes taxa of sub-aerial habitats (Haworth 1976). It represents zone A3 of Pennington *et al.* (1972) and is approximately dated at Cam Loch. This p.a.z. represents a regression to conditions of an incomplete plant cover (favouring *Artemisia*) and increased soil erosion interpreted as due to an increased incidence of freeze-thaw cycles, during the Older Dryas chronozone. At the most northerly site,

Lochan an Smuraich, this zone contains increased percentage values for undoubted *Betula nana* pollen (Pennington 1977).

(vi) *Betula–Juniperus* p.a.z.: 11800–11000 in the Lake District

This represents the upper part of the overall *Betula* zone in the Lake District, is correlated by ^{14}C dating with the Allerød chronozone (figure 2) and appears to be correlative with what the authors regard as Zone II deposits at Glanllynnau (date of 11300 from within the zone,

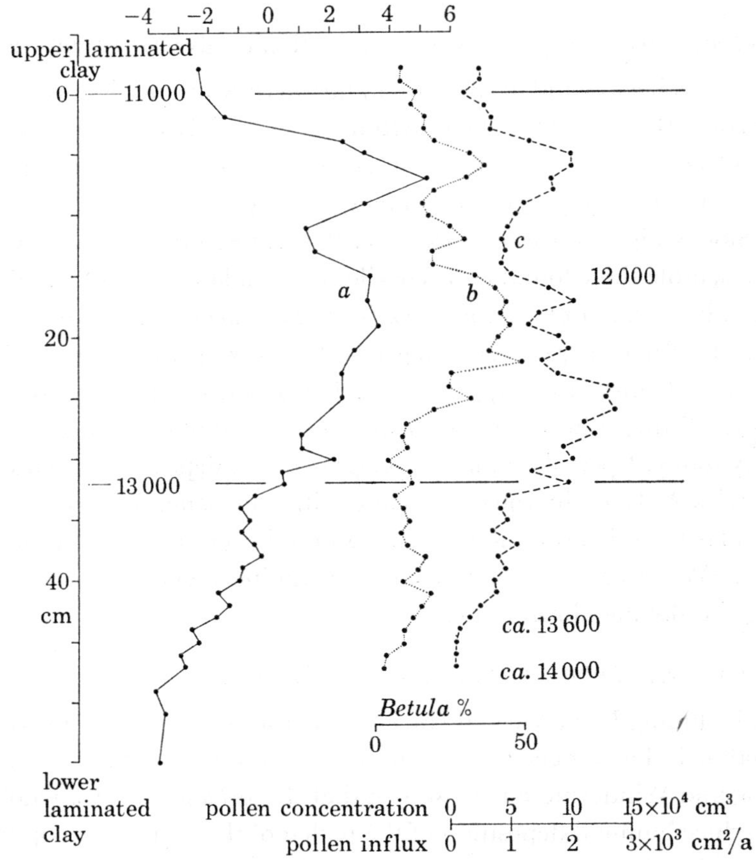

FIGURE 4. Geochemistry compared with pollen stratigraphy in the same profile as in figure 3. (*a*) stratigraphic plot of the sample scores on component 1 of a principal components analysis of the data for distribution within the profile of the elements C, H, N, Fe, Mn, Ca, K and Na; this component includes 68% of total variance in the chemical data; (*b*) *Betula* pollen as a percentage of total pollen; (*c*) pollen concentration, per cm^3. The position of the boundaries of the Windermere Interstadial at 13000 and 11000 is indicated by horizontal lines, which divide the profile into: (1) a pre-interstadial period of progressively ameliorating environment; (2) the interstadial, divided by a regression at *ca.* 12000 and (3) a regression at *ca.* 11000 to an approach to the glacial environment of before 14000.

Simpkins 1974) and Nant Ffrancon (date of 11500 ± 400 from within the zone of *Betula pubescens* macroscopics, Burrows 1975). In both Lake District sites the higher percentages of grass, sedge and *Empetrum* pollen, together with increased values for *Dryopteris* and unidentifiable fern spores, distinguish the spectra of this zone from the earlier *Betula* p.a.z. The fluctuating values for *Betula* and *Juniperus* firstly provide a basis for division of the zone in agreement with Iversen's division of Allerød deposits (Iversen 1954) and secondly suggest an interaction between birch and juniper which represents breakdown of the ecological equilibrium established before

12000. Maximum percentage and absolute values for *Pinus* are found in this zone, and are attributed to an increased representation in the regional pollen rain during Allerød time, when pine was present in the woodlands of southern and eastern England.

A local subzone of Cyperaceae–*Selaginella*, distinguished in the Windermere profile (p. 266), is interpreted as indicative of deterioration in conditions towards the close of this chronozone. At Blea Tarn, higher in the same catchment, intensified erosion of mineral soils at this time led to a complete lithostratigraphic reversion to minerogenic sediment within a correlative *Betula–Empetrum* zone (Pennington 1970, 1973).

(vii) *Empetrum*–Cyperaceae–*Juniperus* p.a.z.: 11800–11000, in northern Scotland

This p.a.z. correlates with zone B of Pennington *et al.* (1972), there called 'interstadial'; ^{14}C dating at Cam Loch identified it with the Allerød chronozone. At no site in this region do *Betula* percentages exceed 10, and at neither Cam Loch nor Lochan an Smuraich do annual deposition rates reach the Greenland values for 'park-tundra' (Pennington 1977), so no local expansion of birch can have taken place. The increased absolute values for *Betula* in this zone are attributed to an increase in the regional pollen rain of NW Europe during Allerød time. Pollen spectra and annual deposition rates resemble in general those of the earlier *Rumex–Empetrum–Juniperus* zone, except for lower values for *Rumex*, and a continuation of *Empetrum*–juniper heath is supposed. Juniper maxima at Loch Tarff (figure 1) and Blackness in the Grampians (table 1*b* (iii)) in the later part of this zone may result from a late-Allerød warmer period such as that interpreted by Iversen (1954) from the *Betula* curve in Denmark.

(*c*) 11000–10000

(i) *Rumex–Artemisia*, *Artemisia–Rumex*, *Artemisia*—Caryophyllaceae p.a.zs: 11000–*ca.* 10500, most parts of the British mainland

Pollen assemblage zones distinguished by maximum values for the profile of *Artemisia* pollen, including *A. norvegica* at Scottish sites but including also much pollen which is not that of *A. norvegica*, have been widely recognized in Britain and correlated with Jessen/Godwin zone III. ^{14}C dating at Blelham Bog, Windermere and Cam Loch has identified the *Artemisia*-dominated pollen zones with the first half of Younger Dryas time, *ca.* 11000–10500, and this must have been the period during which 400–500 paired laminations (varves) were laid down in Windermere from deposition from corrie glaciers in the mountains (figure 5). In almost all profiles the *Artemisia* pollen zones coincide with a distinct lithostratigraphic unit of lower organic content, indicative of soil movements consequent on increased incidence of freeze-thaw cycles, and often identifiable geochemically with the unweathered products of intense periglacial erosion (Pennington 1970, 1973).

It has not so far been possible to find comparable pollen spectra in modern samples. Species of *Artemisia* are associated with dry habitats and do not in general tolerate much snow cover (Iversen 1954); the genus is associated rather with the continuously frost-heaved soils of a continental periglacial environment. The pollen spectra of the Younger Dryas *Artemisia* pollen zones in Britain do not resemble at all closely those of deposits older than 13000, as demonstrated by numerical analysis of data from Blea Tarn and Cam Loch (Pennington & Sackin 1975), but resemble rather the characteristic spectra from unglaciated parts of Europe during the last glacial maximum (Menéndez-Amor & Florschütz 1964; Lang 1970), in the prominence of *Artemisia*, of *Pinus* as a percentage of the total, and in the frequent presence of Chenopodiaceae.

Salix values in the Younger Dryas deposits of Britain are usually much lower than in early Late Devensian deposits.

The absolute increase in pollen deposition of *Artemisia* over much of Britain during Younger Dryas time is surprising, in view of the amount of precipitation as snow which must have been required to re-establish glaciers on the mountains of western Scotland (Sissons 1974), the Lake District (Manley 1959) and North Wales (Seddon 1962), and suggests that strong differentiation with respect to snow cover must have been present in Britain at this time.

In lakes which lie within the limits of Younger Dryas ice (figure 1) most of the Late Devensian deposit consists of coarse minerogenic material, but in some of these lakes a narrow *Artemisia* pollen zone can be found, indicating that some lake sites had become open and were recruiting pollen within the time represented by the *Artemisia* pollen zone.

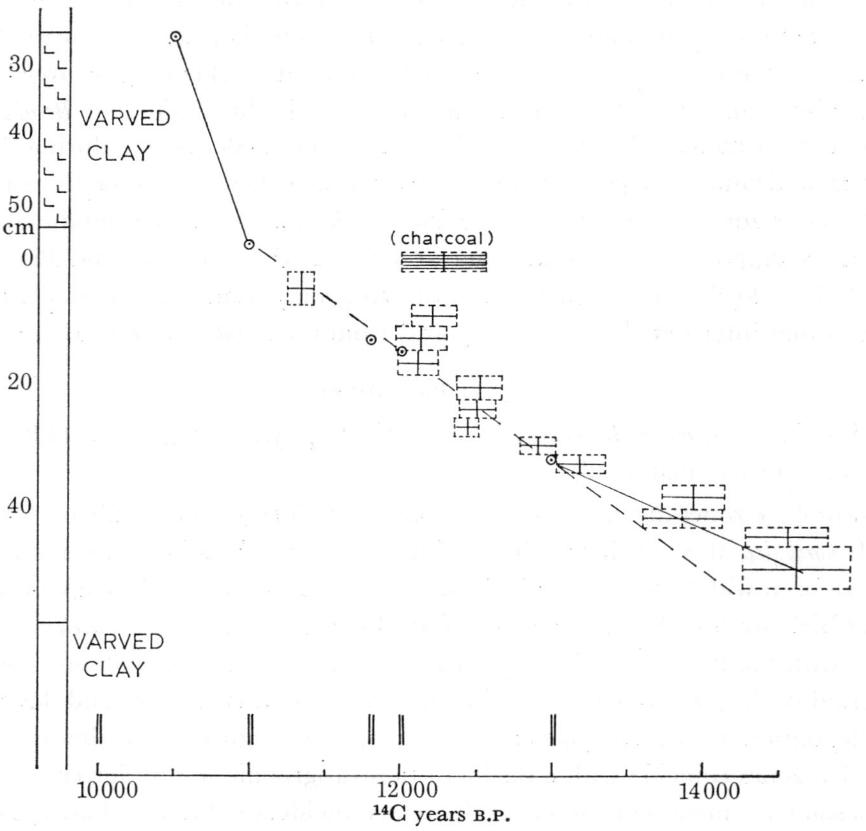

FIGURE 5. ^{14}C dates from the profile in figure 3, plotted to show sample depth and one standard deviation. The dated samples are: SRR-668, −1 to 2 cm, 12272 ± 280, $\delta^{13}C = -26.6‰$; SRR-669, 2–7.5 cm, 11344 ± 90, $\delta^{13}C = -24.1‰$; SRR-670, 7.5–11 cm, 12213 ± 150, $\delta^{13}C = -25.0‰$; SRR-671, 11–15 cm, 12132 ± 175, $\delta^{13}C = -21.1‰$; SRR-672, 15–19 cm, 12112 ± 125, $\delta^{13}C = -26.1‰$; SRR-673, 19–23 cm, 12517 ± 150, $\delta^{13}C = -23.9‰$; SRR-674, 23–26 cm, 12499 ± 120, $\delta^{13}C = -26.6‰$; SRR-675, 26–29 cm, 12441 ± 82, $\delta^{13}C = -26.9‰$; SRR-676, 29–32 cm, 12913 ± 120, $\delta^{13}C = -26.9‰$; SRR-677, 32–35 cm, 13185 ± 170, $\delta^{13}C = -24.7‰$; SRR-678, 35–37 cm, 12567 ± 240, $\delta^{13}C = -24.2‰$; SRR-679, 37–41 cm, 13938 ± 210, $\delta^{13}C = -10.2‰$; SRR-680, 41–44 cm, 13863 ± 270, $\delta^{13}C = -17.0‰$; SRR-681, 44–47 cm, 14557 ± 280, $\delta^{13}C = -25.1‰$; SRR-682, 47–54 cm, 14623 ± 360, $\delta^{13}C = -23.4‰$. The position on the depth–time-scale of chronozone boundaries, determined from the pollen diagram and shown on the time axis by double lines, is indicated by the symbol ⊙. The date of SRR-678 is unacceptable and this sample is not plotted. The sample which contains micro-charcoal, of ^{14}C age greater than the time of deposition in the lake, is hatched. A line fitted to the other samples and the positions of chronozone boundaries, projected backwards, suggests that if deposition was approximately constant there is a hardwater error of up to 500 years in the four lowest dates.

(ii) Gramineae-*Rumex*–*Betula nana* subzone (Birks, 1973), Gramineae–Cyperaceae p.a.z., *Salix*–Cyperaceae p.a.z. (Walker 1975), Gramineae–Cyperaceae p.a.z. (Brown 1977): undated, but correlated Zone III

These as yet undated pollen zones, which in each instance appear on stratigraphy to represent the Younger Dryas chronozone, lie between deposits containing raised percentages of *Betula* which are correlated with the Late Devensian interstadial, and the base of the Flandrian sequence. These pollen zones are of great interest in that they demonstrate how the environment associated with the *Artemisia* pollen zones was not continuous over Britain in Younger Dryas time. Their existence suggests that a different vegetation (associated by Walker with much snow cover) was present in parts of the Highland zone of Britain – Skye, the Grampian Highlands, Bodmin Moor. Birks (1973) discusses the difficulty of precise interpretation of the vegetation represented by this zone on Skye and of the associated climate. The site at Lochan an Smuraich (Pennington, 1977) in the north of Sutherland, represents a link between the vegetation of mainland Scotland and of Skye during this period, in that the percentage rise in *Artemisia* is not accompanied by any absolute rise in Lochan an Smuraich (cf. table 1*b* (ii)).

(iii) Herb-dominated p.a.zs containing much grass, sedge and *Rumex* pollen, *ca.* 10500–10000

These p.a.zs (table 1) are found, throughout western Britain, in sediment of increasing organic content, and are interpreted as representative of the pioneer vegetation which colonized and stabilized soils, as climatic amelioration brought to an end, after 10500, the processes of glacial and periglacial erosion. The ^{14}C dates from Blelham Bog and Cam Loch correlate these p.a.zs with the second half of the Younger Dryas chronozone of Mangerud *et al.* (1974). To some extent the sequence of pollen zones resembles that found at the base of Late Devensian deposits (e.g. in northern Scotland (table 1*a* (ii)) but the succession must have occupied a much shorter period of time. In most regions the period of primarily herbaceous vegetation was followed by successive maxima of *Empetrum* and of *Juniperus*, the latter marking the transition to the Flandrian. In Skye, though juniper subzones were distinguished at some sites, there was a rather rapid transition to the basal *Betula*–*Corylus* zone of the Flandrian.

(*d*) *Synthesis of Late Devensian environmental history, ca.* 14500–10000, *at Windermere; the Windermere Interstadial, ca.* 13000–11000

The site at Low Wray Bay, Windermere, to be more fully described elsewhere, is chosen for a final synthesis for three reasons – (1) it embraces maximum contrast in Late Devensian environments, including both laminated (varved) clays and a woodland biozone, (2) it has been investigated by multidisciplinary Quaternary research over the last 30 years and (3) it has now been more fully dated by ^{14}C than any other Late Devensian site.

Before 14500, coarsely laminated clay was deposited from seasonal melt from valley glaciers fed by a remnant ice cap on the Lake District mountains, and vegetation was so sparse as to have left no records.

ca. 14500–13000; ^{14}C results (figure 5, legend) show that before 13000 there is evidence in δ ^{13}C values for hardwater error due to aquatic photosynthesis, and the true dates probably lie nearer to the backward projection of the linear depth–time-scale fitted to dates (figure 5) between 13000 and 11000. On this, the *Salix herbacea* p.a.z. falls at *ca.* 14000–13600, changes in sediment composition indicative of soil maturation (figure 4, curve *a*) begin at 14000 and continue without regression until *ca.* 12000, and the *Rumex*–Gramineae p.a.z., within which is found evidence for the arrival and establishment of *Betula pubescens*, is dated to *ca.* 13600–13000. Within this period, therefore, the evidence points to continuous climatic amelioration, unaccompanied however by any significant expansion of woody plants. It can be postulated that while summer temperatures were high enough for the moderately thermophilous insect fauna found by Dr Coope (1977), some other factor remained inimical to trees except in particularly favoured (sheltered?) habitats, until *ca.* 13000.

13000–11000; the Windermere Interstadial. This section of the Windermere profile is clearly distinguished on both sediment composition and pollen concentration (figure 4); on percentage composition of pollen spectra it constitutes a woodland biozone or palaeobotanical interstadial, correlated chronostratigraphically with the Bølling–Allerød Late Weichselian Interstadial, and similarly divided by a minor regression between *ca.* 12000 and 11800. The evidence from Blelham Bog (Pennington 1975) confirms that from the Windermere site (1.2 km distant) in showing that maximum rates of annual pollen deposition, interpreted as indicative of maximum pollen productivity of the vegetation, distinguish this interstadial.

Five p.a.zs are distinguished within the Windermere Interstadial in the ^{14}C dated pollen diagram – figure 3. They are:

(1) The *Juniperus* zone, 13000–12500, in which maximum Late Devensian pollen deposition rates are recorded; cf. Watts (1977) for Irish sites. Percentages of juniper indicate tall thickets of the shrub.

(2) The *Betula* zone, 12500–12000, in which pollen deposition rates are almost as high and pollen data suggest a 50 % tree cover of *Betula*; the absolute fall in juniper in this zone may represent the results of unsuccessful competition with birch for restricted habitats favourable for shrubs and trees, or of some other factor as yet unidentified.

(3) The *Betula–Rumex* p.a.z., *ca.* 12000–11800, within which annual pollen deposition by *Betula* is significantly reduced and there is a small percentage and absolute rise in *Rumex*. This zone appears to represent the results of an environmental change (regression) of small amplitude, which however destroyed permanently the ecological equilibrium attained between 12500 and 12000. Within and above this zone, percentages of grass, sedge and herbaceous pollens are higher than in the earlier part of the interstadial, but this is to some extent the result of an absolute decline in deposition rates of *Betula*.

(4) The *Betula–Juniperus* p.a.z., *ca.* 11800– just before 11000, in which values for both *Betula* and *Juniperus* rise again to percentage and absolute maxima, and in which a great variety of herbaceous taxa is recorded. High values for *Dryopteris* and unidentified fern spores in this *Betula–Juniperus* zone suggest that the later interstadial (Allerød) birch woodland was more fern-rich than the earlier, and higher values for grass pollen suggest a more open woodland. Maximum percentages of *Pinus* are found in this p.a.z., though the percentages, which do not exceed 3 (total pollen), do not suggest local presence of pine.

(5) The Cyperaceae–*Selaginella* p.a.z., distinguished by low values for *Betula* and *Juniperus* and by the presence of spores of *Selaginella*, *Lycopodium* spp. and *Cryptogramma*. High percentages of sedge pollen result from constant deposition rates during a period of rapid fall in deposition rates of other taxa (figure 3). It is possible to interpret this pollen zone as indicative of increased snowfall during the period of break-up of Allerød vegetation (cf. Birks 1973, on chionophilous vegetation). The anomalous ^{14}C date, SRR-668, is attributed to the effect of the micro-charcoal common in the sediments of this zone and presumed to have originated with eroding soils; Burrows (1974) noted similar charcoal in correlative sediments at Nant Ffrancon.

11000–10500. The first half of the Younger Dryas chronozone is represented in Windermere (and in the other large lakes of the Lake District) by 400–500 paired microvarves which show strongly graded bedding (Pennington 1947) – indicating that between *ca.* 11000 and 10500 the lake must have been turbid with clay and silt produced by active ice in the high corries of the mountain catchment (Manley 1959). Pollen is extremely sparse in the laminated sediment, which I have explained by demonstrating low rates of pollen deposition at the present time in a Norwegian lake sediment, similarly laminated, where there is a rapid throughput of water from seasonal meltwater from mountain glaciers (Pennington 1973). In Windermere the proportional composition of the sparse pollen is comparable with that found in the correlative solifluxion clay at Blelham Bog (Pennington & Bonny 1970; Pennington 1973, Table 4), but the necessarily laborious analysis was not pursued above the base of this zone.

Conclusions

(*a*) *Late Devensian flora*

The Late Devensian flora was rich and varied and is likely to have included *ca.* 60 % of the species of the present flora. Species assemblages show that (*a*) for all regions of Britain the phytogeographical affinities of the Late Devensian flora were as varied as in the present flora, and (*b*) plants of contrasted modern ranges (e.g. northern, southern, oceanic and steppe) appear to have coexisted in a vegetation different from that of the preceding and following periods. Problems arise in relation to attempted climatic reconstructions based on present geographical ranges of plant species, and raise the question of to what extent the modern distribution of a species is governed by its physiological tolerances, and how far it depends on interaction with other species.

(*b*) *Late Devensian vegetation*

The great diversity in Late Devensian pollen spectra shown by this analysis of results from profiles on a broad north–south transect of western Britain can only be explained in terms of strong regional differentiation of Late Devensian vegetation. Climatic differentiation may be invoked to account for some of these differences in vegetation, but it seems unlikely that any simple explanation in climatic terms can be devised. It must be accepted that the facts of vegetation history in western Britain do not fit the simple model of uniform response by trees to changing temperatures which was developed from work on neighbouring parts of the continental mainland. However, at sites which have been adequately dated by radiocarbon, the profiles can be divided into a series of chronozones whose boundaries coincide with those of recently defined Late Weichselian chronozones, and, like these, appear from interpretation of the pollen data to have been climatically conditioned.

Pollen evidence from western Britain, supported by lithological and geochemical evidence, indicates a continuous and uninterrupted development of plant communities and soils, interpreted as a response to climatic amelioration, from before 14000 B.P. until *ca.* 12000 B.P. Snow-bed communities (postulated by Faegri to be the characteristic periglacial vegetation of the oceanic fringe of Europe), and the tolerant plants of exposed summit-heath today, were followed by pioneer species of immature soils, and by species-rich communities including plants which now demand high base-status. Into these communities dispersed the shrubs and tree birches whose subsequent development varied so greatly with aspect, altitude and soil that it cannot have been controlled primarily by temperature. A widespread and synchronous increase in pollen production by juniper at 13000 is interpreted as the consequence of the general climatic amelioration which has been used to define the lower boundary of the Late Weichselian substage (Mangerud *et al.* 1974). This rapid expansion of juniper defines the base of the Windermere Interstadial. Throughout Bølling time (13000–12000) tree birches must have been present in most of western Britain, dispersed in edaphically favoured and locally sheltered conditions as postulated by Godwin (1975, pp. 488–9), and forming up to 50 % cover in the most favoured areas by the end of the period. In Scotland these favourable areas appear to have been few, and in northern Scotland the *Empetrum–Juniperus* heath, which had succeeded the pioneer vegetation at 13000, persisted. If this was primarily a climatic effect, reflecting cool and cloudy conditions in northern Scotland (cf. Brown 1971) it is difficult to explain the scarcity of *Empetrum* in Skye, and some edaphic factor is almost certainly involved. On the western hills of Skye and the Lake District a species-rich grassland persisted, producing pollen spectra which show no significant expansion of *Betula*, and a comparable vegetation with some juniper is recorded from *ca.* 13000 from Bodmin Moor (Brown 1977).

This highly differentiated interstadial vegetation with a spatially uneven distribution of woody plants appears to have suffered only minor regression during the climatic deterioration of Older Dryas time (*ca.* 12000–11800) – because changes in the pollen spectra are small, e.g. the *Rumex–Artemisia* zone in northern Scotland, and most easily seen in the long and closely sampled profiles from the southern Lake District. In the second part of the interstadial, the Allerød chronozone, 11800–11000, fluctuating curves for *Betula* and *Juniperus* could be explained by the irregularly declining temperatures postulated by Coope & Brophy (1972) on faunal evidence. In parts of Wales and on Bodmin Moor, maximum percentages of birch were reached during Allerød time, as in southern and eastern England. In general, the evidence from western sites suggests an interstadial vegetation type containing varying proportions of *Betula*, *Juniperus* and *Empetrum*, which both developed earlier, and represented a different vegetation type, when compared with the birch-pine p.a.z. of eastern England, which reached its maximum development in Allerød time and correlates on both chronostratigraphy and pollen spectra with the Late Weichselian Zone II. Since this western interstadial has been investigated fully in the sediments of Windermere, it may be termed provisionally the Windermere Interstadial.

The major regression in Late Devensian vegetation in Younger Dryas time, which is characteristic of northern Europe and now explicable in terms of sea temperatures in the North Atlantic (Ruddiman *et al.* 1977), correlates with lithostratigraphic and geomorphological evidence for recrudescence of mountain glaciation in western Highland Britain. Periglacial soil disturbance must have been general outside the limits of glaciation (Watson 1977; Mitchell 1977) and this agrees with vegetational evidence – the widespread occurrence of pollen spectra characterized by *Artemisia* – in suggesting that snow cover was localized to those upland areas of the north and west where the pollen spectra indicate a more snow-tolerant vegetation.

I am greatly indebted to Mr H. C. Gilson, Mr W. Findley and Mr I. Haig for the design, construction and operation of a 12.5 cm corer, which made possible the detailed ^{14}C dating of the Windermere profile, and to Dr D. D. Harkness of S.U.R.R.C., East Kilbride, for his patience in dating these samples, many of which contained only small amounts of organic matter. Mrs J. P. Lishman was responsible for geochemical analyses, and Mr M. J. Sackin, of Leicester University, carried out the computer analysis of the geochemical data. I thank Dr H. J. B. Birks for many discussions of late-glacial problems, and Dr A. P. Brown for access to his as yet unpublished data.

References (Pennington)

Bell, F. G. 1969 *New Phytol.* **68**, 913–922.
Birks, H. J. B. 1973 *Past and present vegetation of the Isle of Skye. A palaeoecological study.* Cambridge University Press.
Birks, H. J. B. & Deacon, J. 1973 *New Phytol.* **72**, 877–902.
Brown, A. P. 1971 *New Phytol.* **70**, 841–849.
Brown, A. P. 1977 *Phil. Trans. R. Soc. Lond.* B **276**, 251–320.
Burrows, C. J. 1974 *New Phytol.* **73**, 1003–1033.
Burrows, C. J. 1975 *New Phytol.* **75**, 167–171.
Conolly, A. P. & Dahl, E. 1970 In *Studies in the vegetational history of the British Isles* (eds D. Walker & R. G. West). London: Cambridge University Press.
Conolly, A. P., Godwin, H. & Megaw, E. M. 1950 *Phil. Trans. R. Soc. Lond.* B **234**, 397–469.
Coope, G. R. 1970 *Rev. Geogr. Phys. Geol. Dynam.* **12** (2), 149–155.
Coope, G. R. 1977 *Phil. Trans. R. Soc. Lond.* B **280**, 313–340 (this volume).
Coope, G. R. & Brophy, J. A. 1972 *Boreas* **1**, 97–142.
Crabtree, K. 1972 *New Phytol.* **71**, 1233–1243.
Davis, M. B., Brubaker, L. B. & Webb, T. III. 1973 In *Quaternary plant ecology* (eds H. J. B. Birks & R. G. West). Oxford: Blackwells.
Degerbøl, M. & Iversen, J. 1945 *Danm. geol. Unders.* II, **73**, 1–62.
Evans, G. H. 1970 *New Phytol.* **69**, 821–874.
Faegri, K. 1933 *Årbok. Mus. Bergen.* no. 7, 255 pp.
Faegri, K. 1953 *Norsk Geogr. Tidsskrift.* **14** (1–4), 61–76.
Fredskild, B. 1973 *Meddr. Grønland* **198** (4), 245 pp.
Godwin, H. 1956 *A history of the British flora.* London: Cambridge University Press.
Godwin, H. 1960 *Proc. R. Soc. Lond.* B **153**, 287–320.
Godwin, H. 1964 *Proc. R. Soc. Lond.* B **160**, 258–275.
Godwin, H. 1975 *A history of the British flora*, 2nd edn. London: Cambridge University Press.
Godwin, H. & Willis, E. H. 1964 *Radiocarbon* **6**, 116–137.
Hammen, T. van der, Maarleveld, G. C., Vogel, J. C. & Zagwijn, W. H. 1967 *Geol. Mijnb.* **46** (3), 79–95.
Harmsworth, R. V. 1968 *Ecol. Monogr.* **38**, 232–241.
Haworth, E. Y. 1976 *New Phytol.* **77**, 225–254.
Iversen, J. 1954 *Danm. geol. Unders.* II, **80**, 87–119.
Iversen, J. 1960 *Danm. geol. Unders.* IV, **4**, 1–32.
Iversen, J. 1973 *Danm. geol. Unders.* II, 7c, 1–125.
Johnson, R. H., Tallis, J. H. & Pearson, M. 1972 *New Phytol.* **71**, 533–544.
Kirk, W. & Godwin, H. 1962–63 *Trans. R. Soc. Edinb.* **65** (11), 225–249.
Lang, G. 1970 *Feddes Repert.* **81**, 315–335.
Livingstone, D. A. 1968 *Ecol. Monogr.* **38**, 87–125.
Mangerud, J., Andersen, S. T., Berglund, B. E. & Donner, J. J. 1974 *Boreas* **3**, 109–126.
Manley, G. 1959 *Liv. and Man. Geol. J.* **2** (2), 188–215.
Menéndez Amor, J. & Florschütz, F. 1964 *Bol. R. Soc. Espanola Hist. Nat.* **62**, 251–255.
Mitchell, G. F. 1977 *Phil. Trans. R. Soc. Lond.* B **280**, 199–209 (this volume).
Mitchell, G. F., Penny, L. F., Shotton, F. W. & West, R. G. 1973 *Geol. Soc. Lond. Spec. Rep.* 4, 99 pp.
Moar, N. T. 1969 *New Phytol.* **68**, 433–467.
Moore, P. D. 1970 *New Phytol.* **69**, 363–375.
Newey, W. W. 1970 *New Phytol.* **69**, 1167–1177.
Pennington, W. 1943 *New Phytol.* **42** (1), 1–27.
Pennington, W. 1947 *Phil. Trans. R. Soc. Lond.* B **233** (596), 137–175.
Pennington, W. 1970 In *Studies in the vegetational history of the British Isles* (eds D. Walker & R. G. West). London: Cambridge University Press.
Pennington, W. 1973 In *Quaternary plant ecology* (eds H. J. B. Birks & R. G. West). Oxford: Blackwells.
Pennington, W. 1975 *Boreas* **4**, 157–171.
Pennington, W. 1977 In *Studies in the Scottish late-glacial environment* (eds J. M. Gray & J. J. Lowe). Pergamon Press. (In the press.)
Pennington, W. & Bonny, A. P. 1970 *Nature, Lond.* **226**, 871–873.
Pennington, W., Haworth, E. Y., Bonny, A. P. & Lishman, J. P. 1972 *Phil. Trans. R. Soc. Lond.* B **264**, 191–294.
Pennington, W. & Sackin, M. J. 1975 *New Phytol.* **75**, 419–453.
Perring, F. H. & Walters, S. M. 1962 *Atlas of the British flora.* London: Thomas Nelson.
Ruddiman, W. F., Sancetta, D. C. & McIntyre, A. 1977 *Phil. Trans. R. Soc. Lond.* B **280**, 119–142 (this volume).
Seddon, B. 1962 *Phil. Trans. R. Soc. Lond.* B **244** (716), 459–481.
Simpkins, K. 1974 *New Phytol.* **73**, 605–618.

Sissons, J. B. 1974 *Scott. J. Geol.* **10**, 311–337.
Walker, M. J. C. 1975 *Pollen Spores* **17** (1), 67–92.
Watson, E. 1977 *Phil. Trans. R. Soc.* B **280**, 183–198 (this volume).
Watts, W. A. 1963 *Irish Geogr.* **IV** (5), 367–376.
Watts, W. A. 1977 *Phil. Trans. R. Soc.* B **280**, 273–293 (this volume).
West, R. G. 1970 *New Phytol.* **69**, 1179–1183.

Discussion

D. M. BERTIE (*Department of Geography, University of Reading, 2 Earley Gate, Whiteknights Road, Reading RG6 2AU*). It is interesting to note that the latest work on the Late Devensian flora of Britain and Ireland appears to be lending support to the results obtained by G. R. Coope with beetles regarding palaeotemperature curves for the British Late-glacial. It is disappointing, therefore, to see that Dr Pennington continues to fix the new British pollen zones to the Scandinavian zones (the dates for the boundaries of the zones are obviously derived from Mangerud *et al.* 1974, p. 119). Dr Pennington appears to have arbitrarily used the dates for the Scandinavian zones for obtaining the boundaries of the new British pollen zones, with rather dubious results. For example, the Older Dryas *Betula–Rumex* zone seems to have no justifiable existence, the zone's boundaries not being marked by emergence or decline of a plant species, other than that it lies between ^{14}C dates 12000–11800 a B.P. It would have been far better to have used Dr Coope's Late-glacial Interstadial ('Windermere' Interstadial), 13000–11000 a B.P., as a framework to work within, thereby bringing the new pollen work into line with current thought on the British Late-glacial. An excellent opportunity for comparative work between plants and beetles in the British Late-glacial has been lost here.

Reference

Mangerud, J., Andersen, B. E., Berglund, B. E. & Donner, J. J. 1974 Quaternary stratigraphy of Norden, a proposal for terminology and classification. *Boreas* **3**, 109–127.

W. PENNINGTON. The zonation of the Windermere Interstadial, based on synthesis of the evidence from pollen and beetles, is discussed by Dr Coope and myself later in this volume (pp. 337–339 and table 1). All forms of evidence agree to indicate a temporary deterioration at *ca.* 12000 B.P. on the Windermere ^{14}C dates.

(*b*) (iv)				(b) (v)			
LAKE DISTRICT		CORNWALL (Brown 1977)		WALES			
Blelham Bog & Windermere		Hawks Tor 3	Bodmin Moor regional p.a.z.	Central Elan valley (Moore 1970)		North Cors Gevallt (Crabtree 1973)	
	10000						
Juniperus Gramineae – herbs *Artemisia–Rumex*			sedge – grass	*Juniperus* *Rumex*	III–IV	*Juniper* *Empetrum*	III–IV
				Artemisia–		*Rumex* *Artemisia*	III
					III		
Rumex–Artemisia				Gramineae			
	11000		11069				
Betula– *Juniperus*			*Betula–* *Empetrum*	*Juniperus–* Cyperaceae	II–III	*Betula–Juniperus*	
		*Betula–*Umbelliferae– *Juniperus*					
Betula–Rumex– Gramineae	11800		11553	*Betula*	II	*Betula–*Gram– *Rumex*	II
	12000	Umbelliferae– *Juniperus*	Grass– sedge – herb			*Betula*	
		12635 ± 300					
Betula *Juniperus*		Cyperaceae– *Ranunculus acris*		*Juniperus–* *Betula nana*	I–II	*Juniperus*	
	13000	13088 ± 300					I
*Rumex–*Gramineae– Rubiaceae–*Artemisia*				*Rumex–* Gramineae	I	*Rumex–*Gramineae– Rubiaceae– *Artemisia* + secondary pollen and *Pinus*	
Salix herbacea– Cyperaceae	13900						
lower laminated clay	14500						

Phil. Trans. R. Soc. Lond. B. **280**, 273–293 (1977) [273]
Printed in Great Britain

The Late Devensian vegetation of Ireland

BY W. A. WATTS, M.R.I.A.
Botany Department, Trinity College, Dublin

[Pullouts 1 and 2]

The vegetation of the Late Devensian period in Ireland is reviewed. New investigations at Ballybetagh, Co. Dublin, Dunshaughlin, Co. Meath, Glenveagh, Co. Donegal and Poulroe, Co. Clare, are reported. A sequence of phases of vegetation development for the Dublin region is described and regional variation elsewhere in Ireland discussed. Pollen influx values for two Late Devensian sites in southeastern Ireland are reported. A case is made that the *Juniperus–Empetrum* phase between 12400 and 12000 B.P. was the warmest phase of the Late Devensian. The reason for absence of birch woodland in late-glacial Ireland is discussed. Evidence for widespread soil erosion at the end of the *Juniperus–Empetrum* phase is presented. The occurrence of a corrie glaciation at Lough Nahanagan in the Wicklow Mountains in the *Artemisia* phase is documented. A radiocarbon chronology for events in the Late Devensian of Ireland is proposed.

1. INTRODUCTION

(*a*) *Review of the literature*

The formal system of stratigraphic nomenclature now in use in Britain and Ireland names the Last Glaciation the Devensian and defines the Late Devensian as the time from 25000 to 10000 years ago (Mitchell, Penny, Shotton & West 1973). The Late Devensian thus includes the time when the ice sheets of the Last Glaciation first accumulated and spread to their greatest extent, their subsequent melting and disappearance and a final phase with fluctuating climate which lasted from about 13000 to 10000 years ago. As nothing is known about the Late Devensian flora and vegetation of Ireland before about 13000 years ago it is convenient to recognize this last phase as the 'Late-glacial period', while recognizing that this is an informal name which lacks exact definition. At 10000 B.P. the Flandrian (Postglacial) period begins.

The first clear identification of a Late-glacial deposit in either Ireland or Britain was the investigation of the bogs at Ballybetagh near Dublin by Jessen & Farrington (1938). At the northwestern bog at Ballybetagh Jessen's excavations showed that a lake deposit which contained bones of the extinct Irish giant deer, *Megaloceros*, lay between two stony layers which he identified as solifluxion earths. Jessen drew attention to the similarity between Ballybetagh and the well-known site at Allerød in Denmark where there was clear evidence of a Late-glacial climatic oscillation. His subsequent investigations showed that Late-glacial deposits are very common in Ireland. In 1949 Jessen's comprehensive account of his work in Ireland was published by the Royal Irish Academy at whose invitation his Irish researches had begun. His splendidly argued and presented paper (Jessen 1949) has been regarded as authoritative, and later studies in Ireland have used his work as a standard against which to measure their conclusions. Jessen concluded that three zones could be identified in the Late-glacial period. It is useful to quote his summary of his tripartite scheme:

Zone I. The older *Salix herbacea* period. The information that is yielded by examination of the plant remains and of the sediments in the lakes suggests that Ireland was covered by an open tundra-like vegetation of a limited number of species and was subject to solifluxion.

Zone II. The Late-glacial *Betula* period or Allerød period. Ireland was an oceanic sector of the sub-arctic Birch region of north-west Europe. There were copses of birch (*Betula pubescens*) and stretches of open country, which in the west were covered by heaths rich in *Empetrum* and elsewhere by a vegetation of grasses and herbs whose character cannot be closely defined.

Zone III. The younger *Salix herbacea* period. The vegetation was clearly more northern in character than in the preceding period, though oceanic influence was maintained. The tree growth was even more restricted but birch copses may have survived in sheltered localities. Solifluxion was a common phenomenon and the country for the most part carried an open tundra-like vegetation with patches of sub-arctic heaths containing, among other plants, *Salix herbacea*, *Dryas octopetala*, with in the northwest much *Empetrum nigrum*.

Parallel with Jessen's studies, his pupil G. F. Mitchell showed that fossils of both *Megaloceros* and the reindeer occur predominantly in the upper part of Zone II and that it could be argued that they had become extinct in Zone III (Mitchell 1941 *a*; Mitchell & Parkes 1949). Both Jessen and Mitchell determined large numbers of Late-glacial plant macrofossils. This work is summarized, and the Irish Late-glacial flora compared with the modern vegetation of south-facing morainic slopes in Swedish Lapland by Mitchell (1954).

In the early work of Jessen and Mitchell only a limited range of pollen types was identified. In particular, *Juniperus*, which proved to be a very important member of the flora, and many herbaceous species were not recognized. As the range of pollen determinations was extended, a simple tripartite scheme for the Late-glacial was found to be inadequate. Zone II proved to be more complex than had been believed, and the pollen zones could not be equated exactly with the sedimentary units. Authors attempted to redefine the pollen zones (Watts 1963) or to interpolate new transitional zones numbered I–II and III–IV (Singh 1970). Craig (1973) abandoned Jessen's zones completely and used a local zonation system only.

A number of pollen diagrams exist in which *Juniperus* and herbs were identified to a modern standard. Watts (1963) published three diagrams of major taxa from sites in western Ireland, while A. G. Smith and his colleagues at Queen's University, Belfast, published several detailed studies (Smith 1961; Morrison & Stephens 1965; Singh 1970). Smith (1970) reviewed Ireland's Late-glacial vegetation and climate. In southwestern Ireland a detailed pollen diagram is available from Muckross, Killarney, Co. Kerry (Vokes 1966). Craig (1973) has investigated two sites in the southeast, Coolteen, Co. Wexford, and Belle Lake, Co. Waterford, making use of absolute pollen counting procedures. These sites and others referred to in the text are located in figure 1.

(*b*) *The present review – some general considerations*

In an effort to make this review as objective as possible, information from Ireland has been considered in isolation, so that the account is uninfluenced by stratigraphic or climatic conclusions reached in other countries. The reason for this approach is the belief that premature emphasis on correlation may obscure real differences between regions. In Ireland the numerous Late-glacial deposits make it possible to choose sites for study where sedimentation has been rapid and a high resolution of stratigraphic detail can be obtained. This is held to be particularly important in the establishment of a satisfactorily detailed account of the regional development of vegetation.

The review presents new site investigations but its objective is not to consider these in great detail, but to use them in conjunction with the published literature to illustrate a number of interesting aspects and problems of Ireland's Late-glacial vegetation (see § 5). The system of pollen zones introduced by Jessen (1949) for the Late-glacial and early Flandrian is abandoned here. As a framework it does not fit, and cannot be adapted to fit, present-day knowledge of the pollen stratigraphy. Unfortunately, the very numerous plant macrofossil records from the period have been referred to one or other of Jessen's zones. This does not mean that the records are unusable, but they cannot be related to modern pollen diagrams. It will be necessary in the future to prepare seed diagrams (Watts & Winter 1966) from the Irish Late-glacial to show the exact relationship of macrofossil records to pollen profiles. In abandoning Jessen's zones, difficult problems of procedure and principle arise in deciding whether, or how, to zone pollen diagrams. The procedures adopted are discussed in the succeeding paragraphs.

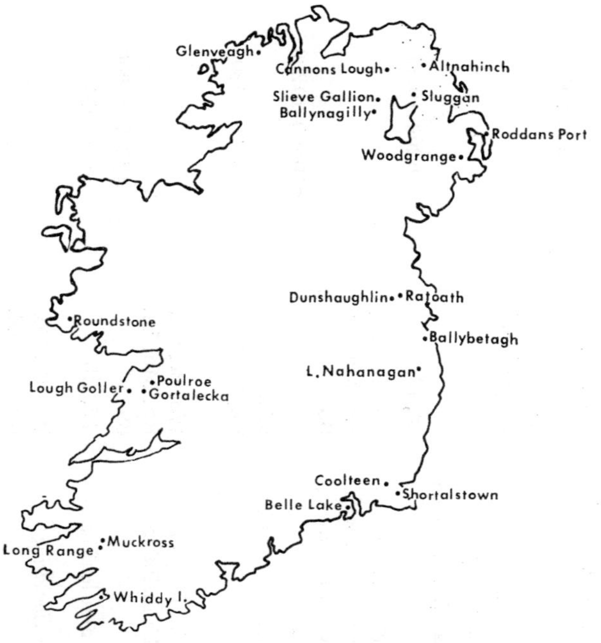

FIGURE 1. Sites referred to in the text.

The diagrams were first divided into local pollen zones. Local zones refer to one site only. Each zone is given a number and a combination of letters which identifies its site. The zones are defined by their pollen composition only. The importance of giving primacy to local pollen zones is that they are defined without giving any consideration to the pollen record at other sites so that they preserve information about local variation. This reminds us that vegetation was not homogeneous in the past and that the investigation of differences between sites may reveal the variability of vegetation cover in the past and its relation, then as now, to factors of history, soil and local climate. Pollen diagrams from one region tend to be very alike and common features are usually more striking than differences. As a second stage local pollen zones which were very homogeneous from site to site were considered to belong to an 'assemblage zone'. Assemblage zones are defined by a characteristic composition of fossil pollen. They can be mapped in time and space (American Commission on Stratigraphic Nomenclature 1961; Cushing 1967). Assemblage zones may begin at different times locally and last for varying periods. They do not provide a basis for time correlation which must depend on radiocarbon-dated profiles. The assemblage zones were named and defined (table 1). For convenience in

TABLE 1. POLLEN ASSEMBLAGE ZONES AND CHRONOLOGY OF THE LATE-GLACIAL PERIOD IN IRELAND

informal names used in text for pollen assemblages	most important pollen types	sites at which identified	approximate duration in radiocarbon years B.P.	chronological subdivision of Late-glacial (Late Weichselian) period in northern Europe (in Pennington 1975)	
the *Corylus* assemblage	*Corylus* (hazel)	Wo, Be, Co, Ro	begins 9000		
the *Betula* assemblage (*Betula* or birch phase)	*Betula* (tree-birch), often with *Salix* (tree-willow) and *Dryopteris* (fern)	Wo, Be, Co, Ro, Mu, Dl, Bb	9500–9000		
the *Juniperus–Filipendula* assemblage	*Juniperus* (juniper), often with *Empetrum* (crowberry) and *Filipendula* (meadow-sweet)	Wo, Be, Co, Mu, Dl, Bb	10000–9500	Flandrian	
the Gramineae–*Rumex* peak	A peak of Gramineae (grass), often with *Rumex* (dock)	Dl, Bb	10200–10000		10000
the *Artemisia* assemblage (*Artemisia* phase)	Variable amounts of *Artemisia* with Caryophyllaceae, often with *Salix* (dwarf willow), *Armeria*, *Koenigia*, *Sedum rosea*	Wo, Be, Co, Ro, Mu, Dl, Bb	10800–10200	Younger *Dryas*	
the Cruciferae peak (crucifer peak)	More than one genus of Cruciferae	Wo, Be, Co, Ro, Dl, Bb	10900–10800		11000
the Gramineae assemblage (Gramineae or grass phase)	Gramineae, often with frequent *Helianthemum* and Umbelliferae	Wo, Be, Co, Ro, Mu, Dl, Bb	12000–10900	Allerød Older *Dryas*	11800 12000
the Hippophaë peak	Hippophaë, pollen and leaf hairs	Dl, Bb	*ca.* 12000		
the *Juniperus–Empetrum* assemblage (*Juniperus–Empetrum* phase, Juniper phase)	*Juniperus*, often with *Empetrum*	Wo, Be, Co, Ro, Mu, Dl, Bb	12400–12000		
the *Rumex–Salix* assemblage	*Rumex*, usually with *Salix* (dwarf willow)	Wo, Be, Co, Ro, Dl, Mu, Bb	13000–12400	Bølling	13000
the pre-*Rumex*-phase flora	Gramineae, much secondary pollen	Wo, Ro	13000	Middle Weichselian	

Wo, Woodgrange; Be, Belle Lake; Co, Coolteen; Ro, Roddans Port; Mu, Muckross; Dl, Dunshaughlin; Bb, Ballybetagh.

discussion the names are as short as possible. This procedure is 'informal' in the sense of the Code of Stratigraphic Nomenclature but the formal naming of a type site and more detailed diagnosis and nomenclature are premature in the present rapidly changing state of knowledge. Formally named assemblage zones may well, because of their rigidity, be less useful than a loosely defined informal scheme. Table 1 contains a tentative chronology for the assemblage zones. These dates should be understood to be no more than very crude estimates of means, the true values of which will vary from site to site by amounts varying from decades to centuries.

In principle it may be doubted whether pollen zone boundaries should be drawn at all. Pollen diagrams are complex records of alternating periods of stability and of change in vegetation. Zone boundaries imply that discrete homogeneous units can be identified within pollen diagrams, but this is not the case, because it ignores transitions which may occupy significant periods of time (Watts 1973).

There is an obvious analogy to a basic problem of vegetation science. Is the nature of vegetation such that its stands can be regarded as discrete units and be classified on the basis of degrees of similarity of their floras, the view of classical plant sociology (Westhoff & van der Maarel 1973) or is each stand continuously variable in space and time, as Gleason (1926) maintained, and therefore unclassifiable? In the author's view (Watts 1973), Gleason's individualistic concept of the plant community is supported by the evidence of pollen analysis. In this strict sense the drawing of zone boundaries whether subjectively or by computer (Gordon & Birks 1974) in vegetation which is 'continuously variable in space and time' is a mistake because it is inconsistent with the character of the scientific evidence.

If one accepts that it may be convenient and practical, even with the reservation that it may be inexact scientifically, to delimit and classify stands (define and correlate pollen zones), then the process of zoning pollen diagrams becomes like the practice of plant sociology. Local pollen zones, or the spectra of which they are composed, become analogous to the relevés or real local records (quadrats) of plant sociology, and the delimitation of assemblage zones analogous to the definition of associations (a synthetic or artificial concept) from the data of relevés assembled in association tables. The necessity to present large quantities of data in a manageable form suggests that local pollen zones, based on actual data and subsequently united into synthetic assemblage zones, may be pragmatically the best way to make the information contained in pollen diagrams accessible in summary form.

2. Pollen diagrams from Ballybetagh and Dunshaughlin (Figures 2, 3)

(a) *The sites*

A re-investigation of Ballybetagh seemed desirable because it is a classic site but lacks a satisfactory modern pollen diagram. Dunshaughlin has already been studied by Mitchell (1940), but also lacks a modern pollen diagram.

Ballybetagh lies at the edge of the Dublin Mountains at an altitude of about 250 m. The three bogs described by Jessen & Farrington (1938, p. 207) overlie boulder clay of the Last Glaciation which is rich in limestone. The 'bogs' are really spring-fed marshes or fens covered by sedges, herbs of shallow water and bryophytes. *Carex paniculata* is a prominent component of the vegetation.

Jessen's studies were carried out mainly on excavations in the northwestern bog. The middle bog of the three, Jessen's southwestern bog, had become famous because upward of 60 giant deer skulls and antlers had been excavated there during the nineteenth century, together with reindeer remains. Jessen considered this bog too disturbed by excavation for giant deer to be satisfactory for study, a very reasonable viewpoint when one considers that the site is only about 300 m long and 50 m wide. However, the author found no difficulty in sampling the middle bog with a Livingstone sampler and in finding undisturbed sediments in its centre. No evidence of nineteenth-century excavation was encountered and it is possible that excavations were limited to a small area or to the bog margin. Cores were obtained from the centre of the northwestern segment of the bog, about 30 m north of a mapped fence line (cores SW-1, SW-2, map in Jessen & Farrington 1938, p. 207). As the early Post-glacial sediments in the middle bog are peats with a strongly locally influenced pollen flora the pollen diagram is completed by a core from the northwestern bog, core NW-1, where the early Post-glacial facies is lacustrine.

Although the altitude of Ballybetagh at 250 m may seem low, it is close to the limit of cultivation at the edge of the Dublin Mountains; heath and bogland predominate from about 300 m. In contrast, Dunshaughlin is a drained lake basin lying at less than 100 m in the Central Plain northwest of Dublin in farming land of high quality over limestone-rich boulder clay. Differences in the pollen diagrams between the two sites probably reflect the poorer soil and more exposed situation of Ballybetagh. Drilling at Dunshaughlin was carried out about 30 m east of map point 29 (map in Mitchell 1940, p. 14).

(*b*) *The pollen sequence*

(i) The *Rumex–Salix* assemblage (dock–dwarf willow)

At the base of the Ballybetagh and Dunshaughlin profiles (figures 2 and 3) pollen of *Rumex* type makes up to 20 % of the pollen rain. *Rumex* type includes *Rumex acetosa*, *R. acetosella* and *Oxyria digyna*, all three of which occurred macroscopically at Ballybetagh (Jessen & Farrington 1938). *Salix*, probably *S. herbacea*, is frequent at the same time. At Ballybetagh a pollen spectrum in silt at the base of the profile has high percentages of grass pollen and relatively little Cyperaceae (sedge). This suggests that the earliest upland flora was largely grasses with dwarf willow, *Rumex* and some diversity of other herbs. The *Rumex–Salix* assemblage is found at all Irish sites that have been investigated in detail. It corresponds more or less with Jessen's Older *Salix herbacea* Zone (Zone I). Much detailed work on the pollen and macrofossil assemblages will be needed before the character of its vegetation can be defined satisfactorily.

(ii) The *Juniperus–Empetrum* (juniper–crowberry) assemblage

Juniperus and *Empetrum* appear at the end of the *Rumex–Salix* assemblage and become abundant immediately. Juniper contributes 60 % of the pollen rain at Ballybetagh, 60 % at Woodgrange (Singh 1970) and 70 % at Belle Lake (Craig 1973). The juniper scrub must have covered the landscape over large areas, much as gorse (*Ulex europaeus*) now does. At sites where it plays a smaller rôle percentages of 20–40 are recorded. *Empetrum* arrived before *Juniperus* at Ballybetagh, simultaneously at Dunshaughlin and Roddans Port (Morrison & Stephens 1965). In western Irish sites, especially at Long Range and Muckross, it arrived clearly before juniper and was already declining as juniper reached a peak. Its percentage values range from high values of up to 20 in southern and western Ireland and at Roddans Port to minor occurrences in other areas of northeast and southeast Ireland.

A small but well defined peak of *Hippophaë* (sea-buckthorn) pollen is present and leaf hairs were observed at both Ballybetagh and Dunshaughlin at the end of the *Juniperus–Empetrum* phase. They are also recorded at Roddans Port (Morrison & Stephens 1965). The peak appears to be stratigraphically significant in the Dublin region, but elsewhere sea-buckthorn has been little recorded. It is recorded as a peak within the *Juniperus–Empetrum* assemblage in table 1, but is judged to be too small a feature to be given separate stratigraphic recognition.

(iii) The Gramineae (grass) assemblage

In this phase Juniper declines to low values and *Hippophaë* disappears. Grass supplies 50 % of the pollen rain. The most conspicuous associated taxa are *Helianthemum* and Umbelliferae. *Betula* is less frequent than in the preceding phase. The grass species involved are not known. There was some open ground, for mineral material continued to arrive in lakes. The assemblage of grass with low-growing genera such as *Helianthemum* and *Galium* suggests predominant short grassland with small herbs and shrubs and some stands of *Betula pubescens*. The birch occurs macroscopically. This was probably the habitat in which giant deer and reindeer flourished.

(iv) The Cruciferae (crucifer) peak

After the grass assemblage there is a brief peak of Cruciferae pollen. There is more than one type of crucifer pollen. At Ballybetagh Jessen identified *Cardaminopsis petraea* from 'zone III' and subsequently Mitchell (1953) added *Draba incana* and *Cochlearia officinalis* to the list. All three species occur relict on cliffs and base rich talus slopes in Irish mountains now. *Cochlearia* also occurs on seashores. The crucifer peak is very strongly developed at Ballybetagh, Dunshaughlin and Roddans Port (18–30 % of the pollen sum) and less strongly but clearly at Woodgrange (about 10 %). Elsewhere it has not been identified, so it is either a local phenomenon or it has not been observed because of too wide sampling. It appears to be a transitional phase from the grass assemblage to the next phase and probably indicates the beginning of the destruction of stable grassland by solifluxion in a deteriorating climate.

(v) The *Artemisia* assemblage

This phase is sharply distinct from all others in the Late-glacial. The most characteristic pollen types are *Artemisia* and Caryophyllaceae but at Ballybetagh, for example, *Sedum rosea*, *Armeria*, *Koenigia* and *Polygonum viviparum* are virtually confined to the zone, and pollen and leaves of *Salix herbacea* are frequent. *Thalictrum* pollen is common (*T. alpinum* was identified by Jessen at Ballybetagh) and at Dunshaughlin pollen of *Saxifraga hypnoides* type and spores of *Selaginella selaginoides* and *Botrychium* are characteristic. Elsewhere, there is a good deal of variation from site to site. *Artemisia* is not always abundant. High peaks occur at some sites with over 20 % of the pollen, at others the values may be as low as 1–2 %. At many sites the Cyperaceae increase in value and Gramineae decrease sharply, but at Ballybetagh both the Gramineae and the Cyperaceae decrease and at Dunshaughlin the Cyperaceae change little from the previous phase. The rôle of *Juniperus* and *Empetrum* in this zone is particularly interesting. They appear to have been under severe pressure from the climate, but to have been able to survive in favoured localities. Both species disappeared completely at Ballybetagh and Coolteen and were very much reduced or disappeared temporarily in the west and southwest. However, at Belle Lake and Dunshaughlin juniper continued to be successful and *Empetrum* survived at Dunshaughlin. There is no evidence to suggest redeposition of pollen from older assemblages.

There is evidence from Lough Nahanagan that mountain glaciers were temporarily re-established at this time and solifluxion seems to have been widespread (see § 5*f*). The phase corresponds approximately with Jessen's Younger *Salix herbacea* zone (Zone III).

(vi) The Gramineae–*Rumex* (grass–dock) peak

At the end of the *Artemisia* phase orderly organic sedimentation began again, and grass pollen was abundant briefly, associated with *Rumex* at Dunshaughlin and several other Irish sites, but not at Ballybetagh. In this brief phase very great changes must have taken place, for the characteristic pollen types of the *Artemisia* phase all disappeared or were greatly reduced and new taxa such as *Dryas*, *Ononis* type and *Trifolium* appeared fleetingly at Ballybetagh.

(vii) The *Juniperus–Filipendula* (juniper–meadowsweet) assemblage

Grassland was invaded in this phase by juniper, *Filipendula* and *Empetrum*. At Dunshaughlin *Filipendula* peaked first, while at Ballybetagh *Empetrum* was first and played a larger rôle, as is suitable for a mountain fringe habitat now with leached soils. Juniper reached high percentage values in this phase and was as abundant once more in the landscape as it had been in the early Late-glacial.

(viii) The *Betula* (birch) assemblage

In this phase tree birches came to dominate the landscape and replace juniper generally, as is particularly well seen at Dunshaughlin. This is the first more or less closed forest of the Post-glacial and is a suitable point to end consideration of the open landscapes of the Late-glacial and earliest Post-glacial period. Shortly afterwards *Corylus*, *Quercus*, *Ulmus* and *Pinus* appeared and initiated the forested Boreal period.

3. Pollen influx studies

Craig (1973) has carried out pollen influx studies at Coolteen and Belle Lake in southeastern Ireland. The pollen sequence is broadly similar to that at Ballybetagh. In the basal *Rumex–Salix* phase pollen influx is very low. In the succeeding juniper-dominated phase it is relatively high but falls again in the grass phase. An interesting result is that the influx of herb pollen of most genera is higher in the juniper than in the grass phase. The high grass values in percentage diagrams appear to be an artefact of the form of calculation used, rather than an expression of high cover of the landscape, for the grass pollen influx itself is less than in the juniper phase. In the *Artemisia* phase pollen influx is again higher than in the grass phase, a quite unexpected result, for most authors have held that the pollen rain of the *Artemisia* phase was probably an expression of sparse open vegetation. Craig's interesting data will require a thorough re-consideration of the Late-glacial plant cover if they are confirmed by further studies at other sites. Table 2 shows a comparison of Craig's data with data from Britain (Pennington 1975) and from the United States (Waddington 1969; Davis 1969; Craig 1972). Craig has shown that the pollen concentration and influx figures at Coolteen in the *Rumex–Salix* phase compare with the Late-glacial herb zone of the northern United States (the treeless vegetation between the forest margin and the ice sheet) but that the grass phase values are comparable to modern prairie. It is, however, well known that lake basins vary greatly in their characteristics as pollen traps, and that influx data may show great variability between different sites (Pennington 1973). This is clear from a comparison of the data from Blelham Bog (Pennington 1975) and from Coolteen. The Blelham values are all much lower, although the pollen concentration values are similar. Even taking this into account, vastly more pollen was sedimented at Coolteen in the *Artemisia* phase than in the comparable phase at Blelham. In view of this variability, caution should be used in making use of these data until more comparative data are available.

TABLE 2. COMPARISON OF POLLEN CONCENTRATION AND INFLUX VALUES FROM SEVERAL SITES

Coolteen (Craig 1973)			Blelham (Pennington 1975)			
local zone	pollen concentration (cm^{-3})	pollen influx ($cm^{-2}\,a^{-1}$)	local zone	pollen concentration (cm^{-3})	pollen influx ($cm^{-2}\,a^{-1}$)	approximate equivalent in Ireland
Co 1 *Rumex* phase	3.2×10^4	2×10^3	Ba	3.2×10^4	1.0×10^3	*Rumex–Salix* phase
Co 2 Juniper phase	21×10^4	50×10^3	Bb	10×10^4	2.5×10^3	*Juniperus–Empetrum* phase
Co 3 grass phase	15×10^4	15×10^3	Bd	7.5×10^4	1.5×10^3	Gramineae phase
Co 4 *Artemisia* phase	15×10^4	35×10^3	Bf	1×10^4	1×10^3	*Artemisia* phase

	Rogers Lake, Connecticut (Davis 1969)		Lake of the Clouds, Minnesota (Craig 1972)	
	pollen concentration (cm^{-3})	pollen influx ($cm^{-2}\,a^{-1}$)	pollen concentration (cm^{-3})	pollen influx ($cm^{-2}\,a^{-1}$)
Late-glacial herb zone	3×10^4	1×10^3	3 to 10×10^4	1.6 to 5×10^3

	Rutz Lake, Minnesota (Waddington 1969)	
	pollen concentration (cm^{-3})	pollen influx ($cm^{-2}\,a^{-1}$)
Post-glacial prairie	10 to 15×10^4	10 to 25×10^3

4. REGIONAL VARIATION IN VEGETATION COVER

While the Ballybetagh and Dunshaughlin pollen diagrams are very similar, there are differences in detail between the two sites which probably reflect the greater altitude of Ballybetagh, its more exposed site, and its poorer soils which have developed from calcareous boulder clay since the Late-glacial. The differences include the greater rôle of *Filipendula* and *Saxifraga hypnoides* type at Dunshaughlin and, at Ballybetagh, the relatively greater abundance of *Helianthemum* as well as the more distinct rôle of *Sedum rosea*, *Armeria*, *Koenigia* and *Saxifraga oppositifolia* in the *Artemisia* assemblage.

In western Ireland at Poulroe on bare limestone of the Burren Region, and at Lough Goller on infertile shales and sandstones, the same general development of vegetation can be seen as in eastern Ireland (figure 4). *Juniperus* plays a much larger rôle on limestone than on shale. *Empetrum* is more abundant at both sites than farther east. At these sites the early Late-glacial juniper peak is not as well developed as in eastern and southern Ireland and juniper is still present in the grass assemblage. Local presence is proved by the occurrence of a juniper needle in the sediments at Poulroe. Poulroe is of some interest because the relationship of macrofossils to the pollen diagram is known. *Betula nana* (dwarf birch) was frequent but *B. pubescens* (tree birch) is represented by several fruits and bract-scales in the grass assemblage. Birch was not abundant in the pollen rain. *Dryas octopetala* leaves are common at this site, as at other sites in the Burren. *Linum catharticum* is a common Late-glacial macrofossil at Poulroe and at Gortalecka (Watts 1963), and *Thalictrum minus* is also present in the Late-glacial of the Burren region.

So far, evidence for Late-glacial *Empetrum* heath in western Ireland has rested on the work of Jessen at Roundstone in western County Galway (Jessen 1949). The author has confirmed by resampling Jessen's site that *Empetrum* was frequent in the pollen rain of the Late-glacial and early Post-glacial even when *Juniperus* is also recorded. More recently pollen studies have been carried out by M. B. Telford at Glenveagh in northwestern Ireland as part of a Ph.D. thesis at Trinity College, Dublin. The Glenveagh site lies at low elevation in a high rainfall region largely covered by blanket-bog. The bedrock is granite.

The outstanding feature of the Glenveagh pollen diagram (figure 4) is the prevalence of *Empetrum* throughout the Late-glacial and early Post-glacial. *Empetrum* heath with grasses and sedges appears to have been the predominant Late-glacial vegetation at all times, including the *Artemisia* phase. The *Artemisia* phase has low percentages of *Artemisia* itself, and spores of *Lycopodium selago* and *Selaginella selaginoides* are frequent.

In the diagram the features which characterize Late-glacial pollen diagrams from eastern and southern Ireland can be distinguished to some extent. There is a *Rumex–Salix* phase and an early *Juniperus* peak. A silt horizon at 772 cm is correlated with a *Rumex* peak. This is the erosion phase referred to below (§ 5*e*). There is a weakly developed grass assemblage, culminating in a high percentage of *Betula* in one sample. Perhaps birch woodland was established locally for a short period. In the early Post-glacial successive peaks of *Rumex* with grass and *Filipendula* with juniper are very similar to the succession in eastern Ireland.

5. Aspects of climate and landscape in the Late Devensian

(*a*) *Reliability of evidence for the nature of the Late Devensian climate*

In the Late-glacial and Flandrian periods plant migration was universal, and is probably still taking place for many species. During migration species were under constant selection pressure to enable them to coexist with others, to occupy new habitats and niches and to adapt to different climates. The plant communities were restructured as new species immigrated and numbers adjustments took place (Watts 1973). It follows that species may not occupy their total potential range now because they are still migrating, they may differ genetically to some extent from their Late-glacial ancestors because of natural selection and they may occur in associations that did not exist in the Late Devensian. These considerations suggest that caution should be exercised in assessing past climates from present plant distributions. More reliable climatic evidence can be obtained from less ambiguous phenomena, such as the temperature of the ocean, the presence or absence of glaciers, geomorphological evidence of permafrost and solifluxion, the productivity of lakes at different times, and tree cover or its absence. The author doubts whether it is possible to state mean July temperatures for the various phases of the Late-glacial with much certainty of being correct, except within wide margins of error.

(*b*) *The main Late Devensian Glaciation*

Ireland was largely covered by ice at the height of glaciation (figure 5). An ice sheet of 400–500 m in thickness covered the Central Plain and through it the higher mountains which themselves bore corrie glaciers stuck out as nunataks. To the east the main ice sheet passed into a large ice mass filling the Irish Sea basin. In the southwest a local ice sheet covered the mountains of Kerry. A considerable area of southern Ireland, perhaps a fifth of the country, was free of ice and there were ice-free areas locally on the north and west coasts (Synge 1969). The ocean surface temperatures west of Ireland were comparable to those of eastern Greenland at present and there may have been floating ice. The Gulf Stream did not influence Ireland at that time, for 17000 years ago polar water extended to northern Spain (Ruddiman & McIntyre 1973). There seems no possibility that Ireland could have contained 'refugia' on its west coast for oceanic plants as suggested by Mitchell & Watts (1970). It is possible that the ice-free areas were isolated from the rest of Europe by ice sheets, although the floor of what is now the Celtic Sea may have provided a land bridge to Britain. Ice-free areas may have been a critical factor in the late survival of abundant large herbivores such as the giant deer, for there is no clear evidence that Palaeolithic hunters ever reached Ireland. Low sea level must have exposed substantial areas of the continental shelf (figure 5) and these were not invaded by the rising sea until after the Late Devensian, for freshwater deposits with Late-glacial pollen floras lie some 60 m below present sea level near Whiddy Island in southwestern Ireland (table 3, Stillman 1968).

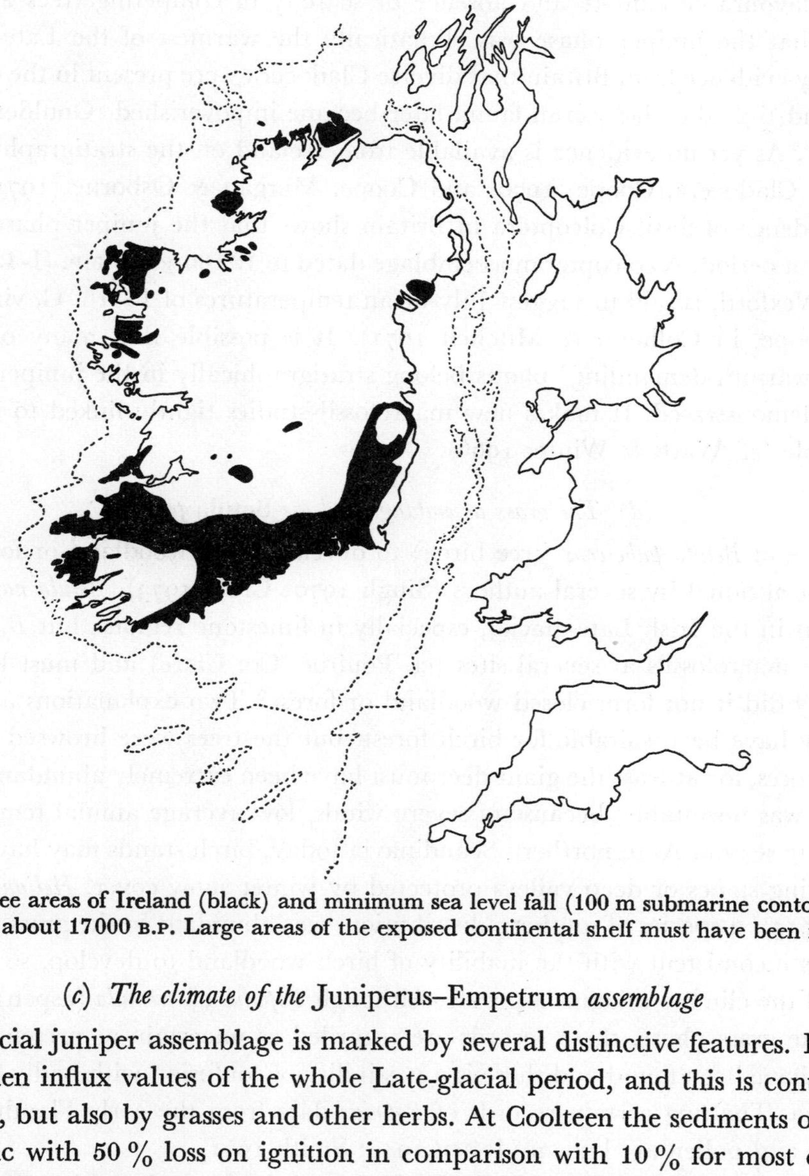

FIGURE 5. Ice-free areas of Ireland (black) and minimum sea level fall (100 m submarine contour shown as dotted line) about 17000 B.P. Large areas of the exposed continental shelf must have been ice-free.

(c) *The climate of the* Juniperus–Empetrum *assemblage*

The Late-glacial juniper assemblage is marked by several distinctive features. It has much the highest pollen influx values of the whole Late-glacial period, and this is contributed not only by juniper, but also by grasses and other herbs. At Coolteen the sediments of the period are very organic with 50 % loss on ignition in comparison with 10 % for most of the Late-glacial. This high productivity may also be observed at Dunshaughlin where whitish grey richly calcareous sediments mark the assemblage stratigraphically. At Roddans Port desmids are found early in the Late-glacial when juniper was frequent (Round, in Morrison & Stephens 1965), and a virtually identical species assemblage is found at Dunshaughlin in the juniper phase. There are numerous species of *Cosmarium* with species of *Staurastrum* and *Euastrum*. It might be argued that the productivity of the lakes and their rich algal floras reflected the interaction of a warming climate with availability of nutrients from unleached till, but the evidence that the upland flora was diverse and that there was a dense vegetation cover suggests that the climate was warm in summer with a long growing season. The success of juniper itself seems to point in this direction. Iversen (1954) explained survival of juniper in the Late-glacial of Denmark by the presence of adequate snow-cover in winter to protect flowering branches in periods of severe climate, with taller growth, free flowering and expansion in a more favourable climate when protection by snow was no longer necessary. In Ireland expansion can be explained by a completely favourable climate and absence or scarcity of competing trees and shrubs. The suggestion that the juniper phase was climatically the warmest of the Late-glacial may be supported by evidence from Britain that diverse Cladocera were present in the early part of the 'Allerød' and that the cladoceran fauna later became impoverished (Goulden 1964; Harmsworth 1968). As yet no evidence is available from Ireland on the stratigraphic occurrence of Late-glacial Cladocera. Coope (1970) and Coope, Morgan & Osborne (1971) have claimed that the evidence of fossil Coleoptera in Britain shows that the juniper phase was an exceptionally warm period. A coleopteran assemblage dated to 12160 ± 180 B.P. (I-4963) at Shortalstown, Co. Wexford, is said to suggest July mean temperatures of 15–16 °C, virtually the same as now (Coope, in Colhoun & Mitchell 1971). It is possible that many older macrofossil records of 'warmth-demanding' plants belong stratigraphically in the juniper phase, but this cannot be demonstrated. It makes new macrofossil studies tightly linked to pollen diagrams very desirable (cf. Watts & Winter 1966).

(*d*) *The grass assemblage and the* Betula *problem*

The failure of *Betula pubescens* (tree birch) to develop birch woodland or forest in the grass phase has been noted by several authors (Singh 1970; Craig 1973). *Betula nana* (dwarf birch) was frequent in the Irish Late-glacial, especially in limestone regions, but *B. pubescens* is also known from macrofossils at several sites (cf. Poulroe, Co. Clare) and must have been widespread. Why did it not form closed woodland or forest? Two explanations are possible. The climate may have been suitable for birch forest, but the trees were browsed very heavily by large herbivores, for at least the giant deer must have been extremely abundant. Alternatively, the climate was unsuitable, because of severe winds, low average annual temperatures and a short growing season. As in northern Scandinavia today, birch stands may have been confined to south-facing slopes or deep valleys protected by winter snow cover. *Helianthemum* cf. *canum* (Mitchell 1953), a species of southern distribution, was abundant in the grass phase. Its abundance seems inconsistent with the inability of birch woodland to develop, so that the overall character of the climate remains a puzzle. Although *Populus* cf. *tremula* (aspen) pollen is recorded from the grass phase, these records are regarded as uncertain at present, because single grains only have been found and there is a possibility of confusion with badly preserved pollen of other taxa. The first certain records of *Populus* date from the early Flandrian where both macrofossils and pollen are known (Jessen 1949; Smith 1961).

The development of vegetation at the beginning of the Late-glacial is strikingly parallel to the Early Flandrian. The Late-glacial may be characterized as an interglacial with deflected development. The juniper phase should have been succeeded by birch and other broad-leaved trees in a steadily improving climate. Instead, a deterioration began at about 12000 B.P. which was not reversed until the Early Flandrian.

(*e*) *Late-glacial erosion*

There is evidence from several sites that erosion characterized the time of transition from the juniper phase to the grass phase. This was already commented on by Watts (1963) and was first noticed by Mitchell (1941 *b*) at Ratoath, Co. Meath, a site close to Dunshaughlin and from which a modern pollen diagram very similar to that from Dunshaughlin is available (Peglar, in Mitchell 1976). Mitchell wrote 'minor climatic oscillations prior to this marked deterioration (Zone III) are possibly indicated by the two grey sandy layers that were traced in the lower chalk-mud over considerable distances'. At many Irish sites, and particularly noticeable in those with predominantly white calcareous lake-muds, a layer of inorganic material, often grey or blue-grey in colour and 2 or 3 cm only in thickness, occurs within a predominantly organic sediment. Such layers have been observed at Muckross, Gortalecka, Lough Goller, Glenveagh and Ratoath. The evidence is not so clear at Dunshaughlin and Ballybetagh where the thick sediments are more varied, but at Dunshaughlin the thick white calcareous muds of the juniper phase are interrupted by a dark silty layer and at Ballybetagh the grass phase is interrupted by somewhat brecciated sediment with large mica flakes. At Belle Lake there is no evidence of erosion, and it has not been observed at other sites with the exception of Coolteen. As it is a relatively minor phenomenon it may well have been disregarded as insignificant in some older studies.

At Coolteen there is very remarkable evidence of erosion. As the juniper phase came to an end sediments which were 50 % organic were succeeded by grey calcareous clays a metre or more in thickness with 10 % organic content. Curves for Na and K are available for the Coolteen profile. These show that if Na and K are expressed as a proportion of the total ash weight their highest values are found in the clays, providing evidence that unleached clay was washed into the basin by erosion.

The inorganic layers at other sites and the thick clays at Coolteen may be contemporary, but the evidence is disputable. At some sites silt layers occur before the end of the juniper phase (Glenveagh, Dunshaughlin), at some they mark the transition to the grass phase (Coolteen, Ratoath, Gortalecka) and at others they lie within the grass phase (Ballybetagh). Craig (1973) has argued that the juniper phase is of different duration at different places and this would explain the apparently variable ages of what might at first sight be a single climatically caused event. However, this cannot be regarded as certain and should not be relied on without further investigation. At Gortalecka, the silt of the erosion phase shows an *Artemisia* peak and at Glenveagh and Dunshaughlin *Juniperus* falls during the erosion phase in favour of *Rumex* and grass respectively.

While unambiguous evidence of a single phase of erosion remains elusive, it is clear that about the time of the end of the juniper phase many lake basins were affected by inflows of inorganic material. This could have any of a number of causes, solifluxion, exceptional precipitation causing sheet erosion, or low water levels with redeposition from shores.

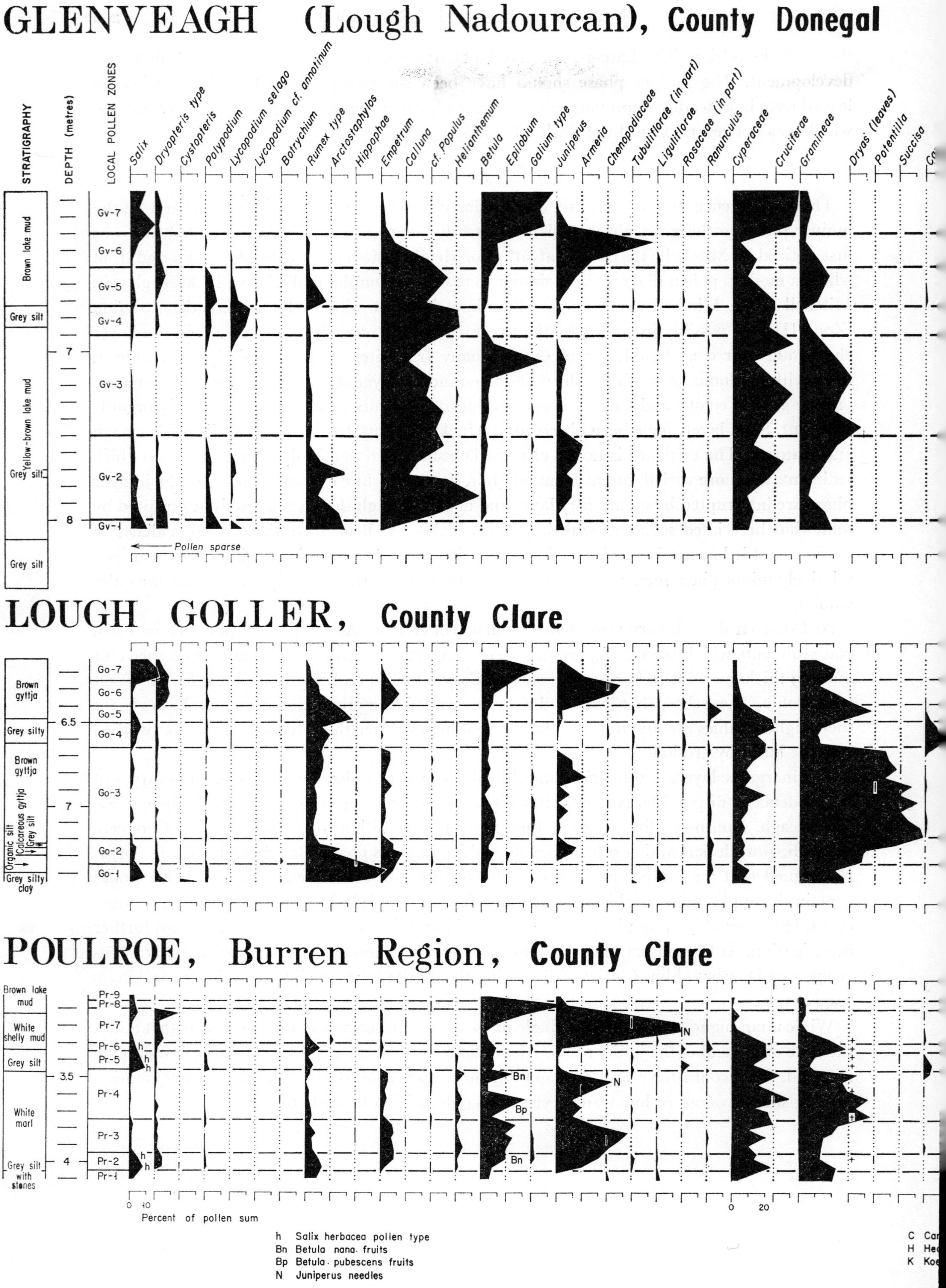

FIGURE 4. Pollen diagrams from Lough Golle

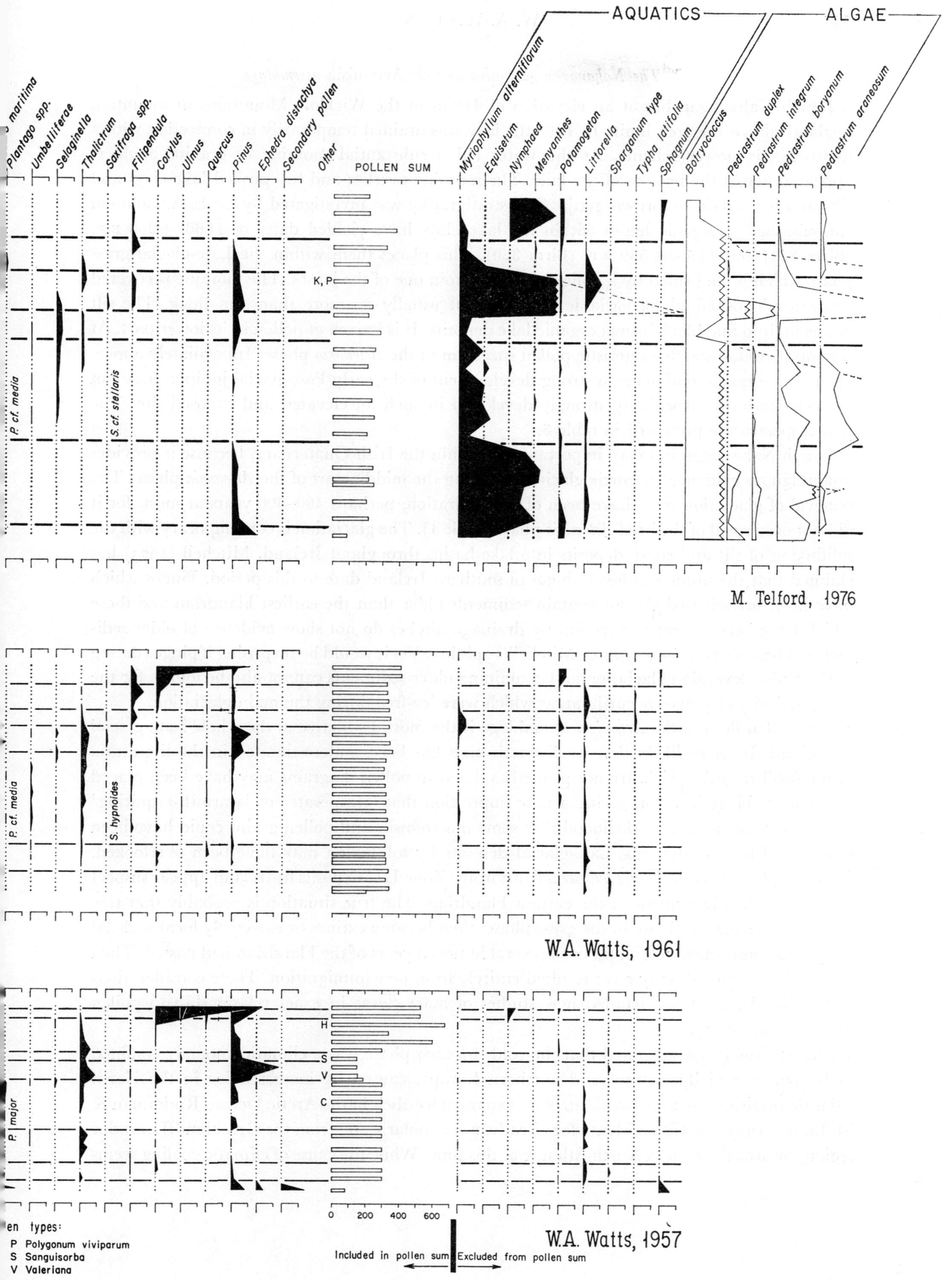

ulroe, Co. Clare, and Glenveagh, Co. Donegal.

(*f*) *The Nahanagan glaciation and the* Artemisia *assemblage*

Lough Nahanagan lies at an elevation of 410 m in the Wicklow Mountains in mountain bogland above the tree limit. In 1968 the lake was drained temporarily in connection with a hydroelectric power scheme. As the water fell a substantial moraine of granite boulders appeared across the bottom of the lake. The moraine overlies and has pushed lake clays and silts and is overlain by brown gyttja. The stratigraphy was investigated by Dr E. A. Colhoun (unpublished). Organic lenses within the lake clays have yielded dates of 11 600 ± 260 B.P. (Birm 321) and 11 500 ± 550 B.P. (Birm 320). This places them within the Late-glacial grass phase which is confirmed by a pollen spectrum from one of the lenses. The moraine is covered by water-deposited silt of variable thickness but usually no more than 1 m thick. The silt passes up into fine black-brown organic lake deposits. It is barren of pollen in its lower part. At the top it contains a characteristic pollen spectrum of the *Artemisia* phase. Immediately above, in the first organic sediments, a strong development of the early Post-glacial juniper peak can be seen, perhaps surprisingly strongly developed in such an elevated and exposed site. The pollen spectra are presented in table 3.

Lough Nahanagan is a very important site within the Irish Quaternary, because it provides unambiguous evidence for corrie glaciation during the middle part of the *Artemisia* phase. The renewal of glaciation must have been of brief duration, perhaps 400–500 years at most, for it did not occupy all of the brief *Artemisia* phase (table 1). The glaciation is contemporary with the solifluxion of silt and stony deposits into lake basins throughout Ireland. Mitchell (1973) has claimed that the numerous fossil pingos of southern Ireland date to this period. Pingos which have been investigated do not contain sediments older than the earliest Flandrian and those which have been exposed in section by drainage ditches do not show evidence of older sediments. These observations support Mitchell's opinion. Yet it would be surprising if pingos dating to the main Devensian glaciation and containing older sediments cannot also be found, for the great majority of pingos occur in areas which were ice-free during the main glaciation.

The pollen flora of the *Artemisia* assemblage is the most distinctive of the whole Late-glacial in Ireland. It seems likely that its distinctiveness has been underestimated in the literature. Macrofossil records which are not properly related to pollen diagrams may have been placed in the assemblage in error, giving a false impression that 'temperate' or 'warmth-requiring' species were present. The likelihood that some macrofossils and pollen grains could have been redeposited from underlying Late-glacial deposits by solifluxion may have been overlooked. For example, reports of *Betula pubescens* fruits from 'Zone III' deposits in Ireland appear suspect because birch is infrequent in the earliest Flandrian. The true situation is probably that tree birch was present until late in the grass phase, then became extinct or extremely local and did not play an important rôle again until several hundred years of the Flandrian had passed. Then its population growth may have resulted entirely from new immigration. These considerations emphasize the need for detailed new studies of macrofloras in exact relationship to pollen diagrams (see § 1*b*).

Mercer (1969) has suggested that the cold *Artemisia* phase was a climatic anomaly confined to Europe, especially western and northern Europe, caused by increased ice in the North Atlantic resulting from the break-up of an extensive ice-shelf in the Arctic Ocean. Ruddiman & McIntyre (1973) provide evidence for a readvance of polar water from the Greenland/Labrador region towards the eastern North Atlantic at this time. While the cause of climatic cooling seems

TABLE 3. POLLEN COUNTS FROM LOUGH NAHANAGAN AND WHIDDY ISLAND

	Betula	*Juniperus*	*Salix*	Gramineae	Cyperaceae	*Rumex*	*Empetrum*	*Filipendula*	*Artemisia*	Other herbs	*Lycopodium selago*	Other pteridophytes	*Pinus*	*Corylus*	Other broad-leaved trees	Pollen sum	pollen assemblage or age
Whiddy Island (depth below sea level)																	
57 m (silty gyttja)	20.0	9.0	3.0	14.0	14.0	12.0	13.0	—	—	3.0	—	12.0	—	—	—	100	*Juniperus–Empetrum* assemblage
Lough Nahanagan (depth below mud-water interface)																	
325 cm (gyttja)	5.0	—	1.6	2.6	0.6	0.6	0.3	—	—	0.6	0.3	6.3	22.0	33.3	26.3	300	Flandrian
327.5 cm (silty gyttja)	23.0	0.8	10.4	10.9	1.7	1.3	1.3	0.4	—	1.3	—	36.3	10.9	1.7	—	230	*Betula* assemblage
330 cm (silty gyttja)	9.5	60.0	4.0	7.1	3.1	1.8	1.8	0.3	0.3	5.5	0.6	4.3	2.8	—	0.3	325	*Juniperus–Filipendula*
332.5 cm (silt)	5.4	0.8	3.4	28.3	11.0	3.0	5.9	1.3	1.3	6.8	14.3	11.0	7.6	—	—	237	Gramineae–*Rumex* peak
335 cm (silt)	3.3	0.8	3.3	4.2	24.4	0.8	1.6	—	7.3	9.8	4.9	24.4	11.4	—	—	123	*Artemisia* assemblage
(from organic lens in silt below Nahanagan moraine)																	
silty gyttja	3.7	16.8	1.1	28.0	17.9	8.6	1.1	—	0.4	6.3	4.9	11.2	—	—	—	268	*Juniperus–Empetrum* assemblage

to be focused on the northwestern North Atlantic, its effect is most strongly expressed on the Atlantic coasts of northwest Europe, and Ireland may have been more affected floristically than areas farther east.

(g) Aquatic plants

The Late-glacial aquatic flora is a rich one and new species have been added to the known list (Mitchell 1954) in recent years. *Elatine hexandra* is known macroscopically from the grass phase at Muckross, Killarney, and has been found elsewhere as pollen. *Ceratophyllum demersum* occurs early in the Late-glacial at Muckross and at Coolteen. At Coolteen *Potamogeton praelongus*, *P. pectinatus*, *P. crispus*, *P. pusillus* and *P. natans* all occur in the earlier part of the Late-glacial as, apparently, does *Typha angustifolia*. *Pilularia globulifera* megaspores were found in the *Artemisia* phase at Belle Lake. At Ballybetagh, pollen of *Polygonum amphibium* is frequent in the early juniper and grass phases.

It is interesting that the aquatics play their largest rôles at two points within the Late-glacial, at the beginning, and at the transition to Flandrian. In each case a warming climate and an abundant release of nutrients from fresh till surfaces may have resulted in high productivity. At the end of the *Artemisia* phase, in which aquatic plants were completely eliminated at many sites, a remarkable burst of growth took place in the aquatics. *Myriophyllum alterniflorum* was particularly successful as was *M. spicatum* but the short-lived success of *Typha latifolia* and, at Ballybetagh, invasion and vigorous growth by *Ceratophyllum demersum* also mark this phase, which was fully developed and, at some sites, completed before the Flandrian expansion of *Juniperus* began. The very rapid response of aquatics to climatic warming during the Late-glacial is also recorded for Denmark by Iversen (1954).

6. Radiocarbon chronology

The number of radiocarbon dates available for the Late-glacial in Ireland is not great. A series is available from Coolteen and Belle Lake in southeastern Ireland (Dresser & McAulay 1974; Craig 1973). Further series are available from Sluggan, Slieve Gallion, Ballynagilly and Altnahinch in northeastern Ireland (Smith, Pearson & Pilcher 1971*a*, *b*, 1973). The northeastern dates are from bog monoliths. Pollen diagrams have not yet been published for these sites. Dates from Roddans Port (Morrison & Stephens 1965) are not used here because of inconsistencies in the long series which apparently arise from complex sources of contamination (cf. discussion by Dresser in Smith, Pearson & Pilcher 1971*b*). The southeastern and northeastern dates are detailed in appendix 1. They are the source from which the chronology given in table 1 is derived. Other dates are referred to in the text as appropriate.

It is desirable that more dates should be obtained from the earliest Late-glacial. We do not know when the *Rumex–Salix* phase began. Dates as old as 13000 may be obtained but there does not seem to be any reason to expect older dates at present, although they are known from Britain (Pennington 1975). The duration of the grass phase is uncertain because D-111 is significantly younger than UB-227F. The older date is preferred because it is consistent with dates from outside Ireland for the beginning of the *Artemisia* phase. Dates for this event should be nearly synchronous because the vegetation changes were caused by a general climatic deterioration related to a readvance of polar water in the North Atlantic (Ruddiman & McIntyre 1973). The dates for the end of the *Artemisia* phase for northeastern Ireland appear young and D-108 is preferred. It is possible that there is a regional difference in age for this event but the clustering of dates from the northeast around 9500 for several events (cf. UB-297, UB-260, UB-411, UB-225F) suggests that there may be contamination from rootlet penetration or movement of humus in some of the bog monoliths. Some of the dates proposed in table 1 are speculative. As yet we do not know anything about the duration of the Crucifer peak or the transitional herb-dominated phases at the beginning of the Flandrian. Both Ballybetagh and Dunshaughlin are unsuitable for radiocarbon dating because of the probable presence of ancient carbon in their calcareous sediments. It will be necessary to obtain further series of dates from sites where the various possibilities of error are more limited, ideally from autochthonous sediments of a large fast-sedimenting lake in a non-calcareous region. It should be remembered (§ 1*b*) that the pollen assemblages which are the basis of stratigraphy are probably non-synchronous and that a uniform chronology is not to be expected. In this sense any outline chronology is an over-simplification that ignores the likelihood of at least small-scale variation in age for dated events from site to site.

7. Discussion

In Britain Late-glacial pollen diagrams differ substantially in detail from those from Ireland (Pennington 1975). The most important difference is that much of Britain seems to have been covered by birch woodland in the Irish 'grass phase' while in the Netherlands, for example, either birch or pine woodland was present throughout the Late-glacial and juniper was much less frequent. The distinctively Irish features are, (1) an open landscape, never forested, at most with local birch woodland patches, (2) a very large rôle for juniper in the vegetation, (3) a very distinctive flora in the '*Artemisia* phase', suggesting that the climatic deterioration from 10900 to 10200 B.P. may have achieved its strongest expression on Atlantic coasts in northwest Europe, (4) the survival of the large herbivores, no doubt correlated with the absence of man.

The early *Juniperus* expansion between 12400 and 12000 B.P. is seen as the period of warmest climate and highest productivity of land and aquatic vegetation in the Late-glacial. The succeeding grass phase probably represents a climatic deterioration with perhaps a shorter growing season and cold winds, but without large-scale solifluxion. It is natural to equate the *Juniperus* expansion with the Bølling oscillation of Denmark and the grass phase with the Allerød (Iversen 1954). In this case the erosion phase (§ 5*e*) becomes 'Zone Ic'. In the author's opinion such a correlation is unjustified. The evidence for a climatic deterioration between 12000 and 11800 B.P. in Britain followed by a recovery has a very slight basis as yet (Pennington 1975). The evidence from Ireland suggests that the climate of the Grass phase was less favourable throughout than the climate of the *Juniperus–Empetrum* assemblage, in other words a permanent deterioration rather than a decline and recovery took place about 12000 B.P. Erosion which took place at about that time may be a series of unconnected local events reflecting a general reduction in vegetation cover.

While the abandonment of Jessen's tripartite zone system is plainly overdue, new investigations will be necessary to improve our understanding of Ireland's Late-glacial flora and climate. In particular new macrofossil studies are needed so that there is no room for doubt about the stratigraphic position of climatically significant species. The identification of the grass species of the grass phase is specially desirable. New information about the supposed 'climatic optimum' of the *Juniperus–Empetrum* assemblage may also come from studies of the Algae, Cladocera and sediments of that time in comparison with those of the grass assemblage.

I would like to thank N. Allott, H. Lamb, J. Shackleton and other final year students at Trinity College for assistance in field work, A. J. Craig and M. Telford for making unpublished data available to me and G. F. Mitchell, F.R.S., H. E. Wright and H. J. B. Birks for assistance in the laboratory and the field and with discussion of the literature.

Appendix 1. Radiocarbon dates from Late Devensian and Early Flandrian deposits in Ireland

(1) *Southeastern Ireland*

D-109	Coolteen	12470 ± 155	upper part of *Rumex–Salix* phase
D-110	Belle Lake	12235 ± 160	transition from *Rumex–Salix* to *Juniperus–Empetrum* phase
I-5039	Coolteen	12390 ± 160	early in *Juniperus–Empetrum* phase
I-5038	Coolteen	12020 ± 180	near end of *Juniperus–Empetrum* phase
D-111	Belle Lake	10590 ± 185	end of Gramineae phase, transition to *Artemisia* phase
D-108	Coolteen	10210 ± 110	beginning of *Juniperus–Filipendula* phase
D-112	Belle Lake	9600 ± 135	early Flandrian *Betula* peak
D-107	Coolteen	9055 ± 95	end of *Betula* phase before *Corylus* rise
D-113	Belle Lake	9100 ± 130	rise of *Corylus* after *Betula* decline

Data from Dresser & McAulay (1974), Craig (1973).

(2) *Northeastern Ireland*

UB-229F	Sluggan	12470 ± 125	boundary of Zone I–II (equivalent to end of *Rumex–Salix* phase)
UB-447	Sluggan	11635 ± 160	lower Zone II
UB-446	Sluggan	10995 ± 160	middle Zone II
UB-227F	Sluggan	10945 ± 145	top of Zone II, just below III
UB-298D	Slieve Gallion	9660 ± 105	end of Zone III
UB-297	Ballynagilly	9595 ± 125	end of Zone III
UB-411	Altnahinch	9555 ± 135	end of Zone III, transitional to next zone
UB-444	Sluggan	9610 ± 130	early Flandrian *Juniperus* rise
UB-281	Slieve Gallion	9215 ± 75	middle of early Flandrian *Juniperus* peak
UB-225F	Sluggan	9475 ± 145	transition from *Juniperus* peak to *Betula* rise in early Flandrian
UB-260	Ballynagilly	9595 ± 80	fall of *Juniperus*, rise of *Betula* in early Flandrian
UB-419	Altnahinch	9045 ± 125	early Flandrian *Betula* maximum
UB-443	Sluggan	9360 ± 150	falling *Betula*, rising *Corylus*
UB-418	Altnahinch	8895 ± 115	beginning of *Corylus* rise

Data from Smith, Pearson & Pilcher (1971*a*, *b*, 1973). Zone numbers after Jessen (1949).

References (Watts)

American Commission on Stratigraphic Nomenclature 1961 Code of Stratigraphic Nomenclature. *Am. Ass. Petroleum Geol. Bull.* **45**, 645–665.

Colhoun, E. A. & Mitchell, G. F. 1971 Interglacial marine formation and late-glacial freshwater formation in Shortalstown townland, Co. Wexford. *Proc. R. Ir. Acad.* B **71**, 211–245.

Coope, G. R. 1970 Climatic interpretations of Late Weichselian Coleoptera from the British Isles. *Rev. Geogr. phys. Geol. dyn.* **12**, 149–155.

Coope, G. R., Morgan, A. & Osborne, P. J. 1971 Fossil Coleoptera as indicators of climatic fluctuations during the last glaciation in Britain. *Palaeogeog., Palaeoclimatol., Palaeoecol.* **10**, 87–101.

Craig, A. J. 1972 Pollen influx to laminated sediments: a pollen diagram from northeastern Minnesota. *Ecology* **53**, 46–57.

Craig, A. J. 1973 Studies in the ecological history of south-east Ireland using pollen influx analysis and other methods. Unpublished Ph.D. Thesis, Trinity College, Dublin. 205 pp.

Cushing, E. J. 1967 Late-Wisconsin Pollen Stratigraphy and the Glacial Sequence in Minnesota. In *Quaternary paleoecology* (eds E. J. Cushing & H. E. Wright), pp. 59–88. New Haven and London: Yale University Press.

Davis, M. B. 1969 Climatic changes in southern Connecticut recorded by pollen deposition at Rogers Lake. *Ecology* **50**, 409–422.

Dresser, P. Q. & McAulay, I. R. 1974 Dublin radiocarbon dates: 2. *Radiocarbon* **16**, 6–9.

Gleason, H. 1926 The individualistic concept of the plant association. *Bull. Torrey bot. Club* **53**, 7–26.

Gordon, A. D. & Birks, H. J. B. 1974 Numerical methods in Quaternary palaeoecology. II. Comparison of pollen diagrams. *New Phytol.* **73**, 221–249.

Goulden, C. E. 1964 The history of the Cladoceran fauna of Esthwaite Water (England) and its limnological significance. *Arch. Hydrobiol.* **60**, 1–52.

Harmsworth, R. V. 1968 The developmental history of Blelham Tarn (England) as shown by animal macrofossils with special reference to the Cladocera. *Ecol. Monogr.* **38**, 223–241.

Iversen, J. 1954 The late-glacial flora of Denmark and its relation to climate and soil. *Danm. geol. Unders.* (Raekke II) Nr. **80**, 87–119.

Jessen, K. 1949 Studies in late Quaternary deposits and flora-history of Ireland. *Proc. R. Ir. Acad.* B **52**, 85–290.

Jessen, K. & Farrington, A. 1938 The bogs at Ballybetagh, near Dublin, with remarks on late-glacial conditions in Ireland. *Proc. R. Ir. Acad.* B **44**, 205–260.

Mercer, J. H. 1969 The Allerød oscillation, a European climatic anomaly? *Arctic, Alpine Res.* **1**, 227–234.

Mitchell, G. F. 1940 Studies in Irish Quaternary Deposits: some lacustrine deposits near Dunshaughlin, Co. Meath. *Proc. R. Ir. Acad.* B **46**, 13–37.

Mitchell, G. F. 1941*a* Studies in Irish Quaternary Deposits, No. 3: the reindeer in Ireland. *Proc. R. Ir. Acad.* B **46**, 183–188.

Mitchell, G. F. 1941*b* Some lacustrine deposits near Ratoath, Co. Meath. *Proc. R. Ir. Acad.* B **46**, 173–182.

Mitchell, G. F. 1953 Further identification of macroscopic plant fossils from Irish Quaternary deposits, especially from a Late-glacial deposit at Mapastown, Co. Louth. (Studies in Irish Quaternary Deposits, No. 8.) *Proc. R. Ir. Acad.* B **52**, 225–281.

Mitchell, G. F. 1954 The Late-glacial Flora of Ireland. *Danm. geol. Unders.* (Raekke II), Nr. **80**, 73–86.

Mitchell, G. F. 1973 Fossil pingos in Camaross Townland, Co. Wexford. *Proc. R. Ir. Acad.* B **73**, 269–282.

Mitchell, G. F. 1976 *The Irish landscape.* London: Collins.

Mitchell, G. F. & Parkes, H. M. 1949 The giant deer in Ireland. *Proc. R. Ir. Acad.* B **52**, 291–314.

Mitchell, G. F., Penny, L. F., Shotton, F. W. & West, R. G. 1973 A correlation of Quaternary deposits in the British Isles. *Geol. Soc. Lond. Spec. Rep.* **4**, 99 pp.

Mitchell, G. F. & Watts, W. A. 1970 The history of the Ericaceae in Ireland during the Quaternary Epoch. In *Studies in the vegetational history of the British Isles* (eds D. Walker & R. G. West), pp. 13–21. Cambridge University Press.

Morrison, M. E. S. & Stephens, N. 1965 A submerged Late-Quaternary deposit at Roddans Port on the northeast coast of Ireland. *Phil. Trans. R. Soc. Lond.* B **249**, 221–255.

Pennington, W. 1973 Absolute pollen frequencies in the sediments of lakes of different morphometry. In *Quaternary plant ecology* (eds H. J. B. Birks & R. G. West), pp. 79–104. Oxford: Blackwell Scientific Publications.

Pennington, W. 1975 A chronostratigraphic comparison of Late-Weichselian and Late-Devensian subdivisions, illustrated by two radiocarbon-dated profiles from western Britain. *Boreas* **4**, 157–171.

Ruddiman, W. F. & McIntyre, A. 1973 Time-transgressive deglacial Retreat of polar waters from the North Atlantic. *Quat. Res.* **3**, 117–130.

Singh, G. 1970 Late-glacial vegetational history of Lecale, Co. Down. *Proc. R. Ir. Acad.* B **69**, 189–216.

Smith, A. G. 1961 Cannons Lough, Kilrea, Co. Derry: stratigraphy and pollen analysis. *Proc. R. Ir. Acad.* B **61**, 369–383.

Smith, A. G. 1970 Late- and post-glacial vegetational and climatic history of Ireland: a review. In *Irish geographical studies* (eds N. Stephens & R. E. Glassock), pp. 65–88. Belfast: Department of Geography, Queen's University.

Smith, A. G., Pearson, G. W. & Pilcher, J. R. 1971*a* Belfast radiocarbon dates: III. *Radiocarbon* **13**, 103–125.

Smith, A. G., Pearson, G. W. & Pilcher, J. R. 1971*b* Belfast radiocarbon dates: IV. *Radiocarbon* **13**, 450–467.

Smith, A. G., Pearson, G. W. & Pilcher, J. R. 1973 Belfast radiocarbon dates: V. *Radiocarbon* **15**, 212–228.

Stillman, C. J. 1968 The post-glacial change in sea level in southwestern Ireland: new evidence from fresh water deposits on the floor of Bantry Bay. *Scient. Proc. R. Dubl. Soc.* A **35**, 125–127.

Synge, F. M. 1969 The Würm ice limit in the West of Ireland, pp. 89–92. In *Quaternary geology and climate.* Publication 1701. Washington D.C.: National Academy of Science.

Vokes, E. 1966 The Late- and Post-glacial vegetational history of Killarney, Co. Kerry in south-west Ireland. Unpublished M.Sc. Thesis, Trinity College, Dublin, 181 pp.

Waddington, J. C. B. 1969 A stratigraphic record of the pollen influx to a lake in the Big Woods of Minnesota. *Geol. Soc. Am. Spec. Pap.* **123**, 263–282.

Watts, W. A. 1963 Late-glacial pollen zones in western Ireland. *Ir. Geogr.* **4**, 367–376.

Watts, W. A. 1973 Rates of change and stability in vegetation in the perspective of long periods of time. In *Quaternary plant ecology* (eds H. J. B. Birks & R. G. West), pp. 195–206. Oxford: Blackwell Scientific Publications.

Watts, W. A. & Winter, T. C. 1966 Plant macrofossils from Kirchner Marsh, Minnesota – a palaeoecological study. *Bull. geol. Soc. Am.* **77**, 1339–1360.

Westhoff, V. & van der Maarel, E. 1973 The Braun–Blanquet approach. In *Handbook of vegetation science. Part V. Ordination and classification of communities* (ed. R. H. Whittaker), pp. 617–726. The Hague: Dr W. Junk.

Phil. Trans. R. Soc. Lond. B. **280**, 295–312 (1977) [295]
Printed in Great Britain

The vertebrates of the Last Cold Stage in Britain and Ireland

By A. J. Stuart
Museum of Zoology, University of Cambridge

Until many more dated records are available it will not be possible to follow in much detail the faunal changes through the Last Cold Stage.

Some of the mammal taxa characteristic of the Devensian were already present in the second half of the Ipswichian, probably in response to decreasing forest cover. The Early and Middle Devensian English faunas include extinct animals (e.g. mammoth, woolly rhinoceros) and living animals whose ranges do not overlap at the present day. Many now live in tundra or tundra and boreal forest (e.g. lemmings, reindeer), others in steppe (e.g. red-cheeked suslik, horse), while others have very southern distributions (lion, spotted hyaena). Some animals (e.g. elk) may have been confined to woodland interstadials, but the fauna as a whole is consistent with treeless herbaceous vegetation. Many taxa (e.g. woolly rhinoceros) failed to reach Ireland.

At some time in the Late Devensian before Zone II several large mammals (e.g. spotted hyaena, mammoth) became locally or totally extinct. The elk failed to reach Ireland in the Late Devensian, and reindeer and giant deer are the only known fauna. The beginning of the Flandrian saw the rapid replacement of the northern and steppe fauna by forest animals.

1. Introduction

On the whole the vertebrates of the Last Cold Stage have received relatively little attention in recent years in comparison with the faunas of earlier stages. Consequently the number of reliably stratified and dated records of fossil vertebrates at present available is not adequate as a basis for any detailed analysis of faunal change during this time. Moreover, the faunal lists from many sites need critical revision.

In view of the emerging complex patterns of climatic and vegetational changes during the Devensian this lack of data is unfortunate, since it seems very probable that the history of the vertebrate fauna was correspondingly complex.

In this paper I have tried to make the best use of such information as we have. The data and conclusions presented here should be regarded as provisional and likely to be much modified by future work.

2. The occurrence of fossil vertebrates

Fossil vertebrate material (mostly mammalian) of Devensian age is known from most parts of the British Isles, and is abundantly represented in museum collections. The finds come from river terrace, lacustrine and cave and fissure deposits. At present, however, only the material from a limited number of sites is of very much value because of the considerable uncertainties in dating most of the finds.

The low terrace sands and gravels of many rivers have yielded bones of mammoth and other animals. In a few cases the fossil vertebrate records are more or less associated with radiocarbon-dated organic horizons which have yielded pollen, macroscopic plant remains and beetles.

Lacustrine deposits with fossil vertebrates are known from the Late Devensian. Records from these deposits are ideal in that the sediments are clearly and simply stratified, contain pollen and are generally dateable by radiocarbon.

Cave and fissure deposits are usually far richer in fossil vertebrates than are the deposits of open sites. Carnivora are much better represented in proportion to herbivores than would be expected from the relative population densities of living animals. This is because carnivorous mammals, e.g. spotted hyaena, bear, commonly used caves as dens and when they died contributed their remains to those of their prey accumulating on the cave floor. As a result of this bias and the overall abundance of remains many records of Devensian vertebrates are available only from cave and fissure deposits. Unfortunately there are serious difficulties in dating these fossil assemblages. Very few caves have been excavated with careful regard to the usually complex stratigraphy, and the fossils from different horizons have therefore been confused. In the few cases where cave deposits have been carefully excavated, however, the age of a faunal assemblage can be approximately estimated from the stratigraphical relationships to other fossiliferous horizons, and by comparison with faunas from other sites. Radiocarbon dates on bone collagen are available from a few sites. These are obviously of value, but inherently less reliable than radiocarbon dates based on plant material.

3. Ipswichian

The faunal history of the preceding Ipswichian stage has recently been described by the author (Stuart 1976). It seems relevant here to mention certain aspects of the faunas of this interglacial as they in some ways anticipate those of the Devensian.

The fauna of zone Ip II included temperate mammals such as a rhinoceros (*Dicerorhinus hemitoechus*), straight-tusked elephant (*Palaeoloxodon antiquus*), hippopotamus (*Hippopotamus amphibius*) and fallow deer (*Dama dama*) in association with regional mixed oak forest vegetation. During the second half of the interglacial the temperate taxa were gradually replaced by taxa usually considered to be characteristic of cold stages. At for example Aveley, Essex, both mammoth (*Mammuthus primigenius*) and horse (*Equus caballus*) were present as early as the beginning of zone Ip III. The fauna of the Lower Brickearth at Crayford, Kent, is considered (Stuart 1976) to be of probable late Ipswichian age, possibly late zone Ip IV. If this is correct then northern vole (*Microtus oeconomus*), arctic lemming (*Dicrostonyx torquatus*), Norway lemming (*Lemmus lemmus*), woolly rhinoceros (*Coelodonta antiquitatis*) and musk ox (*Ovibos moschatus*) were present before the end of the interglacial. The fauna, however, differs from those of Devensian age in the continued presence of temperate mammals such as *D. hemitoechus*, the absence of reindeer (*Rangifer tarandus*), and the occurrence throughout the deposits of abundant temperate and southern molluscs, notably *Corbicula fluminalis*, which suggest that summer temperatures were still high. The early occurrence of such animals as lemmings and woolly rhinoceros may have been related to a regional or local vegetation that was largely open, rather than giving any direct indication of climatic deterioration.

4. Early and Middle Devensian

A list of Early and Middle Devensian mammal taxa and their occurrences at some of the more important sites are given in table 1.

(a) Britain

An Early Devensian fauna is available from terrace deposits of the River Wissey at Wretton, Norfolk (West *et al.* 1974). The vertebrate material is being studied by K. A. Joysey, and only a preliminary faunal list is so far available. The vertebrates recorded (K. A. Joysey, personal communication) include arctic fox (*Alopex lagopus*), mammoth (*Mammuthus primigenius*), horse (*Equus caballus*) (not included in earlier lists), reindeer (*Rangifer tarandus*) and bison (*Bison priscus*). The pollen spectra record a long period of herbaceous vegetation, interrupted by two (pollen-based) interstadials with boreal forest. It should prove possible to relate many of the vertebrate finds to the stratigraphy. The fauna seems consistent with open herb vegetation and includes no characteristically woodland animals.

A skull of *Bison priscus* was recovered from deposits of probable Early Devensian age at Sidgwick Avenue, Cambridge (Lambert, Pearson & Sparks 1963).

Middle Devensian faunas are available from a number of sites, of which the most important described so far is Upton Warren, Worcestershire (Coope, Shotton & Strachan 1961). At this site numerous bones and teeth of large mammals were found in terrace deposits of the River Salwarpe. Remains of small vertebrates were recovered from silt bands, one of which gave a radiocarbon date of 42 000 a B.P. Only a few stickleback remains were, however, actually found in the dated horizon and the large-mammal remains were mostly found in the gravels by workmen. The date can only, therefore, be taken as a general indication of the age of the faunal assemblage. The fauna includes arctic lemming (*Dicrostonyx torquatus*), mammoth, horse, woolly rhinoceros (*Coelodonta antiquitatis*), reindeer and bison. Pollen spectra from the silt bands indicate a treeless herbaceous vegetation.

From the abundance of dung and carcass beetles, Coope *et al.* suggested that the pools, represented by the fossiliferous silt bands, were used as waterholes by large mammals.

A mammoth tusk from Oxbow near Leeds was recovered from organic silts which gave a radiocarbon date of about 38 600 a B.P. (Gaunt, Coope & Franks 1970). The pollen again indicates open herbaceous vegetation.

At Fladbury, Worcestershire, finds of mammoth, woolly rhinoceros and horse from river gravels are broadly associated with a radiocarbon date of about 38 000 a B.P. from a peat horizon within the gravels (Coope 1962).

Many cave deposits, probably covering Early and/or Middle Devensian time, are known from Britain.

The series of deposits in Tornewton Cave, Devon (Sutcliffe & Zeuner 1962; Kowalski 1967) apparently spans the time from the Wolstonian to the Flandrian. Overlying a horizon of Ipswichian age with hippopotamus and other temperate mammals is a series of deposits of probable Devensian age. The Elk Stratum, a cave earth resting on frost-shattered limestone breccia, has yielded arctic lemming, northern vole (*Microtus oeconomus*), spotted hyaena, woolly rhinoceros, horse, reindeer, red deer (*Cervus elaphus*) and elk (*Alces alces*). Man is represented by a few nondescript artefacts.

The fauna of the overlying Reindeer Stratum is rather similar but lacks both elk and red deer. On the other hand bank vole (*Clethrionomys glareolus*) and mole (*Talpa europaea*) are recorded. Man is represented by flint flakes and an incisor.

A preliminary account of the stratigraphy and vertebrate remains of the cave or rock shelter at Picken's Hole, Compton Bishop, Somerset, was given by Tratman (1964). The faunas were

TABLE 1. MAMMALS RECORDED FROM THE EARLY AND MIDDLE DEVENSIAN

	Wretton (Early Devensian)	Upton Warren (Middle Devensian)	Tornewton Cave, Elk Stratum	Tornewton Cave, Reindeer Stratum	Picken's Hole, Layer 5	Picken's Hole, Layer 3	Castlepook Cave, Ireland	other sites	
Insectivora									
1. *Sorex* cf. *araneus* L., common shrew	—	—	—	—	—	—	—	+	Ponders End
2. *Talpa europaea* L., mole	—	—	—	+	—	—	—		
Primates									
3. *Homo sapiens* L., man (bones or artefacts)	—	—	+	+	—	+	—		
Lagomorpha									
4. *Lepus timidus* L., mountain hare	—	—	—	+	+	—	+	+	e.g. Robin Hood Cave
Rodentia									
5. *Spermophilus major* Pallas, red-cheeked suslik	—	—	—	—	—	+	—	+	e.g. Fisherton
6. *Dicrostonyx torquatus* (Pallas), arctic lemming	—	+	+	—	—	—	+	+	e.g. Angel Road
7. *Lemmus lemmus* (L.), Norway lemming	—	—	—	—	—	—	+	+	e.g. Hutton Cave†
8. *Clethrionomys glareolus* (Schreber), bank vole	—	—	—	+	—	—	—		
9. *Arvicola terrestris* L., water vole	—	—	+	+	—	—	—		
10. *Microtus oeconomus* (Pallas) northern vole	—	—	+	+	+	—	—		
11. *Microtus gregalis* (Pallas), tundra vole	—	—	+	+	+	+	—		
12. *Microtus agrestis* (L.), field vole	—	—	+	+	—	—	—		
13. *Microtus* sp., a vole	—	+	—	—	—	—	—		
Carnivora									
14. *Canis lupus* L., wolf	+	—	+	—	+	—	+		
15. *Alopex lagopus* (L.), arctic fox	+	—	—	—	—	+	+		
16. *Vulpes vulpes* (L.), red fox	—	—	+	—	+	+	+		
17. *Ursus arctos* L., brown bear	—	—	—	—	+	+	+		
18. *Ursus maritimus* Phipps, polar bear	—	—	—	—	—	—	—	+	Kew Bridge†

19. *Ursus* sp., a bear	+	—	+	+	—	—	—		
20. *Mustela erminea* L., stoat	—	—	—	—	—	—	+		
21. *Gulo gulo* (L.), glutton	—	—	—	—	—	—	—	+	e.g. Banwell Cave,† Plas Heaton Cave†
22. *Crocuta crocuta* Erxleben, spotted hyaena	—	—	+	+	—	+	+		
23. *Panthera pardus* (L.), leopard	—	—	—	—	—	—	—	+	e.g. Banwell Cave†
24. *Panthera leo* (L.), lion	—	—	—	—	—	+	—	+	e.g. Fisherton
Proboscidea									
25. *Mammuthus primigenius* Blumenbach, mammoth	+	+	—	—	—	+	+	+	e.g. Fladbury, Oxbow
Perissodactyla									
26. *Equus caballus* L., horse	+	+	+	+	—	+	—	+	e.g. Fladbury, Shandon Cave
27. *Coelodonta antiquitatus* Blumenbach, woolly rhinoceros	?	+	+	+	—	+	—	+	e.g. Fladbury
Artiodactyla									
28. *Megaloceros giganteus* (Blumenbach), giant deer	—	—	—	—	—	—	+	+	e.g. Robin Hood Cave
29. *Cervus elaphus* L., red deer	—	—	+	—	+	+	—		
30. *Alces alces* (L.), elk	—	—	+	—	—	—	—		
31. *Rangifer tarandus* (L.), reindeer	+	+	+	+	+	+	+	+	e.g. Fladbury
32. *Bison priscus* Bojanus, a bison	+	+	—	—	—	—	—	+	e.g. Sidgwick Avenue
33. *Bison* and/or *Bos* sp., bison or aurochs	—	—	+	+	+	+	—		
34. *Saiga tartarica* (L.), saiga antelope	—	—	—	—	—	—	—	+	Twickenham†
35. *Ovibos moschatus* L., musk ox	—	—	—	—	—	—	—	+	Ightham†, Fisherton

† Age uncertain.

later revised by the author (see list in Stuart 1974). At this site two fossiliferous cave earths are sandwiched between, and separated by, frost-shattered limestone breccias. The fauna of the earlier horizon, Layer 5, includes wolf (*Canis lupus*), red fox (*Vulpes vulpes*), brown bear (*Ursus arctos*), red deer and reindeer. The fauna of the younger horizon, Layer 3, includes red-cheeked ground squirrel (*Spermophilus major*) (identified by D. F. Mayhew 1975), arctic fox, brown bear, spotted hyaena, lion (*Panthera leo*), mammoth, horse, woolly rhinoceros, red deer, and reindeer. The presence of man is attested by a few indeterminate artefacts and two incisor teeth. The faunas from the sites discussed above are fairly typical and include nearly all the taxa recorded from the Early and Middle Devensian. A few additional sites are, however, also worth mentioning as they have produced records of rare taxa.

The 'Arctic Bed' peats and gravels of the River Lea flood plain terrace of Enfield and Edmonton, North London, have yielded arctic lemming, from Angel Road, Edmonton, mammoth, woolly rhinoceros and horse (Warren 1912; Hinton 1912) and a record of shrew (*Sorex* cf. *araneus*, equals *S. kennardi* Hinton) (Hinton 1911) from Ponders End, Enfield.

In the last century a number of mammalian remains were discovered at Fisherton, near Salisbury, Wiltshire. The finds included mammoth, woolly rhinoceros, reindeer and others, and of particular interest: red-cheeked suslik (identified by D. F. Mayhew 1975) and musk ox (Dawkins & Reynolds 1872–1939; Simons 1962). An Early or Middle Devensian age seems probable for this faunal assemblage.

A single record of saiga antelope (*Saiga tartarica*) is known from River Thames gravels at Twickenham, London (Dawkins & Reynolds 1872–1939). There appears to have been no associated fauna.

The record of polar bear (*Ursus maritimus*) is based solely on an ulna from a railway cutting in low terrace gravels of the River Thames near Kew Bridge, London. The bone was compared in detail with a number of fossil and recent bears by Kurtén (1964) who found it differed from recent polar bear only in its greater size. A number of reindeer remains have also been found at Kew.

Fissures in Lower Greensand at Ightham, Kent, have produced large numbers of vertebrate remains (Newton 1894), probably covering much of Devensian and Flandrian time. The material from different periods appears to have been mixed together when found, but the records of red-cheeked suslik, woolly rhinoceros and others are most probably of Early or Middle Devensian age. Simons (1962) mentioned a vertebra of musk ox from this site.

The caves of Creswell Crags, Derbyshire, have yielded abundant vertebrate remains, apparently in the main of Early or Middle Devensian age. The records of sabre-tooth cat (*Homotherium latidens*) from Robin Hood and Pin Hole Caves, seem out of place, since this genus is not otherwise recorded in Europe later than the Middle Pleistocene. The third English record from Kent's Hole, Devon, could well be of Middle Pleistocene age. The record from Robin Hood Cave is based on the crown of a canine (Dawkins 1877) which was excavated in Dawkins' presence and was directly associated with woolly rhinoceros, mammoth, etc. and man-made flint flakes. A possible explanation is that the sabre-tooth cat remains were collected by man, and are actually much older than the deposits in which they were found.

Remains of glutton (*Gulo gulo*) are known from a very few cave sites in Britain. At Plas Heaton Cave, near Cefn St Asaph, North Wales, it was recorded in association with wolf, bear, horse and reindeer (Dawkins 1871). A Devensian age seems probable, but by no means certain.

Leopard (*Panthera pardus*) is recorded from Robin Hood Cave (Dawkins 1877) and also Banwell and Bleadon Caves, Somerset (Dawkins & Sanford 1866–72).

Some of the finds of cave bear (*Ursus spelaeus*) from south-west England, e.g. Kent's Hole, Torquay, Devon, and Wookey Hole, Mendip, Somerset, are possibly of Devensian age.

Giant deer (*Megaloceros giganteus*) is included in a number of faunal lists from various sites in England, e.g. Robin Hood Cave (Dawkins 1877), although many of these records are suspect because of likely confusion of giant deer remains with these of large red deer or with elk.

Most of the fossil localities are in England, although a number of records, mostly from cave deposits, are known for Wales. The sparse list from Scotland includes arctic lemming, mammoth, woolly rhinoceros, reindeer and giant deer (Delair 1969), some of which could be of Late Devensian age.

(*b*) *Ireland*

Without doubt the most important pre-Late Devensian fossil vertebrate site in Ireland is Castlepook Cave, County Cork. The fauna from this cave was listed by Scharff, Seymour & Newton (1918). The site and fauna have recently been reassessed by A. J. Sutcliffe (in preparation). The mammals believed to come from an earlier set of deposits include arctic lemming, Norway lemming, wolf, red fox, arctic fox, brown bear, spotted hyaena, mammoth, reindeer and giant deer. The fossils were recovered from a series of sands and stalagmite floors, and are probably not contemporaneous. Remains of animals of Flandrian age, including domestic animals, were recovered from later deposits within the cave. Wood mouse (*Apodemus sylvaticus*), an animal characteristic of temperate woodland, was listed by Scharff *et al.* from the earlier deposits, but it seems much more likely to date from the Flandrian and I have therefore not included it in table 1. A radiocarbon date of about 33500 years from collagen of a mammoth femur suggests a Middle Devensian age for the older faunal assemblage.

One animal not recorded from Castlepook Cave, but apparently present in Ireland during the Early or Middle Devensian is horse. Bones of horse in association with remains of hare, bear, mammoth and reindeer were found in stalagmite in a quarry at Shandon Cave, near Dungarven in County Waterford (Brenan 1860; Carte 1860). Lynx (*Lynx lynx*) is recorded from Kilgreany Cave, County Waterford (Savage 1966). It could be either Devensian or Flandrian.

The Devensian mammals of Ireland have been discussed by Savage (1966) and Mitchell (1969).

(*c*) *Distributions and ecology*

Many of the mammals represented as fossils in the Early and Middle Devensian of Britain have very different distributions at the present day.

(i) *Arctic and boreal animals*

The arctic fox (*Alopex lagopus*) has a holarctic circumpolar distribution (figure 1). It is essentially confined to the tundra but its range extends southwards to the northern part of the boreal forest zone in winter (Corbet 1966). It overlaps the northern part of the range of red fox (*Vulpes vulpes*). Voles, and lemmings are an important source of food.

The arctic lemming (*Dicrostonyx torquatus*) also has a holarctic circumpolar distribution confined to the tundra but entirely absent from Fennoscandia (figure 2). The Norway lemming (*Lemmus lemmus*) together with the Siberian and North American forms, often regarded as a different closely-related species *Lemmus sibericus* Kerr, has a similar distribution to the arctic lemming but extends further south into the northern part of the boreal forest zone, and does occur in Fennoscandia (Corbet 1966; Ellerman & Morrison-Scott 1966). Both lemmings burrow extensively and eat grasses, sedges and other available plant food (Banfield 1974; Corbet 1966).

The northern vole (*Microtus oeconomus*) occurs in tundra, boreal forest, wooded steppe and the northern part of the deciduous forest in Eurasia, and also the tundra of northwest North America (Corbet 1966; Banfield 1974). The tundra vole *Microtus gregalis* has a wide distribution in Asia, apparently occurring in tundra, boreal forest and steppe (Ellerman & Morrison-Scott 1966). This, or a closely-related species (*Microtus miurus* Osgood) occurs also in northwest North America (Banfield 1974).

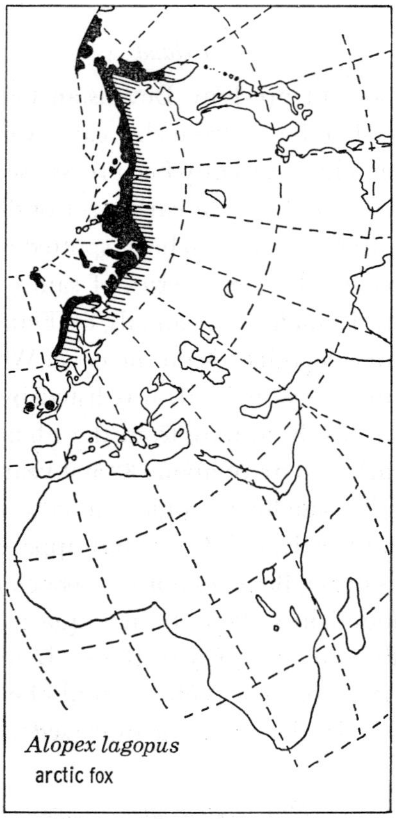

FIGURE 1. Present Old World distribution of arctic fox (after Corbet 1966). Summer range shown in solid black, winter extension hatched. Fossil occurrences in England and Ireland indicated by dots.

The reindeer (*Rangifer tarandus*), a very characteristic animal of Devensian faunas, is nowadays a holarctic circumpolar species, occurring in tundra, boreal forest and mountain regions (figure 3). It has been extensively domesticated in the southern part of the Eurasian range but the distribution of the species has probably been little altered. The tundra populations commonly migrate long distances, up to 300 km, south in winter (Banfield 1974). From the small size of the shed antlers in Tornewton Cave Reindeer Stratum, Sutcliffe & Zeuner (1962) concluded that reindeer were only present at the site in the summer when the females and young shed their antlers. Presumably they migrated south to the Continent for the winter. Reindeer feed on sedges, grasses, other herbs and the shoots of shrubs in summer. Lichens, often dug from under the snow, are very important in the winter diet (Corbet 1966; Banfield 1974).

The musk ox (*Ovibos moschatus*) no longer occurs in Eurasia but is confined to the tundra of North America and Greenland. The food includes grasses, sedges and various shrubs (Banfield 1974).

The polar bear (*Ursus maritimus*) is essentially a marine animal confined to the Arctic Ocean and adjacent coasts. It is seldom reported more than about 100–200 km from the sea (Corbet 1966; Banfield 1974). Very few fossil occurrences are known, although it has been found in Denmark and Scania (Kurtén 1964). The single record from Kew Bridge may represent a solitary wandering individual, perhaps dating from a glacial maximum.

FIGURE 2. Present Old World distribution of arctic lemming (after Banfield 1974). Fossil occurrences in England and Ireland indicated by dots.

The elk, or moose in North America (*Alces alces*), has a holarctic distribution (figure 4) and is mostly confined to boreal forest areas at the present day, although Roman writers record it from North Germany and France about 2000 years ago. It is especially associated with water and marshy forested country. In summer elk feed on herbs, aquatic and riverside plants, and browse on shrubs and deciduous trees. In winter they subsist on shoots and the bark of trees (Corbet 1966). The record of this species from the Elk Stratum of Tornewton Cave (Sutcliffe & Zeuner 1962) therefore seems incompatible with open herbaceous vegetational conditions and suggests that woodland grew in the area at this time.

The glutton or wolverine (*Gulo gulo*) is also primarily found in boreal forest, although migrating northwards onto the tundra in summer (Corbet 1966). It has a circumpolar holarctic distribution.

(ii) *Steppe animals*

The red-cheeked suslik (*Spermophilus major*) has a well-defined steppe distribution at the present day (figure 5). Other species of *Spermophilus* occur in steppe, tundra and mountain areas of Eurasia and North America (Ellerman & Morrison-Scott 1966).

The saiga antelope (*Saiga tartarica*) is found nowadays in the dry steppe and subdesert areas of the southern U.S.S.R. Kahlke (1975) has plotted the recent and fossil occurrences, and his map indicates that during the Last Cold Stage its range was extended westwards across Europe to England, and northwards and eastwards to Siberia and northwest North America.

FIGURE 3. Present Old World distribution of reindeer (modified from Banfield 1974). Fossil occurrences in England and Ireland indicated by dots.

The original distribution of the horse (*Equus caballus*) seems impossible to reconstruct with any precision and it is virtually extinct as a wild animal. It appears to have been primarily a steppe-dweller, although according to some authors (e.g. Brink 1967) there may have been a woodland ecotype living in the forest of Poland and elsewhere to within the last hundred years or so. An engraving made by Upper Palaeolithic man on a bone from Robin Hood Cave (Dawkins 1877) shows a horse with an erect mane, very similar to those depicted in French cave paintings and to the living Mongolian wild horse (*Equus caballus*, or *E. przewalskii* Poliakov).

(iii) *Temperate animals*

The red deer (*Cervus elaphus*) is widely distributed in Eurasia, but its original range is difficult to reconstruct accurately because of interference by man (figure 6). It is or was present in the deciduous woodland of Europe, montane woodland and the transition between steppe and boreal forest in Asia. It is entirely absent from tundra and most of the boreal forest zones. This distribution is paralleled in North America, where the red deer is also known as 'elk' or wapiti (Banfield 1974). It feeds largely by grazing and browsing on the leaves of deciduous trees (Corbet 1966).

FIGURE 4. Present Old World distribution of elk (from data in Ellerman & Morrison-Scott 1966). The fossil occurrences in England are indicated by a dot.

The Early or Middle Devensian records of red deer pose some interesting problems. Perhaps it migrated into England only during woodland interstadials. Alternatively it may have been adapted to the treeless herb vegetation of much of the stage, possible in the form of a distinct ecotype. The survival of red deer on the Scottish moors at the present day gives some insight into the potential adaptability of this species.

The record of mole (*Talpa europaea*) from Tornewton Cave Reindeer Stratum is rather surprising since nowadays it does not occur further north than southern Scandinavia (Brink 1967). Since the animal lives a subterranean existence feeding on soil invertebrates, its occurrence implies that the ground was not permanently frozen at that time.

(iv) *Southern animals*

The lion (*Panthera leo*) is now found only in Africa south of the Sahara, and a small area of India (figure 7). As also shown in the map it formerly occurred over most of Africa except areas of desert and tropical rain forest across southwest Asia to India. According to Herodotus and other classical writers it occurred in northern Greece about 2500 years ago. It is an animal of open or lightly wooded grassland, avoiding dense forest and it preys on a variety of large ungulates (Dorst & Dandelot 1970).

The spotted hyaena (*Crocuta crocuta*), both a predator and scavenger, has a very similar distribution to the lion, but is restricted to Africa south of the Sahara (Dorst & Dandelot 1970).

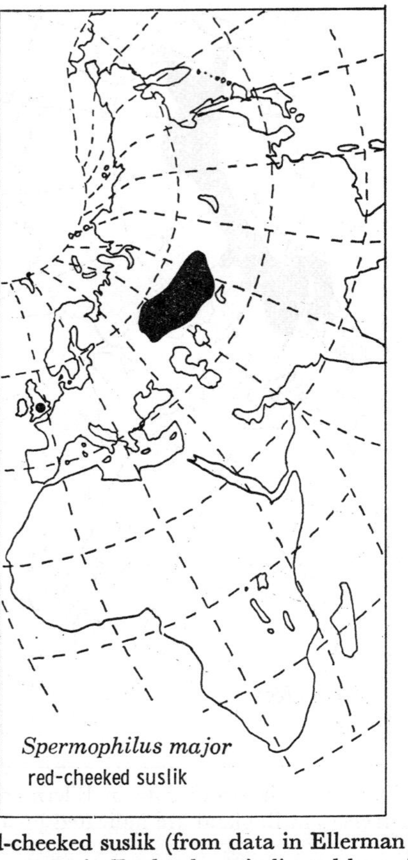

FIGURE 5. Present distribution of red-cheeked suslik (from data in Ellerman & Morrison-Scott 1966). The fossil occurrences in England are indicated by a dot.

(v) *Extinct animals*

There were also four extinct species of large mammals present during the Devensian: mammoth, woolly rhinoceros, giant deer and bison (*Bison priscus*). Of these at least the first two were probably tolerant of extreme cold conditions since we know from finds of carcasses in Siberia and Poland that both animals were protected by long shaggy coats. Nevertheless mammoth, bison and giant deer are also recorded from the temperate parts of interglacials, e.g. at Barrington, Cambridgeshire, they were present in zone IIb of the Ipswichian in association with local deforestation of the river floodplain (Gibbard & Stuart 1975).

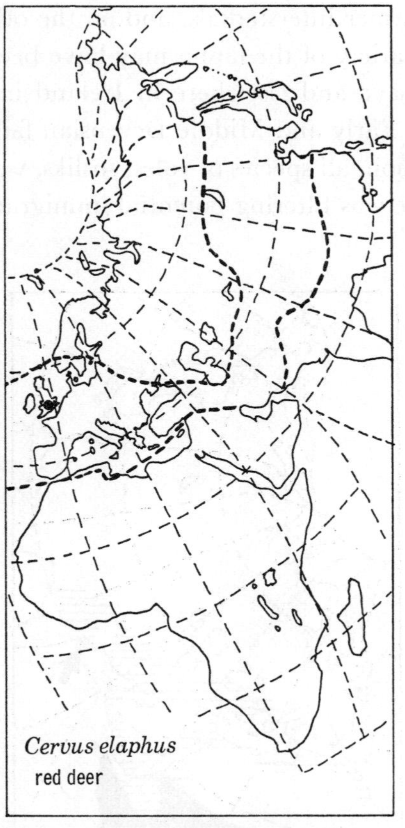

FIGURE 6. Present Old World distribution of red deer (from data in Ellerman & Morrison-Scott 1966). Only the approximate northern and southern limits are shown. The fossil occurrences in England are indicated by a dot.

(*d*) *Faunal history*

The faunas of the Early and Middle Devensian as a whole, and apparently the faunas of particular horizons, therefore comprise a mixture of faunal elements not found anywhere at the present day. Northern animals of tundra and boreal forest are especially well represented as might be expected, but steppe and southern animals were also present. Such animals as elk, mole and perhaps giant deer and red deer may have been confined to woodland episodes or warmer intervals. The rest of the fauna seems to fit in well with the picture of open herbaceous vegetation present throughout much of the Early and Middle Devensian (West 1977, this volume). The particular combination of climatic and vegetational conditions apparently suited animals now confined either to steppe, or tundra as well as the extinct taxa. Both lion and spotted hyena avoid dense forest at the present day and their present southern distribution may be due to the former existence of dense forest in much of Europe, or to the influence of man since both were present in England during the Ipswichian. Similar faunas to those from England are recorded from as far away as Siberia (Klein 1971).

As they are warm-blooded the distributions of mammals are probably not in most cases directly influenced by climate, but they are very much controlled by vegetational conditions. Throughout much of the Devensian one can picture herds of mammoth, woolly rhinoceros, horse, reindeer and bison grazing on the treeless herbaceous vegetation. No doubt many of them underwent considerable seasonal migrations. Woodland taxa such as elk and perhaps red deer may have appeared during warmer interstadials, and on the other hand at times of increased cold and glacial maxima the variety of the fauna may have been reduced.

The faunas of Castlepook Cave and elsewhere in Ireland indicate that a number of taxa present in England during the Early and Middle Devensian failed to cross the land bridge to Ireland. Notable absences are lion, all species of voles, susliks, woolly rhinoceros, bison and red deer. Evidently the land bridge was filtering potential immigrants in some way.

Figure 7. Present distribution of lion, shown in solid black (after Dorst & Dandelot 1970; Ellerman & Morrison-Scott 1966). The isolated population in India is arrowed. The hatched area is intended to give a general idea of the areas in which the lion has become extinct within the last 2500 years. The fossil occurrences in England are indicated by a dot.

5. Late Devensian

(*a*) *Britain*

A few years ago at High Furlong near Blackpool, Lancashire, an elk skeleton was discovered in Zone II lacustrine deposits (Hallam, Edwards, Barnes & Stuart 1973). Radiocarbon dates of about 12200 and 11700 a B.P. were obtained from plant material from the base and top respectively of the Zone II organic muds. A similar find had been made in 1939 at Neasham, County Durham (Trechmann 1939; Blackburn 1952). At both sites the pollen spectra indicate herb vegetation with some birch woods in zone L De II. The High Furlong elk had been hunted, but not butchered, by man. Two barbed points were found with the skeleton and there are numerous injuries on the bones.

Late Devensian muds, probably dating from the end of zone L De II, at Flixton near Star Carr, Yorkshire, yielded three partial skeletons of horse, together with bird bones and a few flint flakes (Clark 1954). The pollen spectra (Godwin, in Clark 1954) indicate open grassy vegetation with perhaps a little tree birch.

Calcareous muds of Late Devensian Zone III age, presenting former pools in the River Lea floodplain, at Nazeing, Essex, yielded frog (*Rana temporaria*), toad (*Bufo bufo* (L.)), common lizard (*Lacerta vivipara* Jacquin), arctic lemming, water vole (*Arvicola terrestris*), northern vole and tundra vole (Allison, Godwin & Warren 1952). The pollen spectra indicate open grassy vegetation. Norway lemming and northern vole are also recorded from muds older than the Zone III horizon.

A faunal assemblage of probable Late Devensian age, including arctic fox, horse and reindeer, was found in a series of sands and limestone breccias banked up against an old limestone sea cliff at Brean Down, Somerset (ApSimon, Donovan & Taylor 1961).

Many English cave deposits containing vertebrate remains of probable Late Devensian age have been excavated, but so far few of the published faunal lists can be relied on to date from a single horizon. One reason for this is the likelihood of contamination by Flandrian material. Animals recorded from Aveline's Hole, Somerset (Davies 1922; Hinton 1924) include varying hare (*Lepus timidus*), pika (*Ochotona pusilla* Pallas), arctic lemming, Norway lemming, brown bear, lynx (*Felis lynx* L.), horse, red deer and reindeer. A number of human skeletons and artefacts were recovered from this site.

D. Bramwell (personal communication) lists a number of animals from Ossom's Cave in the Manifold Valley, Derbyshire. A radiocarbon date of 10590 ± 70 a B.P. (GrN 7000) has been obtained from the collagen of a reindeer bone. A preliminary faunal list from the same horizon as the dated bone includes eagle owl (*Bubo bubo*), grouse (*Lagopus lagopus*), ptarmigan (*Lagopus mutus*), black grouse (*Lyrurus tetrix*), corncrake (*Crex crex*), jackdaw (*Corvus monedula*), mistle thrush (*Turdus viscivorus*), arctic lemming, Norway lemming, various voles, horse, reindeer and red deer. Birds of probable Late Devensian age from a number of English cave sites were discussed by Bramwell (1960).

(*b*) *Ireland*

Many finds of giant deer and reindeer have been made in the zone L De II marls beneath the peat bogs in Ireland, e.g. at Ballybetagh near Dublin. The occurrences were discussed in detail by Mitchell (1941) and Mitchell & Parkes (1949). There is no indication of the presence of man at any of these sites or indeed anywhere in Ireland before the Flandrian.

According to Jessen (1949) the Late Devensian Zone II vegetation of much of Ireland was almost treeless herb communities with *Empetrum* heaths in the west. Mitchell & Parkes suggested that the absence of giant deer and reindeer finds in the west was due to these animals avoiding the areas of heath vegetation. Giant deer is also recorded from Zone II deposits at Ballaugh, Isle of Man (Mitchell 1958).

(*c*) *Faunal history*

The most striking feature of faunas dating from the later part of the Late Devensian is the absence of many of the characteristic larger mammals of the Early and Middle Devensian, i.e. mammoth, woolly rhinoceros, bison, musk ox, lion and spotted hyaena. The first three appear to have been extinct in Europe by this time and mammoth and woolly rhinoceros were totally extinct by the end of the Last Cold Stage. Other large mammals, e.g. horse, reindeer and giant deer, survived throughout all or much of the Late Devensian. The small-mammal fauna was little affected, and the pika, so prominent in Late Devensian cave deposits in Britain, has not so far been recorded from earlier in the stage. The much better known and dated Late Weichselian faunas from North Germany and Denmark (Degerbøl 1964) and numerous other continental sites similarly lack mammoth etc.

The almost total glaciation of Ireland at the beginning of the Late Devensian presumably exterminated most, if not all, of the Irish vertebrate fauna. Giant deer and reindeer therefore must have repopulated Ireland via Britain after the ice had retreated.

The extinction of many large mammals in Britain and Europe towards the end of the Last Cold Stage is best viewed in a wider context as part of the general phenomenon of the accelerated late Pleistocene extinction of large mammals in Eurasia, North America and elsewhere. Similar extinctions did not occur for example during the preceding (Wolstonian) cold stage (Stuart 1974). The hypothesis for man directly or indirectly exterminating many large mammals has been advocated by a number of authors, while others favour climatic and vegetational changes (see Martin & Wright 1967).

The Late Devensian fauna as a whole still had a major northern component including ptarmigan (now found in tundra and mountain areas of Europe including the Scottish Highlands), lemmings, northern vole, tundra vole and reindeer.

Steppe animals are represented by the pika, which has a somewhat similar distribution to the red-cheeked suslik (figure 5), and horse.

In England the northward spread of birch woodland in Zone II was accompanied by the immigration of elk. It was probably absent from the open herbaceous vegetation of Zone III, but returned with the trees at the beginning of the Flandrian, e.g. at Star Carr, Yorkshire (Clark 1954).

In Ireland the many finds of giant deer and reindeer from Zone II comprise all that is known of the Late Devensian fauna, although some of the cave finds are no doubt also of this age. The elk evidently failed to reach Ireland, suggesting that as in the Middle Devensian environmental conditions in the region of the land bridge were in some way unsuitable for elk, i.e. it was acting as a filter. Giant deer is not known from deposits later than Zone II. Presumably it could not adapt to the harsher climatic and vegetational conditions of Zone III.

The climatic amelioration and spread of forests at the beginning of the Flandrian saw the rapid disappearance of reindeer, lemmings, northern vole and tundra vole. Horse is recorded from the early Flandrian at Thatcham, Berkshire (Wymer 1962) and pika (*Ochotona* sp.) appears to have survived until Zone V (Fl Ib) at Nazeing, Essex, although the faunal remains may have been reworked from Late Devensian deposits (Allison, Godwin & Warren 1952).

I thank Mr D. Bramwell and Dr A. J. Sutcliffe for unpublished information on Ossum's Cave and Castlepook Cave respectively. I am also grateful to Professor G. F. Mitchell, F.R.S. for drawing my attention to the record of horse from Shandon Cave.

References (Stuart)

Allison, J., Godwin, H. & Warren, S. H. 1952 Late-glacial deposits at Nazeing in the Lea Valley, North London. *Phil. Trans. R. Soc. Lond.* B **236**, 169–240.

ApSimon, A. M., Donovan, D. T. & Taylor, H. 1961 The stratigraphy and archaeology of the late-glacial and post-glacial deposits at Brean Down, Somerset. *Proc. speleol. Soc.* **9**, 67–136.

Banfield, A. W. F. 1974 *The mammals of Canada.* Toronto: University of Toronto Press.

Blackburn, K. B. 1952 The dating of a deposit containing an elk skeleton at Neasham near Darlington, County Durham. *New Phytol.* **51**, 364–377.

Bramwell, D. 1960 Some research into bird distribution in Britain during the late-glacial and post-glacial periods. *Bird Rep. Merseyside Nat. Ass.* 1959–60, 51–58.

Brenan, E. 1860 Notice of the discovery of extinct elephant and other animal remains, occurring in a fossil state under limestone, at Shandon, near Dungarvan, County of Waterford. *J. R. Dublin Soc.* **2**, 344–350.

Brink, F. H. van den. 1967 *A field guide to the mammals of Britain and Europe.* London: Collins.

Carte, A. 1860 Description of the fossil discovery by Mr Brenan at Shandon near Dungarvan. *J. R. Dublin Soc.* **2**, 351–357.

Clark, J. G. D. C. 1954 *Excavations at Star Carr.* Cambridge University Press.

Coope, G. R. 1962 A Pleistocene coleopterous fauna with arctic affinities from Fladbury, Worcestershire. *Q. Jl geol. Soc. Lond.* **118**, 103–123.

Coope, G. R., Shotton, F. W. & Strachan, I. 1961 A late Pleistocene fauna and flora from Upton Warren, Worcestershire. *Q. Jl geol. Soc. Lond.* **118**, 103–123.

Coope, G. R., Shotton, F. W. & Strachan, I. 1961 A late Pleistocene fauna and flora from Upton Warren, Worcestershire. *Phil. Trans. R. Soc. Lond.* B **244**, 379–421.

Corbet, G. B. 1966 *The terrestrial mammals of western Europe.* London: Foulis.

Davies, J. A. 1922 Aveline's Hole, Burrington Combe. *Proc. speleol. Soc.* **1**, 61–68.

Dawkins, W. B. 1871 On the discovery of the glutton (*Gulo luscus*) in Britain. *Q. Jl geol. Soc. Lond.* **27**, 406–410.

Dawkins, W. B. 1877 On the mammal-fauna of the caves of Creswell Crags. *Q. Jl geol. Soc. Lond.* **33**, 589–612.

Dawkins, W. B. & Reynolds, S. H. 1872–1939 The British Pleistocene Mammalia. 3. The British Pleistocene Artiodactyla. *Palaeontogr. Soc.* (*Monogr*).

Dawkins, W. B. & Sanford, W. A. 1866–72 The British Pleistocene Mammalia. A. The British Pleistocene Felidae. *Palaeontogr. Soc.* (*Monogr*).

Degerbøl, M. 1964 Some remarks on late and post-glacial vertebrate fauna and its ecological relations in northern Europe. *J. Anim. Ecol.* **33** (Supplement), 71–85.

Delair, J. B. 1969 North of the hippopotamus belt: a brief review of Scottish fossil mammals. *Bull. Mammal Soc. Br. Isl.* **31**, 16–21.

Dorst, J. & Dandelot, P. 1970 *A field guide to the larger mammals of Africa.* London: Collins.

Ellerman, J. R. & Morrison-Scott, T. C. S. 1966 *Checklist of palaearctic and Indian Mammals 1758–1946.* British Museum (Natural History).

Gaunt, G. D., Coope, G. R. & Franks, J. W. 1970 Quaternary deposits at Oxbow opencast coal site in the Aire Valley, Yorkshire. *Proc. Yorks. geol. Soc.* **38**, 175–200.

Gibbard, P. L. & Stuart, A. J. 1975 Flora and vertebrate fauna of the Barrington Beds. *Geol. Mag.* **112**, 493–501.

Hallam, J. S., Edwards, B. J. N., Barnes, B. & Stuart, A. J. 1973 The remains of a Late Glacial elk associated with barbed points from High Furlong near Blackpool, Lancashire. *Proc. prehist. Soc.* **39**, 100–128.

Hinton, M. A. C. 1911 The British fossil shrews. *Geol. Mag.* **8**, 529–539.

Hinton, M. A. C. 1912 Note on the lemming remains from the Arctic Bed at Angel Road. *Q. Jl geol. Soc. Lond.* **68**, 249.

Hinton, M. A. C. 1924 Notes on the rodent remains from Aveline's Hole. *Proc. speleol. Soc.* **2**, 34–37.

Jessen, K. 1949 Studies in the late-Quaternary and flora history of Ireland. *Proc. R. Ir. Acad.* B **52**, 85–290.

Kahlke, H. D. 1975 Der Saiga-Fund von Bottrop/Westfalen. *Quartär* **26**, 135-146

Klein, R. R. 1971 The Pleistocene prehistory of Siberia. *Quat. Res.* **1**, 133–161.

Kowalski, K. 1967 *Lagurus lagurus* (Pallas 1773) and *Cricetus cricetus* (L. 1758) (Rodentia, Mammalia) in the Pleistocene of England. *Acta. zool. cracov.* **12**, 111–122.

Kurtén, B. 1964 The evolution of the polar bear, *Ursus maritimus* Phipps. *Acta zool. fenn.* **108**, 1–26.

Lambert, C. A., Pearson, R. G. & Sparks, B. W. 1963 A flora and fauna from Late-Pleistocene deposits at Sidgwick Avenue, Cambridge. *Proc. Linn. Soc. Lond.* **174**, 13–29.

Martin, P. S. & Wright, H. E. 1967 (eds). *Pleistocene extinctions, the search for a cause.* Yale: University Press.

Mayhew, D. F. 1975 *The Quaternary history of some British Rodents and lagomorphs.* Unpublished Ph.D. thesis, University of Cambridge.

Mitchell, G. F. 1941 Studies in Irish Quaternary deposits. 3. The Reindeer in Ireland. *Proc. R. Ir. Acad.* B **46**, 183–188.

Mitchell, G. F. 1958 A late-glacial deposit near Ballaugh, Isle of Man. *New Phytol.* **57**, 256–263.

Mitchell, G. F. 1969 Pleistocene mammals in Ireland. *Bull. Mammal Soc. Br. Isl.* **31**, 21–25.

Mitchell, G. F. & Parkes, H. M. 1949 The giant deer in Ireland. *Proc. R. Ir. Acad.* **52** B, 291–314.

Newton, E. T. 1894 The vertebrate fauna collected by Mr Lewis Abbott from the fissure near Ightham, Kent. *Q. Jl geol. Soc. Lond.* **50**, 188–210.

Savage, R. J. G. 1966 Irish Pleistocene mammals. *Ir. Nat. J.* **15**, 117–130.

Scharff, R. F., Seymour, H. J. & Newton, E. T. 1918 The exploration of Castlepook Cave, County Cork. *Proc. R. Ir. Acad.* B **34**, 33–72.

Simons, J. W. 1962 New records of musk ox from Plumstead, Kent, and Cosgrove, Northants. *Lond. Nat.* **41**, 42–53.

Stuart, A. J. 1974 Pleistocene history of the British vertebrate fauna. *Biol. Rev.* **49**, 225–266.

Stuart, A. J. 1976 The history of the mammal fauna during the Ipswichian/Last interglacial in England. *Phil. Trans. R. Soc. Lond.* B **276**, 221–250.

Sutcliffe, A. J. & Zeuner, F. E. 1962 Excavations in the Torbryan Caves, Devonshire. 1. Tornewton Cave. *Proc. Devon Archaeol. Exploration Soc.* **5**, 127–145.

Tratman, E. K. 1964 Picken's Hole, Crook Peak, Somerset. A Pleistocene site. *Proc. speleol. Soc.* **10**, 1–4.

Trechmann, C. T. 1939 A skeleton of elk (*Cervus alces*) from Neasham, near Darlington. *Proc. Yorks. geol. Soc.* **24**, 100–102.

Warren, S. H. 1912 On a late glacial stage in the valley of the River Lea subsequent to the epoch of river-drift Man. *Q. Jl geol. Soc. Lond.* **68**, 213–229.

West, R. G., Dickson, C. A., Catt, J. A., Weir, A. H. & Sparks, B. W. 1974 Late Pleistocene deposits at Wretton, Norfolk. II. Devensian deposits. *Phil. Trans. R. Soc. Lond.* B **267**, 337–420.

Wymer, J. 1962 Excavations at the Maglemosian sites at Thatcham, Berkshire, England. *Proc. prehist. Soc.* **28**, 329–361.

Phil. Trans. R. Soc. Lond. B. **280**, 313–340 (1977) [**313**]
Printed in Great Britain

Fossil coleopteran assemblages as sensitive indicators of climatic changes during the Devensian (Last) cold stage

By G. R. Coope
Department of Geological Sciences, The University of Birmingham

[1 Pullout]

Coleoptera are abundant fossils in Quaternary deposits laid down under freshwater or terrestrial conditions. They display a remarkable degree of evolutionary stability and reasons are adduced for believing that this morphological constancy is associated with physiological constancy. Thus whole communities of species have been assembled in the past, drawn together by common ecological preferences, so that the species composition of fossil assemblages resembles that of modern faunas. Marked changes in the geographical distribution of Coleoptera during the last glacial–interglacial cycle conform to an orderly pattern of climatic fluctuations. The Coleoptera contribute most information about the Devensian climates during warmer interstadial periods because during the colder episodes conditions in Britain became more or less intolerable to insect life and the fossil content of the sediments approaches zero. The term interstadial is here used for an interlude of milder climate in an otherwise cold period which either does not attain temperatures equivalent to those of the present day or which attains temperatures as warm, or even warmer than those of today but which does not last long enough for floral and faunal equilibrium to become established.

During the Chelford Interstadial, at the limit of acceptable radiocarbon dating but possibly about 60000 years (a) ago, the climate in central Britain was rather cooler than now with a moderate degree of continentality.

The Upton Warren Interstadial complex, between about 45000 and 25000 a ago, reached its thermal maximum at about 43000 a before present when temperatures were rather higher than those of the present day and the climate was moderately oceanic. This episode may have been as short in duration as 1000 a. After this the interstadial is characterized by a period of much lower temperatures, with a greatly increased degree of climatic continentality, lasting for about 15000 a.

Few insect faunas are known from the period of maximum ice expansion but the scant evidence supports an interpretation of a climate of arctic severity.

During the closing phases of the Devensian cold period there is faunal evidence for only one major climatic oscillation – here called the Windermere Interstadial. The sharp rise in the thermal environment at its beginning took place rather before 13000 a ago but later than 14000 a ago. Thermal maximum was attained almost immediately with temperatures during the summer at or above their present day level. Moderate oceanicity of the climate at this time means that winter temperatures were not much lower than those of the present day. At least during the earliest parts of this interstadial a temperate insect fauna was associated with a flora almost entirely dominated by herbs. The decline of the Windermere Interstadial from its thermal maximum seems to have been more or less synchronous from southern to northern England and to have taken place at about 12200 a ago. A cool temperate phase then ensued for over one thousand years with summers about 3 °C cooler than during the thermal maximum. This episode corresponds in time to the Allerød oscillation.

The Loch Lomond Stadial between 11000 and 10000 a ago saw the return of arctic faunas to the British Isles even as far south as Cornwall. The presence of Asiatic species, though not abundant, suggests that the climate at this time may have been rather continental.

The timing and intensity of the climatic changes during deglaciation show close parallelism to the changes in oceanic circulation in the eastern Atlantic now being interpreted from cores of ocean bottom sediments.

1. Introduction

The Coleoptera are characterized by a number of features, which taken together, combine to make them one of the most valuable components of the terrestrial biota in the interpretation of Quaternary environments. They have for the most part robust exoskeletons that survive as excellent fossils that are complicated enough in their structural details to permit many of them to be identified specifically. The fossils are often abundant, so that several thousand specimens can be obtained from a few kilograms of organic silt. In terms of numbers of species the order Coleoptera is one of the largest of the insects (in Britain alone there are over 3800 named species). Because they are easily caught and need little other than drying for their preservation, they have been, and still are, avidly collected, so that their taxonomy, geographical distributions and ecological preferences are relatively well known, particularly in the more inhabited parts of the northern hemisphere. From the point of view of the Quaternary palaeoecologist trying to reconstruct and understand past environments, this latter point is of the greatest importance since we are entirely dependent for our basic ecological data upon the skills of our neontological colleagues. It is thus both stimulating and humbling to read of the wealth of ecological information available and yet to realize how little we really know about the factors controlling the ecological fastidiousness of a species in its natural environment. For a particular species these environmental preferences are often quite narrow but, taken as a whole, the order Coleoptera contains species that are adapted to almost all terrestrial and freshwater habitats and trophic levels in the ecosystem.

The Quaternary fossil record of the Coleoptera has become extensive during the last two decades, especially in western Europe and North America. In spite of the fact that about 2000 species have been recognized, only one or two from the lower Pleistocene of Alaska (Matthews 1974) show any evidence of evolutionary change, and even in these cases the differences are of the same order of magnitude as we would expect today to differentiate geographical races of a single species. We have to look back into deposits that probably date from the late Miocene, namely the Beaufort formation in arctic Canada, before we see the first signs of evolution at the species level (Matthews 1976). Unfortunately the fossil evidence is restricted so far to temperate and polar regions of the northern hemisphere and we know very little of the fossil history of the Coleoptera in the tropics. However, it can be demonstrated that wherever fossils are available the Coleoptera show a remarkable degree of morphological stability throughout the Quaternary. Such is the precision of the preservation of these fossils that often exactly the same criteria can be used to differentiate between related species by both the modern coleopterist and the palaeontologist. It is not at all uncommon to find in Quaternary deposits the abdomens of beetles from which can be dissected the male genitalia which can be shown to match their modern counterparts, even to the intricacies of their internal sclerites. There can thus be little doubt that our Quaternary fossil Coleoptera represent exactly the same species as defined by present-day coleopterists.

However, although morphological constancy can be demonstrated directly, physiological constancy can not. Since the physiological make-up of a species determines its ecological tolerances or preferences, it is important to know to what extent physiological evolution might have taken place without any associated morphological change. Fortunately this is not as intractable a problem as it might seem at first sight. There is a substantial and growing body of evidence to show that the communities of species that occurred in the past are similar to those that can be found today. In other words species of beetle to a large extent kept the same company in the past as they do today. Furthermore, when it is possible to find independent evidence of the local environment or climate, such as for example the presence of a particular host plant represented by its macrofossils or the presence of ice-wedge casts indicating permafrost conditions formed contemporaneously with the insect bearing horizon, the species of beetles that occur as associated fossils are the same as those that would be expected under similar circumstances today. Certainly ecologically 'mixed' assemblages do occur occasionally, but they are relatively rare and usually represent transitional faunas characteristic of periods of rapid environmental change. If gradual physiological evolution had taken place on a large scale hidden under the mantle of morphological stability, it might have been expected that, with increasing age, our fossil assemblages should have included increasingly bizarre assortments of species of contrasting environmental requirements and climatic tolerances. In fact no such progressive deviation from present-day population structures is found in our fossil assemblages even as far back as the Cromerian interglacial. This suggests that, for the most part, species of Coleoptera have remained as physiologically stable as their morphological constancy should have led us to expect.

There are, however, a few species that appear to be exceptions to this general pattern of ecological conformity; that is they persistently occur in our fossil assemblages in company with others with which they have not been found living today. Such species require special consideration. Of course these strange associations might be construed as indicating environmental conditions for which no analogue exists at the present day, but since these odd species make up such a small proportion of the total fauna (almost always less than 5 %) and since it is usually the same few species that offend against ecological conformity, it is more probable that the oddness of their associations is a property of the species concerned rather than of the environment in which they lived. It is not intended here to give the impression that all Quaternary environments have equivalent conditions to be found somewhere on the Earth today, since in a large number of cases this is almost certainly not true, but merely to emphasize that the occurrence of these few nonconformist species in our fossil assemblages is inadequate grounds for this sort of environmental special pleading.

Two examples of these misfit species will be given here. In the latter half of the Upton Warren (Mid-Devensian) Interstadial *Timarcha goettingensis* (L.) occurs at a number of localities in association with many species that are today characteristic of tundra environments. This species is widely distributed in Europe south of latitude 60° N and although it is a plant-feeding species it is unlikely that this northern limit is determined by the availability of some preferred host plant since the species is polyphagous. The second example comes from the Windermere (Late-glacial) Interstadial (see p. 326). *Hypnoidus rivularis* Gyll. is a circumpolar boreal species that in Europe has its southern limit of continuous distribution across Fennoscandia at about 60° N. This northern species has been found as a fossil at a number of sites but always in company with a suite of other species of more temperate preferences. It seems likely that species such as these have either changed their environmental requirements since the Devensian by physiological evolution, or else their present-day geographical range does not include the whole area potentially available to the species. Alternatively it is possible that, in the past, a gene pool of greater diversity was available to the species, only part of which has survived to be handed down to their present-day representatives. Whichever explanation eventually accounts satisfactorily for these anomalously occurring species, their existence highlights the danger of using a limited number of indicator species as important clues to the determination of Quaternary environments. Since evolutionary stability seems to characterize the vast majority of species of Coleoptera during the Quaternary, it is the fossil assemblage viewed as a whole that provides the best evidence upon which to base our palaeoecological interpretations.

The remarkable degree of evolutionary stability shown by the Coleoptera during the Quaternary might at first sight seem unexpected in the light of the instability of the Quaternary environment. It might have been expected that successive ice advances and recessions with their associated climatic effects well south of the glacial limits, would have reduced population sizes, isolated them in restricted refuges and imposed upon them intense selection pressures – a certain recipe for rapid speciation. Yet in the northern latitudes most influenced by the vicissitudes of the ice age climates, rapid speciation evidently did not occur. There must, therefore, be something wrong with our view of the effects of the climatic oscillations on the fauna; some essential ingredient missing from our recipe. The fossil record provides abundant evidence that the geographical ranges of even the most sedentary species have changed on an enormous scale even within the last glacial/interglacial cycle. These rates of population movement were clearly adequate to enable species to keep pace with the changes in the location of suitable living conditions, so that the actual environment in which a species lived remained constant in spite of the drastic changes in the Quaternary climate that took place at any fixed geographical location. Thus the fact that insect species have remained stable during the Quaternary requires no special reconciliation with these environmental fluctuations.

It is of course possible that many species actually owe their constancy to the instability of the Quaternary environment. The hypothesis runs as follows. Many low temperature stenothermic species, that were widespread in Europe during the Last Glacial period, have today very disjunct distributions, being found in high latitudes in a zone of more or less continuous distribution down to sea level and also as isolated populations at high altitudes in the mountains of central Europe. This boreo-montane distribution pattern is largely due to the restriction of acceptable thermal environments during the climatic amelioration of the Post-glacial period. The northern and southern populations must have been isolated from one another for at least 10000 a, that is since the Loch Lomond Stadial (Younger Dryas times) which was the most recent period when such species were widespread throughout the lowlands of western Europe, and they may well have been isolated for much longer than this. It is evident that mere geographical isolation for this length of time has not led to any apparent speciation in spite of the fact that many of these isolated populations must be relatively small. However, large-scale climatic oscillations whose period is measured in a few tens of thousands of years, mean that these mountain top or high latitude evolutionary traps are easily escaped from, if climatic deterioration takes place before the marooned populations have become genetically isolated from one another. Under such conditions geographical barriers to genetic mixing are continuously being broken down or reformed in different contexts. The large-scale movements of insect populations that are so characteristic a feature of the fluctuating Quaternary environment would have repeatedly caused the merging or fragmentation of subpopulations; deployments and regroupings that would have ensured that the gene pool was kept well stirred. Speciation under such effervescent circumstances must have been well nigh impossible. It was this ability of insect populations to move rapidly, and far, in response to changes in the environment that was left out of our initial consideration of the response of the fauna to the vicissitudes of the Quaternary climate (p. 316). Certainly this readiness to move and to track environmental changes across the surface of the Earth coupled with the frequency of major climatic oscillations must be an important factor in maintaining specific stability of insects during the Quaternary.

The hypothesis outlined above is most relevant for the species that occur on the continental masses in temperate or polar latitudes where the opportunities and need to move in response to climatic changes are most pronounced. Conversely it should be expected that those species which live in situations where the option of large-scale population movement is not available, such as for example on oceanic islands or in cave systems, should show tendencies for more rapid evolution. There is of course abundant evidence that this is so. Similarly in those parts of the world where climatic changes associated with glacial/interglacial cycles would be expected to be minimal, namely in tropical latitudes, the necessity for population movement was no doubt correspondingly diminished with the consequent less frequent disruption of geographical barriers to genetic mixing. Under these circumstances evolution might have been more rapid than in the middle and polar latitudes. Paradoxically, therefore, it appears that Quaternary climatic instability could be the cause of evolutionary stability of insects during this period and conversely for the species of continental masses, long periods of climatic stability might be expected to lead to relatively rapid evolution.

2. The climatic significance of Quaternary coleopteran assemblages

The evolutionary stability of the Coleoptera during the Quaternary both in their morphology and their physiology means that the properties of present-day species can be used directly in the interpretation of fossil assemblages. This actualistic approach can be justified on the grounds that, in almost all cases, fossil assemblages when treated in this way make ecological sense; the best possible justification for a hypothesis, namely that it works.

All modern ecologists stress the fundamental rôle that climate plays in determining the geographical range, and often also the abundance, of insect species. Thus we find in that valuable compendium of papers on insect abundance published by the Royal Entomological Society (Southwood 1968), in discussing the processes that modify the size of insect populations Richards & Southwood (1968) state, 'First there is climate which sets the arena in which everything else interacts.' Later in the same volume Henson (1968) points out 'From the earliest stages in the development of our interest in insect populations, there has been a continued awareness that climate and its short-term expression which we call weather, have profound impact on the numbers of insects present at any one time'. This paper gives a useful account of reviews of biometeorological works. Finally Wilson (1968) states 'no one would doubt that climate is a major factor in determining limits of distribution'. Statements such as these serve to emphasize the importance that almost all insect ecologists attach to climate as one, if not the most potent, arbiter of insect distribution.

There are, of course, many factors other than climate that play important parts in the control of abundance and distribution, but these vary in their significance from species to species and place to place. They do not possess that all-pervasive quality that typifies the effects of climate. Thus in the central parts of the range of a species where environmental conditions are most favourable to its survival, population control is achieved largely by biotic factors such as competition for food or living space or by predation and parasitism (figure 1). Factors such as these are almost always density dependent in their effectiveness. Even under these more or less favourable conditions, variations in the physical environment may often exert a limiting influence on the population which may also be density dependent since refuges from adverse circumstances may be limited and easily filled to capacity. Also climate may control food supply and thus indirectly influence density dependent processes. On the other hand at the fringes of the range of a species conditions become increasingly unfavourable and the physical environment, chiefly in the form of climate, begins to exert a limiting influence that becomes increasingly independent of density. Fringe populations are characterized by being transitory; subject to successive eliminations by hostile climatic events and recolonizations from the main body of the population. Climatic factors can thus be expected to be among the most important controlling influences on the limits of the geographical ranges of insect species.

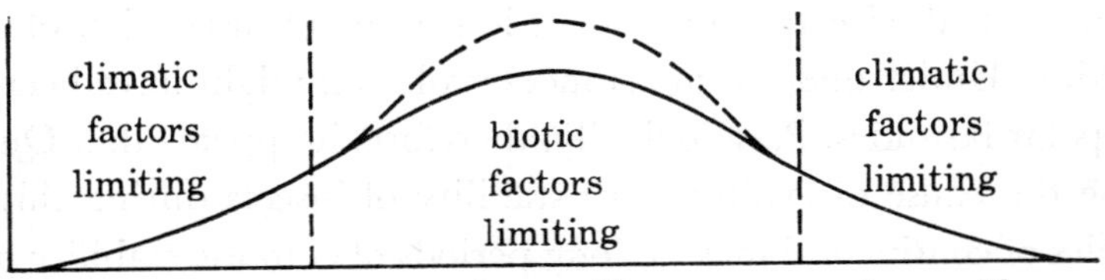

FIGURE 1. Hypothetical relation between the major regulating factors on a species throughout its geographical range (from Price 1975).

In the interpretation of the climatic significance of fossil insect assemblages, greatest attention is thus paid to the limits of the modern geographical distributions of the species of which it is composed. Some measure of the ancient climate can then be gained by comparing the ranges of a large number of species and the general 'fashion' of distribution determined. From some sites over 200 named species of fossil Coleoptera have been used in the interpretation of the climatic regime of a single time horizon. At localities where thick successions of fossiliferous deposits occur, spanning a considerable period of time, a sequence of insect faunas may be obtained. The varying fortunes of each species can then be plotted against time on a graph akin to those produced by palynologists. In the interpretation of the climatic significance of such diagrams, greater attention is paid to the simple presence of a species rather than to its abundance. This is because abundance is largely a reflection of the local environmental conditions and may well be determined by biotic factors, while the range of a species in time, just as it is in space, is likely to be governed chiefly by climatic factors. In this respect, therefore, beetle diagrams are dealt with rather differently from pollen diagrams.

The climatic factors governing insect distributions are chiefly those of the thermal environment and of moisture. Although these factors are known to interact, purely for convenience they will be dealt with here as separate entities.

The part played by moisture in determining the composition of fossil insect assemblages is difficult to assess beyond the immediately local environment. Most of our fossiliferous deposits were laid down in water, accumulating in shallow ponds or peat bogs, and the sediment has usually remained waterlogged throughout the year. Many of the fossil insect assemblages are thus dominated by aquatic or semi-aquatic species. The existence of these shallow ponds, that must have survived for much of the summer and never dried out completely, suggests that for much of the time during the Devensian glaciation adequate precipitation was available to maintain them. This precipitation need not have fallen in the summer since for much of the time winter temperatures were well below freezing and thus any precipitation would fall as snow and accumulate. The whole year's precipitation would then be available for biological and physical processes during the summer as the accumulated winter snows melted. There is ample evidence from fossil insect assemblages that, away from the marshy hollows, much of the country was dry with a patchy vegetation cover and well drained soils. There is almost no evidence of the steppe species that we might expect had these areas been truly arid during this period.

The appropriateness of the thermal environment is of fundamental importance in determining the suitability of habitats for poikilothermic animals. This may be viewed in two ways. Thermal intensity, measured as temperature, will set existence limits within which a species can survive. Furthermore, animals exhibit decided temperature preferences, well within existence limits, that determine behaviour patterns such as breeding and feeding, essential for the survival of the population. The availability of adequate heat, measured in day degrees, is also of great importance, contributing to the survival limits of a species as well as, for instance, to the rate of development and thus duration of life cycle and the rate at which a hibernating animal consumes its fat reserves, a factor that may be of importance to the survival of insect species in areas of oceanic climate where winters are relatively warm and thermally unstable. These examples are quoted in order to indicate the intricacy of the controls exerted by the thermal environment upon insect populations; they are in no way intended to be comprehensive but merely to hint at the complexity of the subject. Andrewartha & Birch (1954) and Price (1975) provide numerous case-histories and detailed discussions based upon data from both laboratory and field experiments.

When dealing with the vast majority of modern insect species, however, it must be admitted that we do not know the actual factors governing their distributions or abundance. In practice therefore we must fall back upon the simple expedient of investigating the modern geographical ranges of the species that make up our fossil assemblages and interpreting these in terms of the known environmental parameters that circumscribe these ranges. Usually these environmental parameters are expressed in terms of average figures of temperatures or rainfall and it must thus be emphasized that these are not the actual factors experienced by the organisms and influencing their population dynamics. They are, however, valuable as approximations that enable environments to be compared objectively.

In the interpretation of the climatic significance of a Quaternary insect assemblage, greatest attention is paid to the limits of the present-day geographical distributions of its component species. Assuming that the limits of these ranges are governed, in most cases, by climatic factors, and the justification for this assumption has been outlined above, species whose ranges are restricted to narrow climatic zones are considered to be stenothermic, while those with widespread distributions are interpreted as eurytherms. Naturally, in making climatic inferences, greatest attention must be given to the stenothermic species. In those cases where the range of a species is not governed by climate the use of this assumption should pick them out as recurring misfits in the overall distributional style of the bulk of the fauna.

This procedure does seem empirically justifiable. At least in Europe, where both the distributions of insects and climatic parameters are fairly well known, it can be shown that for many species their ranges do in fact correspond to well known climatic zones. This correspondence can well be illustrated by comparing the distributional limits of species of Coleoptera in Fennoscandia (a region chosen solely because it is one of the few places for which detailed data are available) in the *Catalogus Coleopterorum Fennoscandiae et Daniae* (Lindroth, ed. 1960) and the climatic parameters for the same area obtainable from numerous standard atlases. Such a comparison shows that the most consistent correlation is between summer temperatures and the northern limits of the ranges of those species which, whilst occurring in the southern areas of Fennoscandia, do not extend into the far north. The southern limits of the more boreal species are rather more ragged, breaking up into isolated patches associated, no doubt, with colder local climates. Even in such cases, however, the main areas of distribution have limits that also correspond to a large extent with summer temperature parameters. It is thus apparent that in latitudes such as north and central Europe the Coleoptera are best equipped to provide information about the thermal environment of the summers. Species with restricted ranges in western Europe, or conversely in the east palaearctic region, may also imply oceanicity or continentality of climate respectively with the added implication of winter temperatures. In our palaeoclimatic reconstructions we have used the July average temperatures of the regions that would best suit each fossil assemblage if they were living today. The July average temperature in this context is only intended to be used as a crude assessment of summer warmth.

3. Climatic changes during the Devensian cold stage

Numerous assemblages of fossil Coleoptera are now known that date from the various phases of the Devensian cold period and, though not all of them are fully investigated, over 50 fossiliferous localities have been discovered. Two recent reviews deal with the climatic changes during the Devensian as inferred from assemblages of fossil Coleoptera (Coope, Morgan & Osborne 1971; Coope 1975). Although this paper presents new information, particularly concerning the episodes of climatic amelioration during this period, there has not been any reason to alter the inferred curve for variations in the summer warmth (figure 2).

In the discussion that follows the term interstadial is used for an interlude of climatic amelioration which, either does not reach the degree of summer warmth characteristic of the present day, or for a period during which the intensity of summer warmth may have reached or even exceeded that of the present day but which did not last long enough for floral and faunal equilibrium to become established.

(*a*) *The Chelford Interstadial*

Sites that date from this period are still remarkably rare in spite of 20 years search since the discovery of the original locality at Chelford, Cheshire (Simpson & West 1958). Only one other deposit can be ascribed with certainty to this interstadial, namely the lens in the Four Ashes gravel pit that underlay the mid-Devensian horizons (Morgan 1973). A puzzling lens of organic silt at Wretton, Norfolk (locality of WG of West *et al.* 1974) has a pollen spectrum not unlike that of Chelford suggesting that woodland existed at that time. On the other hand an assemblage of Coleoptera from the same horizon suggests truly arctic conditions with an open treeless landscape (Coope 1974). This anomalous association has been further discussed (Coope 1975).

The only new information of the faunas of this interstadial arises from a re-evaluation of the original fossils from Chelford (Coope 1959) in the light of improved comparative material taxonomy and, by no means least, personal expertise. The emendations and additions to the original faunal list are as follows. *Bembidion harpaloides* Serv. was incorrectly named and is in fact *Trechus rivularis* Gyll. The numerous fossils of *Hydroporus*, many of them thought at the time to be unidentifiable, have been investigated by R. B. Angus. He reports that only three species can be named with certainty; *H. longicornis* Sharp, *H. melanarius* Sturm and *H. memnonius* Nic. These are all species of acid pools with *Sphagnum*, though *H. longicornis* seems to require trickles

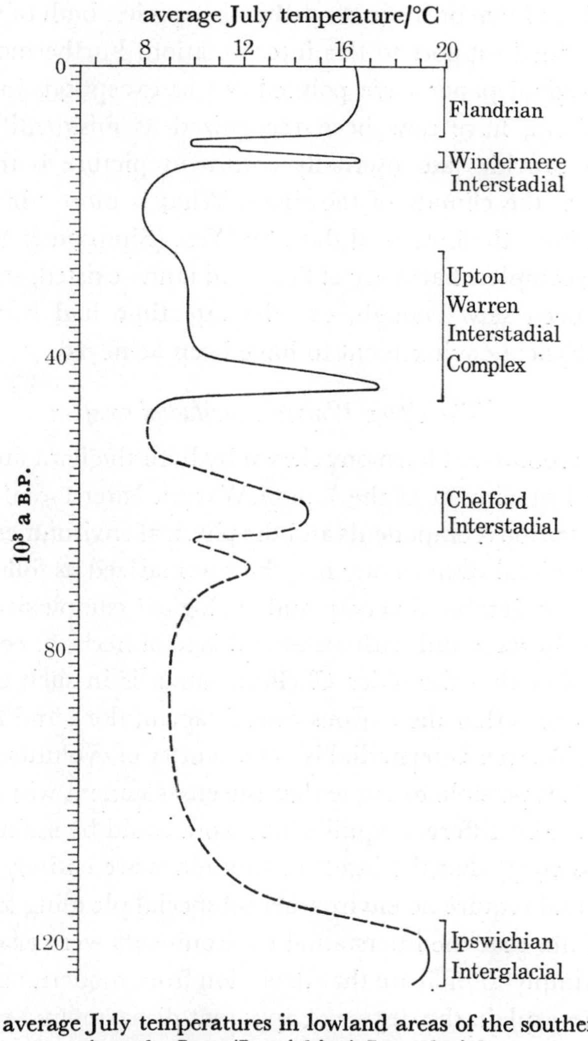

FIGURE 2. Variations in the average July temperatures in lowland areas of the southern and central British Isles since the Last (Ipswichian) Interglacial.

of water, such as those found at mossy flushes. Until its occurrence in this habitat was discovered (Foster 1968) the species was considered to be one of our rarest water beetles. The small staphylinid species *Boreaphilus henningianus* Sahlb. must be added to the Chelford faunal list. It is widespread in boreal Europe but absent from the present-day fauna of the British Isles. There are two emendations necessary among the moss feeding byrrhid beetles. *Simplocaria metallica* Sturm must be added. This is a very rare species today that is widely distributed in boreo-alpine parts of Europe but is absent from Britain. *Morychus aeneus* F. replaces *Cytilus* sp. This species is found all over Europe as far north as 65° N. The identification of two weevils also needs correcting; *Limnobaris pilistriata* Steph. must replace *L. t-album* L. and *Otiorrhynchus scaber* L. is almost certainly the correct identification for some badly preserved fossils originally named *O. ligneus* Oliv.

These alterations and additions to the original faunal list from Chelford do not alter the overall interpretation of the local environmental conditions. The climatic inferences are, however, considerably reinforced. Originally the fossil insect assemblage was interpreted as indicating a climatic regime similar to that in southern Finland between 60 and 65° N. The addition to the faunal list of two or more 'non-British' species, both of which live today in this part of Finland, gives added support to this interpretation. Furthermore, two out of the four species which in the original paper were pointed out as exceptions in that they did not live today in southern Finland, have now been recognized as misidentifications. The Chelford insect assemblage thus provides an internally consistent picture both of the local environmental conditions and of the climate of the times. What is more, this picture is remarkably similar to that derived from the botanical data by West (Simpson & West 1958). Clearly, for this interstadial at least, complete harmony of flora and fauna existed, suggesting that either the climatic changes had been slow enough, or adequate time had been available, for biotal equilibrium with the physical environment to have been achieved.

(*b*) *The Upton Warren Interstadial complex*

In contrast to the environmental harmony shown by both the flora and fauna from Chelford, the biota of the thermal maximum of the Upton Warren Interstadial (figure 3) suggest that equilibrium between its various components and the physical environment had not been reached during this period. This biotal disharmony may be summarized as follows. A thoroughly temperate insect fauna of considerable diversity and ecological complexity was associated with a flora completely lacking in trees and with an assemblage of herbs of contrasting distributional styles (Bell 1969). The fact that the older Chelford fauna is in such complete environmental accord with the flora suggests that the curious assemblage of flora and fauna from the thermal maximum of the Upton Warren Interstadial is not a matter of evolution in the tolerances of the species concerned. Nor is it possible to argue that the environment was so different from that of the present day that a totally different equilibrium biota could be maintained since it has been shown (Coope & Angus 1975) that the insect communities are entirely meaningful in terms of present-day conditions and require no environmental special pleading for their occurrence. It is not intended here to claim that the interstadial environments were exactly analogous to those of the present day but simply to indicate that deviation from modern environmental conditions is inadequate by itself to explain the degree of apparent disharmony between the flora and the fauna. The reasons for this curious concourse of species must, I believe, be sought for in the historical context of this interstadial; that is in the sequence of environmental changes that came both immediately before and after its thermal maximum.

The climatic changes of the Mid-Devensian (Weichselian) in western Europe inferred from fossil coleopteran assemblages have recently been revised (Coope 1975) and only a brief summary will be given here. The Upton Warren Interstadial complex spans about 20000 a between 45000 and 25000 B.P. (figure 3). Before the thermal maximum of the interstadial, the insect fauna was dominated by boreal or boreo-montane species, many of which (about 20 % of the fauna) do not now live in the British Isles. There can be little doubt that the climate at this time was of arctic severity but the absence of any exclusively eastern species from the faunas of this period may indicate that the climate was not markedly continental though the winters were no doubt severe.

The thermal maximum of the interstadial apparently follows this phase of arctic climate with dramatic suddenness. Thus in a stratigraphical sequence at Tattershall, Lincolnshire (Girling 1974) a temperate insect assemblage replaces the boreal fauna within a few centimetres of organic silt. Furthermore, the radiocarbon dates for these horizons emphasize their proximity

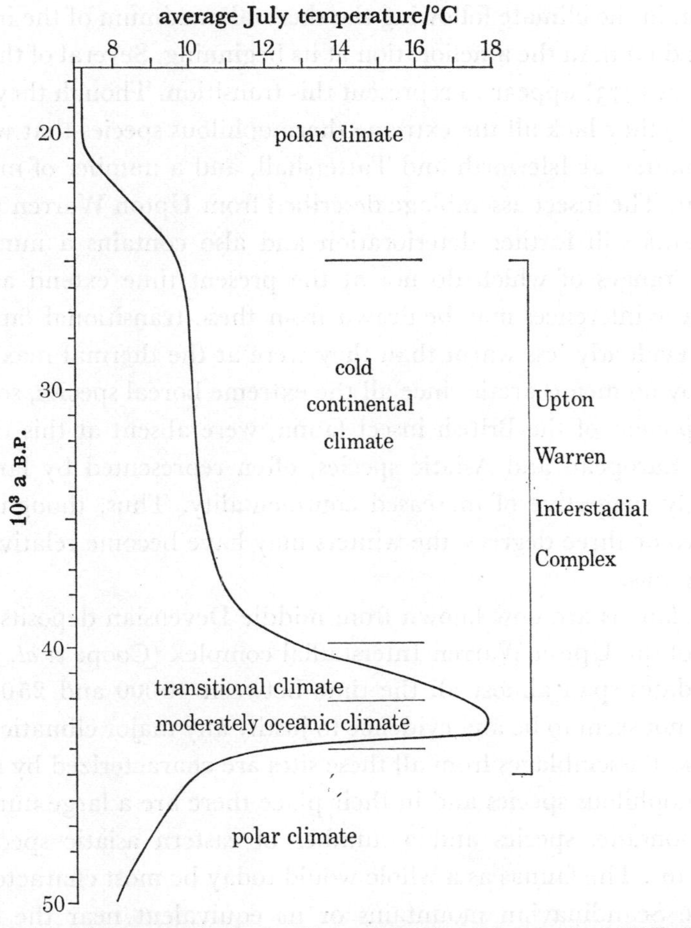

FIGURE 3. Climatic changes in lowland central England during the Middle Devensian.

in time; 44300 ± 1600 B.P. (Birm-408) for the stratum with the boreal fauna and 43000 ± 1200 B.P. (Birm-341) for the layer above with the temperate insect assemblage. Some idea of the intensity of this episode of climatic amelioration may be gained from the ecological study of a similar temperate insect assemblage to that from Tattershall, obtained from the flood plain terrace of the Thames at Isleworth (Coope & Angus 1975). This deposit gave an almost precisely comparable radiocarbon date to that from Tattershall, namely 43140 ± 1400 B.P. (Birm-319). Out of a total of 248 named species, only one has a northern geographical distribution today and it, *Deronectes griseostriatus* DeG., ranges as far south as the northern parts of the British Isles. None of the boreal species that were so important an element of the preceding faunas survive into this episode and in their place there are a large number of species that are widespread in central and southern Europe but which barely reach the extreme south of

England. The bulk of the Isleworth fauna (95 %) can be found today in north central Europe; say in the region of the north German plain. The insect fauna thus indicates that at its thermal maximum the summer temperatures were somewhat *warmer* than those of the present day in southern England. The absence of the far eastern species that characterize the next phase of the interstadial and the presence of exclusively west European species suggests that the climate of this period was moderately oceanic with winter temperatures only a little below those in southern Britain today.

The deterioration in the climate following the thermal maximum of the interstadial seems to have been more gradual than the amelioration at its beginning. Several of the faunas from Four Ashes (Anne Morgan 1973) appear to represent this transition. Though they contain a number of temperate species, they lack all the extreme thermophilous species that were so conspicuous an element in the faunas at Isleworth and Tattershall, and a number of moderately northern species were present. The insect assemblage described from Upton Warren (Coope, Shotton & Strachan 1961) shows still further deterioration and also contains a number of exclusively eastern species the ranges of which do not at the present time extend as far westwards as Europe. Two climatic inferences may be drawn from these transitional faunas. First, though the summers were evidently less warm than they were at the thermal maximum of the interstadial, they were by no means arctic since all the extreme boreal species, soon to become such an important component of the British insect fauna, were absent at this time. Secondly, the occurrence of east European and Asiatic species, often represented by large numbers of individuals, is strongly suggestive of increased continentality. Thus, though the summer temperatures fell by two or three degrees, the winters may have become relatively colder, but it is difficult to quantify this.

Numerous insect faunas are now known from middle Devensian deposits that post-date the temperate episode of the Upton Warren Interstadial complex (Coope *et al.* 1971). Collectively their radiocarbon dates span almost all the time between 40000 and 25000 B.P. and at the moment there does not seem to be any evidence to justify any major climatic oscillations during this period. The insect assemblages from all these sites are characterized by almost total lack of the relatively thermophilous species and in their place there are a large number of exclusively boreal, or boreo-montane, species and a number of eastern asiatic species that are often numerically abundant. The faunas as a whole would today be most characteristic of the Lower Alpine zone in the Scandinavian mountains or its equivalent near the forest limit in the Siberian arctic. Average July temperatures must thus have been at or just below 10 °C. The abundance, in these fossil assemblages, of east palaearctic species may be indicative of exceedingly cold winters with average February temperatures below -20 °C.

(*c*) *The Devensian ice maximum*

Very little evidence is available from the British Isles of the flora and fauna at the time when the ice front reached its most southerly extent, that is from the northwest of Norfolk in the east, across the English Midlands to south Wales in the west. A restricted insect fauna was obtained from a mossy silt beneath about 30 m of Devensian till on the Holderness coast of East Yorkshire (Penny, Coope & Catt 1969). Radiocarbon dates on the moss 18500 ± 400 B.P. (I-3372) and 18240 ± 250 B.P. (Birm-108) indicate that the insect fauna must have only just pre-dated the phase of maximum ice advance. Unfortunately, only a few species are represented but they indicate a harsh climate with summer temperatures probably several degrees below 10 °C.

Though arguing from absence of evidence is precarious, and doubly so in this case because of the smallness of the fauna, the lack of the exclusively Asiatic species, that were so common in the latter parts of the Upton Warren Interstadial complex, may indicate that the climate was less continental than it was at that time. If this is so, increased precipitation may have been largely responsible for the advance of the ice sheets into lowland Britain at this time, rather than a deterioration in the thermal environment. Certainly, for the 15000 years that followed the thermal maximum of the interstadial, the temperatures in the English Midlands would seem to have been adequate to sustain lowland ice sheets, yet throughout this period they were evidently absent.

(*d*) *The closing phases of the Devensian*

The disappearance of the ice sheets from the British Isles, after their maximum extent between 17000 and 18000 radiocarbon years ago, must have occurred with great rapidity. The whole process may well have taken place within 5000 years (Sissons & Walker 1974). Evidently the climatic amelioration was not constant or gradual but was characterized by relatively sudden changes reflected in the response of the flora and fauna as well as in geomorphological evidence of the readvance of the ice sheets. In so far as the evidence from the British Isles is concerned there is information to support only one large scale climatic oscillation towards the close of the Devensian period. During the thermal maximum of this oscillation it is possible that the ice sheets disappeared totally from Britain to be regenerated anew in the Western Highlands of Scotland (Sissons 1974) and to build corrie glaciers in the mountains further south as the climate subsequently deteriorated. This ice advance is called the Loch Lomond Readvance and the colder episode should be termed the Loch Lomond Stadial. It is approximately equivalent in time to Pollen Zone III and to Younger Dryas times of continental authors.

Before discussing the insect faunas of this period it is necessary to comment upon the naming of the episode of milder climate that immediately preceded the Loch Lomond Stadial. Much of this period falls within the time that has been rather loosely allocated to the Late-glacial and the period of climatic amelioration has been informally referred to as the Late-glacial Interstadial. However, regardless of the utility of these terms, they have strictly no validity according to the internationally accepted rules of stratigraphical classification. These rules require the establishment of stratotypes, i.e. standard sections wherein the lower boundary of the stratigraphical unit can be defined. In the British Isles the Late-glacial Interstadial has most frequently been equated with the Allerød oscillation which has its type section in Denmark and with which a great many continental sequences have been correlated. At first little evidence was found in Britain for the earlier Bølling Interstadial, also with its type section in Denmark, but recently abundant evidence has been recognized (Pennington & Bonny 1970; Pennington 1975) for a marked climatic warming more than a thousand years before the generally accepted date for the beginning of the Allerød. Numerous coleopteran assemblages provide independent evidence for this early warming and indicate that the climate was as warm during the summers as it is today, even before the conventional beginning of the Bølling oscillation (Vogel & Zagwin 1967), and apparently earlier than the date selected for the beginning of the Bølling chronozone in Fennoscandia, namely 13000 B.P. (Mangerud, Anderson, Berglund & Donner 1974). The Late-glacial Interstadial of the British Isles cannot, therefore, be satisfactorily equated with either the Allerød or Bølling interstadials as at present interpreted by continental authors, nor can it be accommodated in a combination of both.

It thus seems desirable that a stratotype should be selected to represent the deposits of the Late-glacial Interstadial in the British Isles.† Since we are concerned with the stratigraphical expression of a climatic event it is necessary to choose either a terrestrial or freshwater deposit. Three sedimentary environments are available, peat bog accumulations, kettle hole or small pond infillings and lake bottom sediments. Peat bogs with their unique biota and likelihood of breaks in sedimentation are, for these reasons, unsuitable as representative sections. Kettle hole sequences, though often extremely rich in flora and fauna, suffer from the complications inherent in the hydroseral development of all shallow ponds and also they may have the added complication of spasmodic melting out of buried ice. Limnic deposits, on the other hand, are much more likely to occur as continuous sedimentary sequences and those that accumulate near the lake margins are likely to reflect the terrestrial environment as well as that of the lake itself. As sources of fossil coleopteran assemblages, however, lake sediments are far from ideal since most of the beetles must have been incorporated in the deposit by accidental derivation from the surrounding land surface and are therefore sparsely distributed in the sediment. Representative faunas can thus be obtained only from relatively large samples. To some extent this problem can be overcome if limnic deposits are investigated in the vicinity of incurrent streams that are likely to carry terrestrial plant and animal macrofossils into the lake.

As a result of discussions with a number of workers in the field and particularly with Dr Pennington, it is suggested that the stratotype for the Late-glacial Interstadial in Britain should be selected from the deposits in Windermere (see Discussion). Perhaps the best locality is Low Wray Bay at the north end of the lake where minor streams have brought in much terrigenous material. This is also the site of numerous detailed investigations of both the lithostratigraphy and biostratigraphy carried out during the past 30 years.

In the account that follows of the climatic implications of coleopteran assemblages during the closing phases of the Devensian, the faunas will be considered under three headings; pre-interstadial, the Windermere Interstadial and the Loch Lomond Stadial.

(i) *Pre-interstadial*

No fossils have yet been found in the lower laminated clays in Windermere that predate the interstadial sediments. Four other sites in the British Isles have, however, yielded insect faunas that can be ascribed to the period immediately before the Interstadial.

The Late-glacial deposits at Hawks Tor on Bodmin Moor, Cornwall, have for long been the object of biostratigraphical investigations (Connolly, Godwin & Megaw 1950). Recently a reinvestigation of the stratigraphy and palynology of the site has been carried out by Brown (1977) during which he discovered a sequence of organic silts and gravels whose stratigraphy suggested that they predated the interstadial sequence. From samples of this deposit (section No. 4 of Brown's account) a sparse insect fauna of 17 taxa was obtained which contained a number of boreal species such as *Helophorus sibiricus* Mtsch., *Pycnoglypta lurida* Gyll., *Boreaphilus henningianus* Sahlb. and *Olophrum boreale* Payk., none of which live in the British Isles today. There can be little doubt that the climate at this time was very cold and probably too cold for the growth of trees. Unfortunately the sequence is extensively penetrated by roots from the overlying Flandrian deposits, so any radiocarbon date for this horizon would be unreliable.

At Glanllynnau 4.5 km west of Criccieth, North Wales, a sequence of Late-glacial deposits are immediately underlain by fossil-bearing grey silty clays and are exposed on the coast. The

† See discussion by Coope & Pennington, pp. 337–339.

insect fauna from this site has been described by Coope & Brophy (1972). The assemblage of Coleoptera from the pre-interstadial horizons is sparse but characterized by a variety of low temperature stenothermic species: these include *Bembidion fellmanni* Mnh., *Bembidion hasti* Sahlb., *Amara alpina* Payk., *Deronectes griseostriatus* DeG., *Helophorus sibiricus* Mtsch., *Helophorus obscurellus* Popp., *Helophorus splendidus* Sahlb., *Boreaphilus henningianus* Sahlb., *Simplocaria metallica* Sturm, *Syncalypta cyclolepidia* Munst. and *Phytonomus obovatus* Cki. This is by far the richest arctic assemblage known from pre-interstadial deposits and indicates very open country with no trees and an extremely cold and probably continental climate. A ^{14}C date of 14468 ± 300 B.P. (Birm-212) was obtained from a moss rich band at the base of the succession. Above this dated horizon were 120 cm of silty clay containing throughout the arctic–alpine assemblage of Coleoptera described above. A second radiocarbon date was obtained from seeds of terrestrial plants washed out of a detritus mud with a temperate insect fauna, that lay 20 cm above the highest stratum that contained arctic beetles. This upper date was 12556 ± 230 B.P. (Birm-176). The type of material dated precludes the possibility of 'hard water error' and prolonged pre-treatment of the sample was carried out to remove any possibility of contamination by modern humate. These two dates set time limits for the age of the arctic–alpine fauna from below the interstadial deposits of Glanllynnau.

At Glen Ballyre near Kirkmichael, on the northwest coast of the Isle of Man, a sediment filled kettle hole is exposed in section at the top of the cliff (Mitchell 1965; Dickson, Dickson & Mitchell 1970). Gravelly silts and clays underlie organic deposits that can be correlated with the Windermere Interstadial. From the lowest 30 cm of these deposits (that is well below the samples analysed by Dickson *et al.* 1970) an assemblage of northern Coleoptera was obtained that included *Bembidion hasti* Sahlb., *Bembidion lapponicum* Zett., *Agabus arcticus* Payk. and *Syncalypta cyclolepidia* Munst. One of the most abundant of the water beetles was *Coelambus mongolicus* Jak. that today is restricted to eastern Asia. A moss layer (chiefly *Drepanocladus*) at the base of this sequence gave radiocarbon dates of 18900 ± 280 B.P. (Birm-213) and 18700 ± 500 B.P. (Birm-270). A detritus mud with temperate insects about 5 cm above the topmost strata that contained the northern species gave a radiocarbon date of 12645 ± 280 B.P. (Birm-412). Here again the two radiocarbon dates set limits for the age of the pre-interstadial fauna from Glen Ballyre.

A rather enigmatic assemblage of fossil coleoptera from Colney Heath, Hertfordshire, has been described by Pearson (1962) from a peat erratic embedded in gravels stratigraphically below the level of a pollen bearing sequence characteristic of the Allerød climatic oscillation. A radiocarbon age of 13560 ± 210 B.P. (Q-385) was obtained from wood contained within the erratic. The assemblage of Coleoptera contains a curious assortment of relatively northern and southern species but since the fossils were not preserved it has not been possible to verify their determinations. However, the unmistakable *Diacheila arctica* Gyll., *Miscodera arctica* Payk. and large numbers of *Notaris aethiops* F., indicate that the climate must have been considerably colder than it is at the present time. However, the presence of relatively southern species, ranging barely north of 60° N, such as *Carabus monilis* F., *Bembidion minimum* F., *Aphodius plagiatus* L. and *Dorytomus longimanus* Forst., suggest a climate not much colder than the present day. It is an intriguing possibility that this peat erratic spanned the period of abrupt change from arctic to temperate climatic conditions that must have occurred at about the same time as the radiocarbon date from the Colney Heath erratic.

A single sample from the base of a deep depression beneath the Late-glacial and Flandrian deposits at Stafford (Morgan, A.V. 1973, Fig. 30) provided an insect fauna that included

Helophorus obscurellus Popp., and was indicative of a climate of arctic severity (Anne Morgan 1970). A radio-carbon date of 13490 ± 375 B.P. (Birm-150) was obtained from this sample which, since it was obtained from about 15.5 m depth, is unlikely to be influenced by modern humate. This is the most recent date obtained for a pre-interstadial arctic–alpine insect fauna.

(ii) *The Windermere Interstadial*

The occurrences of various species of Coleoptera through the interstadial sediments at Low Wray Bay, Windermere, are given in figure 4. Sampling datum O is taken at the base of the upper laminated clay, a level that may be correlated with the beginning of the Loch Lomond Stadial (= Pollen Zone III or Younger Dryas times). Direct comparison may therefore be made between the pollen spectrum (Pennington 1977) and the Coleoptera diagram. No effort has been made to indicate the abundance of species because of the smallness of the fossil assemblage. Enough evidence is, however, available to permit this spectrum to be compared profitably with other more completely sampled sequences and also to provide information on the climatic changes during the Windermere Interstadial.

Even at the base of the interstadial deposits there are none of the obligate arctic–alpine species that were so characteristic an element of the pre-interstadial faunas. In their place is a suite of species that would be quite at home in the Windermere area today and which includes a number of species that are either absent from the more northerly parts of Europe or else very rare there. Such species are indicated at the foot of figure 4. These relatively southern species are not present in the upper part of the interstadial sequence and in their place we have the first incoming of the more northern species heralding the truly arctic faunas of the ensuing Loch Lomond Stadial. The faunal change within the latter part of the interstadial seems to correspond with the layer of clay with rock fragments that occurred at about 15–20 cm below sampling datum 0. The change in the fauna and lithology suggest a sharp decline in the thermal environment at this time, but after it there is little evidence of any climatic recovery and conditions must have remained very much colder than during the early part of the interstadial.

The pattern of climatic events inferred from the interstadial Coleoptera from Windermere has now been recognized at a number of widely scattered localities in the British Isles (Coope 1975). The rapidity with which an entirely temperate assemblage of beetles replaced the arctic–alpine fauna at the start of the interstadial at both Glanllynnau and Glen Ballyre, where sedimentation throughout this time seems to have been complete, suggests that the initial climatic amelioration must have been sudden and intense. The precise dating of this event is, however, difficult because of the largely inorganic nature of the associated sediments. Even where the deposits are organic the plant debris is frequently the remains of aquatic plants, notably of *Chara*, with the consequent risk of 'hard water error'. In a Late-glacial context in North Jutland radiocarbon dates based on bulk samples of pond bottom detritus were greater by 1700 a than those obtained from twigs at the same stratigraphical position (Shotton 1972). Only two dates on early interstadial temperate faunas so far seem free of the imputation of 'hard water error'. At Hawks Tor section No. 3 (Brown 1977) where the Late-glacial sediments rest on bedrock a date of 13088 ± 300 B.P. (Q-979) was obtained and, though this was from a bulk sample of a water lain deposit, it is unlikely to be afflicted with 'hard water error' for two reasons. First, the site is in a hollow on the Bodmin Moor granite providing a thoroughly acid substrate, and second, there are no calcareous drifts as far south as this in the British Isles and thus no possibility of a base rich veneer over the granite. The second date 12940 ± 250 (Q-643) was obtained from wood out of a Late-glacial sequence at Roberthill, near Lockerbie, Dumfriesshire (Bishop & Coope 1977). In both these cases temperate conditions were evidently already established for sufficient time for the relatively thermophilous insect species to have spread as far west as Cornwall and as far north as southwest Scotland. It is likely therefore that the climatic amelioration at the start of the Windermere Interstadial took place somewhat earlier than 13000 B.P. but probably not much earlier than 13500 B.P. if the Stafford date is reliable.

The decline from the thermal maximum of the interstadial seems also to have been sudden and on the evidence of the assemblages of fossil Coleoptera, there would seem to have been a diminution of about 3 °C in the average July temperatures compared with those of the thermal maximum. At Hawks Tor, though no dates actually were measured at the change in fauna, the best fit by intercalation between known dates is about 12300 B.P. At Church Stretton this event can be dated at 12135 ± 200 B.P. (Birm-158). At Red Moss near Harwich, Lancashire (Ashworth 1972) the thermophilous Coleoptera had been replaced by 'colder' species by 12160 ± 140 B.P. (Birm-127). In the Glen Ballyre section this same faunal change can be dated at about 12150 B.P. (GRO. 1616) (Joachim 1974). A reinvestigation of the Coleoptera from St Bees, Cumbria, also shows a sudden change from a rich, temperate fauna to an assemblage of thoroughly northern affinities, slightly before 12200 B.P., after adjustments had been made for 'hard water error' (Williams & Johnson 1976). In the interstadial deposits of Windermere the loss of the thermophilous element from the fauna and the incoming of northern species took place sometime shortly before 12000 B.P. In southwest Scotland a cool temperate insect fauna had already replaced the more temperate assemblage by 12290 ± 250 (Q-816) at Redkirk Point (Bishop & Coope 1977). There thus seems to be abundant evidence from radiocarbon dates and from changes in the beetle fauna that there was a marked deterioration in the thermal environment about two centuries before the date 12000 B.P. selected for the upper limit of the Bølling chronozone in Norden (Mangerud *et al.* 1974). In spite of this minor discrepancy in timing, it seems likely that this climatic deterioration is the same as that ascribed to the Older Dryas period of continental authors.

If the episode of climatic cooling mentioned above is to be equated with the Older Dryas period, there is very little evidence from the Coleoptera to support an inference of a period of climatic warning after this, that would be equated with the Allerød Interstadial. In general the assemblages of Coleoptera that date from between 12200 and 11000 B.P. are monotonously similar and suggest a cool temperate climate throughout. There is evidence now accumulating to suggest that during this period the gradient in average July temperatures from the Midland Valley of Scotland to the south of England was about 4 °C. Three faunal sequences suggest that this climatic plateau may not have been entirely flat but may have had a minor episode of slightly warmer conditions. The Church Stretton fauna (Osborne 1972) shows a small-scale increase in the proportion of temperate species shortly after 12000 B.P., but one of the important indications of this change was the southern species *Berosus affinis* Brulle that has now been redetermined as *Berosus luridus* L. (Osborne, personal communication), a species that ranges today as far north as Lapland. Even before this adjustment to the faunal list Osborne had commented (1972, p. 351) 'Despite a rise in the thermophile graph . . . it seems unlikely that a marked rise in summer temperatures occurred'. At Red Moss (Ashworth 1972) there is a diminution in the northern element in the fauna but no evidence of the return of temperate species at about the time normally attributed to the Allerød oscillation. Finally, in the Windermere sequence, there is evidence from the sedimentology and entomology that the period of sudden climatic deterioration shortly before 12000 B.P. may have been rather colder than the period that immediately followed it. Against this evidence of a small-scale climatic oscillation, the Coleoptera from Glen Ballyre and St Bees show no significant changes in the climate between about 12000 and 11000 B.P. Preliminary work on Coleoptera of this period from Sproughton, Suffolk, also show no evidence of any significant oscillation during this period (M. Bryan, personal communication). Thus the Coleoptera provide scant evidence for a climatic oscillation equivalent in time to the Allerød and what evidence there is seems to be most marked in the northern half of the British Isles. In comparison to the total amplitude of the change in thermal environment during the Windermere Interstadial, the climatic oscillation during the equivalent time to the Allerød was very small indeed.

A brief comparison will be made here between the records of palynology and Coleoptera from the Windermere Interstadial. A taxonomically diverse insect fauna that indicates temperate conditions, with summer warmth at least equivalent to that of the present day, is characteristic of the earliest parts of this Interstadial. In these early stages this fauna is associated with herb-dominated pollen assemblage zones with Graminae, Cyperaceae and *Rumex* the chief groups present. The temperate fauna shows gradual increase in diversity throughout this period and there is no acceleration of this process to mark the incoming of *Juniperus* in the pollen diagrams. This relationship holds good for Hawks Tor, Church Stretton, Glanllynnau, Glen Ballyre, St Bees, Windermere and Roberthill, in fact for all localities where pollen and beetle data are available. In general the pollen assemblage zone dominated by *Betula*, that occurs during the second part of the Windermere Interstadial at many localities in England and southern Scotland, is accompanied by a beetle assemblage that lacks all the more southern species that were so characteristic a feature of the faunas from the early part of the Interstadial. Complimentarily a number of boreal species are present that were absent from faunas of the first half of the Interstadial. When interstadial conditions came to an end, the fall in the proportion of *Betula* pollen and the associated increase in curves for open country herbaceous plants is accompanied by a great increase in the numbers of northern beetle species as well as in their specific abundance. Radiocarbon dates for these floral and faunal changes marking the termination of the Windermere Interstadial tend to cluster around 11000 B.P.

(iii) *The Loch Lomond Stadial*

The assemblages of Coleoptera from this period are remarkably consistent and include a large number of boreal or boreomontane species. Even the most southerly site, known so far, Hawks Tor, has yielded a thoroughly northern fauna from this period including such species as *Olophrum boreale* Payk., *Acidota quadrata* Zett. and *Boreaphilus henningianus* Sahlb. This fauna was obtained from section No. 2 (Brown 1977) midway between two samples dated 10884 ± 210 B.P. (Q-1016) and 9654 ± 190 B.P. (Q-1017). Osborne (1971) reports a thoroughly arctic beetle assemblage from the Croydon area, that was associated with wood fragments that gave a radiocarbon date of 10130 ± 120 B.P. (Birm-101). A similar insect fauna has recently been obtained at Sproughton, Suffolk, associated with a radiocarbon date of 10880 ± 250 B.P. (Birm-779). These three faunas serve to illustrate the degree to which arctic species extended their ranges to the southernmost part of England during the period of the Loch Lomond Stadial. Numerous arctic assemblages are now known from localities in the English Midlands, the North of England, North Wales, the Isle of Man and Ireland but so far only one locality in Scotland has yielded an assemblage of Coleoptera that dates from this period. At Bigholm Burn, Dumfriesshire, a small fauna dominated by *Olophrum boreale* was extracted from a peaty lens in gravels overlying peats of Windermere Interstadial age. This is the most northerly insect assemblage that can be dated to the Loch Lomond Stadial and suggests that conditions approaching polar desert must have existed in Scotland at this time (Bishop & Coope 1977).

A curious feature of these faunas is the presence in the British Isles at this time of species that have since become apparently extinct in Europe. Thus *Pterostichus magus* Mnh. is a Siberian species that occurred as a fossil in the low terrace of the Thames at Farmoor near Oxford, in deposits that gave a radiocarbon date of 10600 ± 240 B.P. (Birm-590). *Helophorus jacutus* Popp. has been recovered from 'Pollen Zone III' deposits at Comberton near Ludlow. Its range today is exclusively east Siberian. Finally *Tachinus jacuticus* Popp. occurs as a fossil in the upper parts of the Late-glacial deposits of the Isle of Man. This is an eastern Asiatic and North American species that today lives no nearer to the British Isles than 7000 km. The occurrence of these species suggests that the climate of the times was continental but because this eastern element in the fauna is less, both in terms of numbers of species and their abundance, than in the faunas from the colder parts of the Upton Warren Interstadial, it is probable that the climatic continentality was less extreme during the Loch Lomond Stadial.

(iv) *Summary of climatic changes during the closing phases of the Devensian*

Figure 5 is an attempt to show the pattern of change in summer warmth for various latitudes in the British Isles during the closing phases of the Devensian. It is based on assemblages of Coleoptera from over 25 widely scattered localities that cover most of the British Isles with the

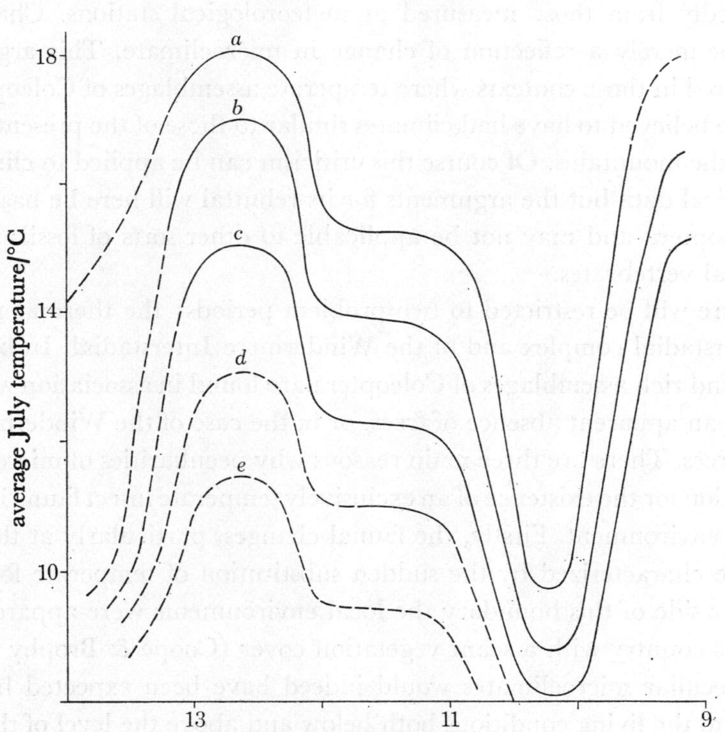

FIGURE 5. Variations in the average July temperatures, near sea level, during the closing phases of the Devensian (Last) cold stage (*a*) for southern England, (*b*) for central England and north Wales, (*c*) for northern England and southwest Scotland, (*d*) for central Scotland and (*e*) a hypothetical extrapolation for the north of Scotland. Dashed lines are used where these are conjectural.

exception of Scotland north of the Midland Valley. The estimates of the average July temperatures are based wherever possible on sites near to present-day sea level. For localities at higher altitudes a normal lapse rate of 6 °C per 1000 m has been assumed.

It is evident that during the period under consideration there is only one large-scale climatic oscillation. The initial amelioration seems to have been somewhat earlier in the south of England than in the Midlands and North. The thermal maximum seems to have been reached sometime shortly before 13000 B.P. and probably lasted for about 1000 radiocarbon years. The decline from the thermal maximum seems to have taken place in two steps. The first of these, a drop of almost 3 °C in the average July temperature, took place almost synchronously at all latitudes. The second step in the climatic deterioration must have taken place at about 11000 B.P. and may have involved a decrease of over 4 °C in the average July temperature. At all localities in the British Isles, where insect faunas of Loch Lomond Stadial age are known, they suggest a climatic régime over the whole area equivalent, at its thermal minimum, to that of the arctic tundra today. The recovery from glacial conditions was apparently very sudden and by 9500 B.P. the thermal environment of the British Isles was fully as warm as it is at the present time.

4. General discussion

(a) *Microclimate versus macroclimate*

The interpretation of ancient climates on the basis of assemblages of fossil Coleoptera has been criticized on the grounds that the beetles actually live in microenvironments whose parameters differ markedly from those measured at meteorological stations. Changes in beetle faunas might thus be merely a reflection of change in microclimate. This argument is most frequently encountered in those contexts where temperate assemblages of Coleoptera are found in situations hitherto believed to have had climates similar to those of the present-day tundra or the alpine zones of the mountains. Of course this criticism can be applied to climatic inference based on any biological data but the arguments for its rebuttal will here be based on information about the Coleoptera and may not be applicable to other sorts of fossils such as plants, molluscs or terrestrial vertebrates.

The discussion here will be restricted to two problem periods; the thermal maxima of the Upton Warren Interstadial complex and of the Windermere Interstadial. In both these cases entirely temperate and rich assemblages of Coleoptera are found in association with herbaceous vegetation and with an apparent absence of trees, or in the case of the Windermere deposits, a marked scarcity of trees. There are three main reasons why peculiarities of microclimate are an inadequate explanation for the existence of an exclusively temperate insect fauna in what appears to be a fully glacial environment. Firstly, the faunal changes, particularly at the beginning of the interstadials, are characterized by the sudden substitution of temperate for arctic–alpine species and on either side of this boundary the local environments were apparently identical; namely open treeless country with a scant vegetation cover (Coope & Brophy 1972). In such exposed habitats, peculiar microclimates would indeed have been expected but they would have been common to the living conditions both below and above the level of the change from an arctic to a temperate fauna. Secondly, a number of diverse and fairly sizeable ecological communities are present at these times and all of them include numerous thermophilous species (species intolerant of colder summers than those of the present day). Peculiar microclimates would have had to be almost universal, in all ecological contexts, everywhere hostile to arctic–

alpine species and permissive to thermophilous species. Thirdly, and lastly, if the presence of the numerous thermophilous Coleoptera is to be accounted for by invoking the peculiar qualities of the arctic style microclimate, which can indeed involve high temperatures (Corbet 1972), these same species would be present in the arctic today and conversely, modern arctic species would be expected in these interstadial faunas. In fact neither of these conditions is fulfilled.

Special pleading that the Coleoptera are merely indicative of microclimates, is thus unconvincing. It is certainly true that microclimates will frequently determine the local distribution of a species but the occurrence of acceptable microclimates is ultimately dependent on suitable macroclimates. For this reason gross geographical ranges have been used in this climatic interpretation. We are left with the inescapable, if rather commonplace, conclusion that a temperate European style fossil fauna implies a temperate European style climate.

(*b*) *Patterns of climatic change during the Devensian*

A number of features that are common to all the climatic changes outlined above, emerge from this study. Thus the amplitude of the oscillations in summer warmth is similar in the two interstadials that are adequately known; the Upton Warren and Windermere Interstadials. At its beginning each interstadial is characterized by a sudden increase in warmth to an early thermal maximum, after which the deterioration seems to have been more gradual. The duration of the period of thermal maximum seems to have been relatively short, of the order of a thousand years or so, but in spite of this a rich and thoroughly temperate insect fauna was able to colonize at least the southern half of the British Isles. At the thermal maxima of both interstadials the average degree of summer warmth was as high or even slightly higher than it is in central and southern Britain today.

The incoming of eastern Asiatic species, often specifically in great abundance, seems to be a feature of the colder phases, particularly the long period of more or less arctic conditions that followed the thermal maximum of the Upton Warren Interstadial. During this period and also during the time of the Loch Lomond Stadial, the insect faunas were rich in arctic–alpine species suggesting that the average July temperature was at or below 10 °C. The diversity of these faunas and the abundance of some species, however, is strongly suggestive of temperatures not much below this level. In this context the eastern Asiatic species are of particular interest. They are well adapted to surviving winters of extreme harshness, and though many can tolerate high summer temperatures they are usually species that have ranges that include the Siberian arctic or the high cold steppes further south. Thus, though they can tolerate high summer temperatures they are also able to accept much lower levels of summer warmth. Furthermore, their arctic–alpine companions in these fossil faunas preclude the possibility of persistent high summer temperatures at this time.

An attempt is made in figure 6 to illustrate the likely seasonal variation in mean monthly temperatures for the three main climatic types found in the interstadials of the Devensian. The figures are estimates for central England. Little need be added to what has already been said about the interludes of temperate climate (figure 6*c*). So far only two short intervals are known from the Devensian that have climatic régimes of this type; the thermal maxima of the Upton Warren and the Windermere interstadials. If the deposits at Chelford represent the thermal maximum of that interstadial, then this phase was somewhat cooler than the other two interstadials. The climate was also rather more continental with greater difference in seasonal temperatures (figure 6*b*). The more continental style climates (figure 6*a*) probably characterized much of the Middle Devensian, the phase immediately before the Windermere Interstadial and also the period of the Loch Lomond Stadial. During these periods the strong seasonal variation in temperature may have been associated with fairly large diurnal temperature range distinguishing our glacial climate from that of arctic Siberia at the present day.

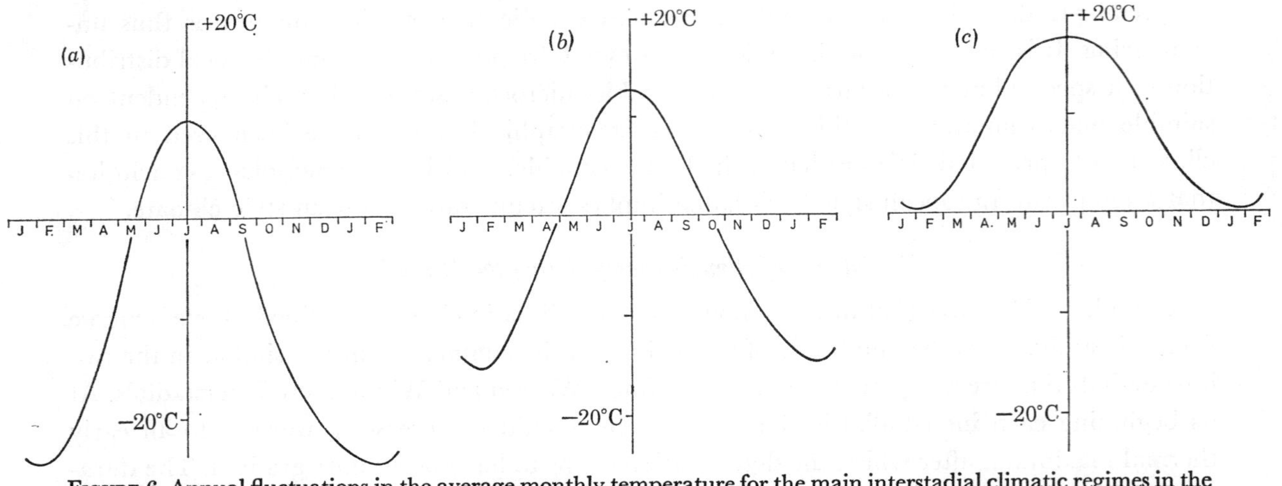

FIGURE 6. Annual fluctuations in the average monthly temperature for the main interstadial climatic regimes in the British Isles. (*a*) During the latter part of the Upton Warren Interstadial complex, (*b*) during the Chelford Interstadial, (*c*) during the thermal maxima of the Upton Warren and Windermere interstadials. Note: The climate in much of England during the Loch Lomond Stadial probably resembled (*a*).

Support for this interpretation of a continental style climate in the British Isles during the colder episodes of the Devensian comes from an entirely independent quarter. Deposits that date from the long cold tail of the Upton Warren Interstadial, that is between about 40000 and 25000 B.P., are often riven by thermal contraction cracks now represented either by ice-wedge casts or by sand wedges. Similar structures can be found penetrating deposits that can be dated with certainty to the Loch Lomond Stadial, 11000–10000 B.P., almost at sea level in the Isle of Man. These cracks are one of the few recognized features that indicate the existence of permafrost at the time of their formation (French 1976, p. 136). Furthermore, these cracks are frequently developed in gravels (including those from the Isle of Man) suggesting that average annual temperatures must have been below −6 to −8 °C (Williams 1975). If these cracks were formed more or less at the same time as the insect-bearing deposits, it is possible to associate these average annual temperatures with the figure of about 10 °C for the average July temperature inferred from the Coleoptera. Though precise contemporaneity cannot be established with certainty there can be little doubt that some of the cracks cannot have been far removed in time from the insect faunas with their abundant continental species. If some of these wedges are indeed sand wedges, and some of them certainly resemble these rather than ice-wedge casts, there is evidence that the cold period that characterized much of the mid-Devensian was arid as well.

No thermal contraction cracks have yet been found that suggest contemporaneity with the thermal maxima of the interstadials. What is more some deposits that yield temperate insect faunas also contain large molluscs, such as *Anodonta* indicating that at such times, precipitation was adequate to maintain shallow ponds not only throughout the whole year but for several years in succession. At least during the thermal maxima of the Upton Warren and Windermere interstadials the climate appears to have been moderately oceanic and not much different from that of northwest Europe of the present time.

The rapid switches in the terrestrial climate of the British Isles from cold-continental conditions on the one hand to temperate-oceanic conditions on the other may be a reflection of the changes in the north Atlantic circulation described by Ruddiman & McIntyre (1973). Certainly with respect to the Windermere Interstadial both the timing and intensity of the climatic change on land resemble the changes in the temperature of the ocean waters of the west coast of Britain. Similarly the return of polar water to the western coasts of Britain corresponds with the timing of the ice advances of the Loch Lomond Stadial (W. F. Ruddiman, personal communication). If this relation between the climate of the British Isles and the circulatory régime of the adjacent ocean is a general one, we might expect a northward excursion of the polar front in the eastern Atlantic around 43000 B.P. and corresponding to the thermal maximum of the Upton Warren Interstadial. Such is the shortness of this episode that it may be close to the resolving power of ocean core analysis.

In spite of the intensity of the thermal maximum of the Upton Warren Interstadial it is seemingly short and hemmed in on either side by climates of near arctic severity. It is unlikely, therefore, to be reflected by a return to present-day figures either in the sea-level changes or isotopic fluctuations in ocean water that are both indications of gross ice volume. A short sharp interstadial is unlikely to inflict the same damage on the world's ice sheets as a cooler but much longer interstadial interval.

There can be little doubt therefore that the rapid and intense climatic changes in western Europe, and particularly in the British Isles where they are likely to be at their most extreme, are dependent on the interaction of the oceanic and terrestrial geometry upon the circulatory patterns of the hydrosphere and atmosphere in this particular area. It is not to be expected that, in their intensity, timing and duration, those climatic changes have necessarily any hemisphere wide or global equivalents.

References (Coope)

Andrewartha, H. G. & Birch, L. C. 1954 *The distribution and abundance of animals.* Chicago: The University of Chicago Press.

Ashworth, A. C. 1972 A Late-glacial insect fauna from Red Moss, Lancashire, England. *Entomol. scand.* **3**, 211–224.

Bell, F. G. 1969 The occurrence of southern, steppe and halophyte elements in Weichselian (Late-glacial) floras from southern Britain. *New Phytol.* **68**, 913–922.

Bishop, W. W. & Coope, G. R. 1977 Stratigraphical and faunal evidence for Lateglacial and Early Flandrian environments in south-west Scotland. In *Studies in the Scottish Lateglacial Environment* (eds J. M. Gray & J. J. Lowe). Oxford: Pergamon Press Ltd.

Brown, A. 1977 Late Devensian and Flandrian vegetational history of Bodmin Moor, Cornwall. *Phil. Trans. R. Soc. Lond.* B **276**, 251–320.

Connolly, A. P., Godwin, H. & Megaw, E. M. 1950 Studies in Post-glacial history of the British vegetation. XI. Late-glacial deposits in Cornwall. *Phil. Trans. R. Soc. Lond.* B **234**, 397–469.

Coope, G. R. 1959 A late Pleistocene insect fauna from Chelford, Cheshire. *Proc. R. Soc. Lond.* B **151**, 70–86.

Coope, G. R. 1974 Report on the Coleoptera from Wretton. (In West, R. G., Dickson, C. A., Catt, J. A., Weir, A. H. & Sparks, B. W.: Late Pleistocene deposits of Wretton, Norfolk. II. Devensian deposits.) *Phil. Trans. R. Soc. Lond.* B **267**, 414–418.

Coope, G. R. 1975 Climatic fluctuations in northwest Europe since the Last Interglacial indicated by fossil assemblages of Coleoptera. *Geol. J.* special issue No. 6, 153–168.

Coope, G. R. & Angus, R. B. 1975 An ecological study of a temperate interlude in the middle of the Last Glaciation, based on fossil coleoptera from Isleworth, Middlesex. *J. Anim. Ecol.* **44**, 365–391.

Coope, G. R. & Brophy, J. A. 1972 Late Glacial environmental changes indicated by a coleopteran succession from North Wales. *Boreas* **1**, 97–142.

Coope, G. R., Morgan, Anne & Osborne, P. J. 1971 Fossil Coleoptera as indicators of climatic fluctuations during the Last Glaciation in Britain. *Palaeogeogr., Palaeoclimatol., Palaeoecol.* **10**, 87–101.

Coope, G. R., Shotton, F. W. & Strachan, I. 1961 A Pleistocene fauna and flora from Upton Warren, Worcestershire. *Phil. Trans. R. Soc. Lond.* B **244**, 379–421.

Corbet, P. S. 1972 The microclimate of arctic plants and animals on land and in fresh water. *Acta Arctica* **18**, 1–43.

Dickson, C. A., Dickson, J. H. and Mitchell, G. F. 1970 The Late-Weichselian flora of the Isle of Man. *Phil. Trans. R. Soc. Lond.* **258**, 31–79.

Foster, G. 1968 *Hydrophorus longicornis* (Col. Dytiscidae) in East Sussex with notes upon its known distribution. *Ent. Mon. Mag.* **104**, 139–146.

French, H. M. 1976 *The Periglacial environment*. London and New York: Longman.

Girling, M. A. 1974 Evidence from Lincolnshire of the age and intensity of the mid-Devensian temperate episode. *Nature, Lond.* **250**, 270.

Henson, W. R. 1968 Some recent changes in the approach to studies of climatic effects on insect populations. *Insect abundance*. Symposium Number 4. London: Royal Entomological Society.

Joachim, M. J. 1974 A Late-glacial coleoptera fauna from Glen Ballyre, Isle of Man. Unpublished M.Sc. thesis, University of Birmingham.

Lindroth, C. H. (ed.) 1960 *Catalogus Coleopterorum Fennoscandiae et Daniae* Lund.

Mangerud, J., Anderson, S. T., Berglund, B. E. & Donner, J. J. 1974 Quaternary stratigraphy of Norden, a proposal for terminology and classification. *Boreas* **3**, 109–128.

Matthews, J. V. 1974 Quaternary environments at Cape Deceit (Seward Peninsula, Alaska): evolution of a tundra ecosystem. *Bull. Geol. Soc. Am.* **85**, 1353–1384.

Matthews, J. V. 1976 Evolution of the subgenus *Cyphelophorus* (Genus *Helophorus*, Hydrophilidae, Coleoptera): description of two new fossil species and discussion of *Helophorus tuberculatus* Gyll. *Can. J. Zool.* **54**, 652–673.

Mitchell, G. F. 1965 The Quaternary deposits of the Ballaugh and Kirkmichael districts, Isle of Man. *Q. Jl geol. Soc. Lond.* **121**, 359–381.

Morgan, Anne 1970 Weicheselian insect faunas of the English Midlands. Unpublished Ph.D. thesis, University of Birmingham.

Morgan, Anne 1973 Late Pleistocene environmental changes indicated by fossil insect faunas of the English Midlands. *Boreas* **2**, 173–212.

Morgan, A. V. 1973 The Pleistocene geology of the area north and west of Wolverhampton, Staffordshire, England. *Phil. Trans. R. Soc. Lond.* B **265**, 233–297.

Osborne, P. J. 1971 The insect faunas of the organic deposits within the Wandle Gravels. *Proc. Croydon Nat. Hist. Sci. Soc.* **14**, 162–175.

Osborne, P. J. 1972 Insect faunas of Late Devensian and Flandrian age from Church Stretton, Shropshire. *Phil. Trans. R. Soc. Lond.* B **263**, 327–367.

Pearson, R. G. 1962 The Coleoptera from a detritus mud erratic of Full-glacial age at Colney Heath near St Albans. *Proc. Linn. Soc. Lond.* **173**, 37–55.

Pennington, W. 1975 A chronostratigraphic comparison of Late-Weichselian and Late-Devensian subdivisions, illustrated by two radiocarbon dated profiles from western Britain. *Boreas* **4**, 157–171.

Pennington, W. 1977 The Late Devensian flora and vegetation of Britain. *Phil. Trans. R. Soc. Lond.* B **280**, 247–271 (this volume).

Pennington, W. & Bonny, A. P. 1970 Absolute pollen diagram from the British late-glacial. *Nature, Lond.* **226**, 871–873.

Penny, L. F., Coope, G. R. & Catt, J. A. 1969 Age and insect fauna of the Dimlington silts, East Yorkshire. *Nature, Lond.* **224**, 65–67.

Price, P. W. 1975 *Insect ecology*. New York, London, Sydney and Toronto: J. Wiley & Sons.

Richards, O. W. & Southwood, T. R. E. 1968 The abundance of insects: introduction. *Insect abundance*. Symposium Number 4. London: Royal Entomological Society.

Ruddiman, W. F. & McIntyre, A. 1973 Time-transgressive deglacial retreat of polar waters from the North Atlantic. *Quat. Res.* **3**, 117–130.

Simpson, I. M. & West, R. G. 1958 On the stratigraphy and palaeobotany of a late-Pleistocene organic deposit at Chelford, Cheshire. *New Phytol.* **57**, 239–250.

Sissons, J. B. 1974 The Quaternary in Scotland: a review. *Scot. J. Geol.* **10**, 311–337.

Sissons, J. B. & Walker, M. J. C. 1974 Lateglacial site in central Grampian Highlands. *Nature, Lond.* **249**, 822–824.

Shotton, F. W. 1972 An example of hard-water error in radiocarbon dating of vegetable matter. *Nature, Lond.* **240**, 460–461.

Southwood, T. R. E. (ed.) 1968 *Insect abundance*. Symposium Number 4. London: Royal Entomological Society.

Vogel, J. C. & Zagwijn, W. H. 1967 Groningen radiocarbon dates: VI. *Am. J. Sci. Radiocarbon Suppl.* **9**, 63–106.

West, R. G., Dickson, C. A., Catt, L. A., Weir, A. H. & Sparks, B. W. 1974 Late Pleistocene deposits at Wretton, Norfolk. II. Devensian deposits. *Phil. Trans. R. Soc. Lond.* B **267**, 337–420.

Williams, R. B. G. 1975 The British climate during the Last Glaciation; an interpretation based on periglacial phenomena. *Geol. J.* special issue No. 6, 95–117.

Williams, R. E. G. & Johnson, A. S. 1976 Birmingham University radiocarbon dates: X. *Am. J. Sci. Radiocarbon Suppl.* 18.

Wilson, F. 1968 *Insect abundance:* Prospect. Insect Abundance. Symposium Number 4. London: Royal Entomological Society.

Discussion

By G. R. Coope and Winifred Pennington.

The Windermere Interstadial of the Late Devensian

We propose to use the Windermere section as a standard reference section because:

(i) A large lake has several favourable characteristics, the most important being the *continuity* of the sedimentary sequence,

(ii) Lake sediments reflect the terrestrial as well as the lacustrine environment,

Table 1. Correlation table for the lithostratigraphy, biostratigraphy and chronostratigraphy of the Late Devensian Deposits at Low Wray Bay, Windermere

lithology	fossils: plants	fossils: animals – Coleoptera	fossils: animals – others	^{14}C dates	interpretation: environment	interpretation: climate	duration of interstadial climate: W.P.	duration of interstadial climate: G.R.C.
rock fragments increase upper laminated clay	*Artemisia* p.a.z. pollen very sparse	sparse fauna with characteristic arctic–alpine species	Trichoptera larvae		seasonal melt from mountain glaciers in Lake District corries	very cold		
				(11 000)				
becoming less organic maximum carbon organic detritus silt	Cyperaceae *Betula*–*Juniperus* p.a.z. tree birch macroscopics	fauna less rich, without 'southern' species woodland species present	Cladocera decline to 0 Chironomidae larvae *Cristatella* statoblasts	SRR-669 11 344 ±90	increased snowfall; declining temperatures; increased soil erosion re-establishment of *Betula*–*Juniperus* woods	cool temperate becoming colder towards the top	↑	↑
paler silt with rock fragments	*Betula*–*Rumex* p.a.z.	*Helophorus glacialis*	Cladocera reduced	SRR-670 12 213 ±150	increased soil erosion woodland reduced	sudden cooling		
maximum carbon organic detritus silt	*Betula* p.a.z. *Juniperus* p.a.z.	sparse fauna at base, becoming richer towards the top species present that are absent from arctic Europe		SRR-672 12112 ±125 SRR-673 12517 ±150 SRR-675 12441 ±82 SRR-676 12913 ±120	maximum environmental diversity of Windermere Interstadial (woodlands established)	temperate throughout	↓	
				(13 000)				
somewhat organic silt carbon content increasing	*Rumex*-Gramineae p.a.z. (tree birches present)		Cladocera suddenly increase	SRR-677 13185 ±170 SRR-679 13938† ±210 SRR-680 13863† ±270	phase of pioneering flora and fauna; vegetation rich in herbs, including species which do not grow in the arctic today; although tree birches are present, no development of woodland	plant evidence inconclusive on temperature, but some factor unfavourable to trees / temperate throughout on insect evidence		
slightly organic silt	*Salix herbacea*-Cyperaceae-*Lycopodium selago* p.a.z. moss stems	very sparse fauna; no alpine species		SRR-681 14557 ±280 SRR-682 14623 ±360	on plants, compare with Middle Alpine zone of NW Europe / insect fauna lacks any alpine species and suggests early pioneer fauna	plant evidence suggests alpine conditions / insects inconclusive but no evidence to support alpine conditions		↓
clay	no plant fossils							
lower laminated clay					seasonal melt from mountain glaciers	very cold		

† ^{14}C dates with evidence of 'hard-water error'.

(iii) The position of Windermere is central in the British Isles, and the lake received glacial drainage, both in the closing phases of the main Devensian glaciation and during the period between 11000 and 10000 ^{14}C a B.P. which is correlated with the Loch Lomond Readvance of Sissons (1974),

(iv) The interstadial sequence, most fully developed in the marginal parts of the lake (Pennington 1943), is best known from Low Wray Bay, where the sediments contain both pollen and the macroscopic remains of plants and animals which have been washed into the lake by small streams.

Detailed analyses of this section, the results of which are presented in our papers, show that the lower boundary of interstadial sediments may be drawn at different horizons, depending on the criteria that are adopted, whether they be lithostratigraphic, palaeoentomological, or palaeobotanical. Table 1 indicates the position of the lower boundary of the Windermere Interstadial on each criterion. If the notion of interstadial is based on a climatic interpretation, two alternative views (those of W.P. and G.R.C.) are given on the right-hand side of the table.

Description of stratotype

1. *Lithostratigraphy*

We interpret the upper boundary of the Lower Laminated Clay (Pennington 1943) as representative of the termination of the main Devensian glaciation of northwest England, and the commencement of interstadial sedimentation. 'Interstadial' is here used in its lithostratigraphic sense, and the climatic change represented is from a glacial to a non-glacial environment.

Above the Lower Laminated Clay the lowest sediments are almost entirely minerogenic with extremely sparse fragments of plant and animal tissue. The change to biogenic sediment is very gradual and there is no clearly discernible boundary between clay and organic silt. At its most organic the silt contains *ca.* 5% carbon. At about 15 cm below the top of the organic silt is a persistent horizon of grey silt containing rock fragments. The base of the Upper Laminated Clay (Pennington 1943) defines the commencement of Loch Lomond Stadial sedimentation in this area.

2. *Biostratigraphy*

The biostratigraphy is summarized in table 1. The environmental interpretation based on both floral and faunal biostratigraphy yields a consistent ecological picture throughout most of the time represented by non-glacial sediments.

On the palaeobotanical definition of 'interstadial' – namely an environment capable of supporting a vegetation of woody plants (trees and shrubs), only part of these non-glacial sediments can be described as interstadial – see table 1. The usefulness of the palaeobotanical definition is that comparison between British sites indicates the presence of a generally distinctive vegetation of woody plants between the ^{14}C dates which delimit the woodland biozone in Windermere, so the particular environment required by a vegetation of woody plants appears to have been restricted to the period between *ca.* 13000 and 11000 ^{14}C a B.P.

Below this woodland biozone the plant remains define two botanical subzones within a herbaceous biozone. Pollen data equate the lower of these with present alpine vegetation in northern Europe. In the upper zone the assemblage of herbs identified includes species which do not grow in the Arctic today, and both pollen and macroscopic evidence shows that tree birches were present but did not expand into woodland. This is interpreted as indicative of conditions not generally favourable to trees.

Only within the herbaceous biozone is there any discrepancy between the interpretations based on botanical and entomological data. Throughout the two divisions of the herbaceous biozone the fauna shows a continuous development, but even at the base no obligate alpine species are present. A number of insect species occur which, though widespread in Europe today, are absent from arctic regions and the alpine zones of the mountains, suggesting that, at least during the *Rumex*–Gramineae pollen zone, temperate climatic conditions prevailed. Comparable temperate insect assemblages, often much richer than that from Windermere, are now known from a number of localities in Britain always in the same palynological context – namely well before the commencement of the woodland biozone. Whatever factor was unsuitable for trees during the herbaceous biozone, it did not inhibit the development of an ecologically complex temperate insect fauna.

No contemporary biota match the fossil assemblage of plants and insects from these lower zones. Both pollen and insect data suggest primarily pioneer communities. This may be an important reason why it is difficult to match these phases of colonization with existing biota which are in equilibrium with contemporary climates.

References

Pennington, W. 1943 Lake sediments: the bottom deposits of the north basin of Windermere with special reference to the diatom succession. *New Phytol.* **42**, 1–27.
Sissons, J. B. 1974 The Quaternary of Scotland: a review. *Scot. J. Geol.* **10**, 311–337.

G. F. Mitchell, F.R.S. and R. G. West, F.R.S. It is important to provide a defined lower boundary for the proposed Windermere Interstadial. The difficulty is that such a boundary may be defined either lithostratigraphically or biostratigraphically, and if biostratigraphically on the basis of pollen or beetles. There is the usual difficulty of definition which confronts the Pleistocene stratigrapher when there is a plethora of information. One solution is to draw an arbitrary line at a radiocarbon age, say 14000 B.P., a time when pollen content increases and *Betula* and *Juniperus* increase their representation (see Tutin, this volume, p. 337). Alternatively, a pollen-based interstadial could be defined as beginning at 13000 B.P. at the base of the *Juniperus* pollen frequency rise, or a beetle-based interstadial could be defined as beginning earlier than 14000 B.P. (see Coope, (this volume, p. 337).

The decision to place the definition of the base of the interstadial must be an arbitrary imposition on the observed lithostratigraphical–biostratigraphical sequence. We suggest 14000 B.P. is a position which has some significance for the evolution of the climate in the late-Devensian, with observed changes in pollen and beetles in the Windermere sequence, and that this age should mark the base of the Windermere Interstadial. We note that this level corresponds closely with the boundary between the *Salix herbacea*-Cyperaceae and *Rumex*-Gramineae-Rubiaceae-*Artemisia* pollen assemblage biozones at Windermere (Tutin, this volume, p. 255).

In Ireland the corresponding Woodgrange Interstadial opens at about 14000 B.P. at the base of the Irish *Rumex*-Gramineae pollen assemblage biozone (see Mitchell 1976, pp. 68–73).

A. V. Morgan (*Department of Earth Sciences, University of Waterloo, Waterloo, Ontario, Canada*). Dr Coope has shown a number of remarkable shifts in the distribution of beetle species from their known fossil localities in the British Isles to their present ranges in northern, central or southern Europe, and even beyond, into Asia. I would like to point out that changes of similar magnitude have taken place in North America. Dr Anne Morgan, working on beetle faunas from Toronto, has shown that many of the species found in Early Wisconsin deposits at Scarborough occur today along the northern treeline of Canada. Several of these species now live in the extreme northwest of the continent, e.g. *Diacheila polita* (one of the species mentioned in Dr Coope's lecture) is present in Alaska and occurs at its closest point (as far as is known) in the Mackenzie Delta some 4000 km from the fossil locality. *Helophorus sibiricus* has a similar distribution except that its present occurrence seems centred on the west coast of Alaska over 4800 km from Scarborough. Unfortunately one is faced with the dilemma of poor collecting records in areas nearer Ontario (northern Ungava/Labrador for example), absence of readily available collections (there are no easily accessible Victorian collections in Canada) and a dearth of knowledge about the present ecology of the much larger North American fauna. Distributions of other species of Coleoptera, *Asaphidian yukonense* and *Carphorborus andersoni*, indicate that these northwestern species occupied sites in southern Ontario and Wisconsin until about 10000 a B.P.

The distribution of relict permafrost features in North America does not appear to have the same widespread areal extent as those described in western and central Europe. Permafrost features believed to have been formed contemporaneously with the maximum Wisconsin ice advance (or with the retreating ice) have been described from Washington, Montana, North Dakota, Wisconsin, Illinois, Indiana and possibly the New England States in the United States, and from southwestern Ontario, Quebec, Nova Scotia and Newfoundland in Canada. The permafrost indicators are usually ice-wedge casts, although associated polygonal ground has been found in North Dakota, Indiana and southwestern Ontario. The patchy nature of the relict structures suggests that permafrost was present in restricted areas, possibly due to localized climatic conditions or to favourable lithologic and hydrologic regimes. The disappearance of permafrost in southern Ontario (43° N) is roughly coincident with that in the central English Midlands (53° N) about 13500 a B.P. Permafrost conditions prevailed later in Gaspé, Quebec, and in southwestern Newfoundland, apparently disappearing about 10000–10500 a B.P.

J. D. PEACOCK (*Institute of Geological Sciences, Murchison House, West Mains Road, Edinburgh*). At the Institute of Geological Sciences we have been examining the late-glacial raised marine deposits in the Clyde area, and have now obtained radiocarbon-dated, overlapping sections for the interval from a little before 12000 B.P. to 10000 B.P. It is clear from the lithological and faunal data that water temperatures were generally much lower than at present in the area, and that conditions were arctic during the Younger Dryas period (*ca.* 11000–10000 B.P.). However, there is evidence for a short-lived amelioration about 11000 B.P. when low-boreal Mollusca and foraminifera immigrated into the district, though winter temperatures probably remained low. Similar temperature variations can be inferred from the records of a poorly dated borehole from the Sea of the Hebrides (Binns, Harland & Hughes 1974), and from undated late-glacial sections described 100 a ago near Greenock. Since these temperature changes appear to have been general, it is unlikely that they were influenced by any relict cover of glacier ice in the Highlands. The form of our temperature curve thus differs from that of Dr Coope, and one wonders if there is any evidence from deposits on land for a more temperate interval at 11000 B.P.

Reference

Binns, P. E., Harland, R. & Hughes, M. J. 1974 Glacial and post-glacial sedimentation in the Sea of the Hebrides. *Nature, Lond.* **248**, 751–754.

Phil. Trans. R. Soc. Lond. B. **280**, 341–350 (1977) [341]
Printed in Great Britain

GENERAL DISCUSSION

Climatic analysis

By H. H. Lamb
Climatic Research Unit, School of Environmental Sciences, University of East Anglia

In opening the proceedings of this meeting Mr J. S. Sawyer of the U.K. Meteorological Office uttered the view that it is really premature as yet to ask a meteorologist to interpret the events of the last glaciation. Nevertheless, that is just what I have been asked to do and, though one must tread warily, I believe it must be attempted. Indeed, it *has* been attempted in a number of scientific papers in recent years and these two days' proceedings have surely helped us to see a number of things more clearly.

I have no time to do justice to all the points that appear interesting.

A meteorologist looks first for the large-scale patterns, which are probably the least difficult to discern and which provide the framework into which all else must fit.

Most meteorologists, oceanographers and perhaps all who are neither geologists nor palaeobotanists, will also probably echo Professor W. A. Watts's reservations about the many names adopted for the various cold and warm stages. One must acknowledge the need which has called the multiplicity of naming systems into being, but the outsider can only accept them reluctantly as marking a provisional stage until the dating is firm and the correspondence of the events known by different names in different parts of the world has been established. This huge, and growing, vocabulary is formidable to those outside the debates about the field evidence and tends to deter other scientists who might contribute to interpreting the processes of climatic change in the Quaternary. From this point of view, and to such audiences, dates and numbered stages, or one single series of names, are much to be preferred.

Data

This meeting has revealed an abundance of data about the last glaciation which would have astounded any previous generation of scientists. With many new types and techniques of observation, measurement and analysis have come global coverage, knowledge of many more fluctuations, and possibilities of corroboration from quite independent lines of evidence. Discrepancies in the apparent climatic indications, where they occur, provide insights into different types of impact and response rates. It seems likely that the most reliable approximation to absolute values of the elements of past climatic regimes will generally come from physical science, particularly from isotope measurements. The records of past ice volume and palaeotemperatures indicated by the oxygen isotope measurements on material from ocean-bed cores (Shackleton's record of the last 2 Ma) and from the Greenland and Antarctic ice sheets, together with the CLIMAP world map of ocean surface temperatures at the last glacial maximum, seen at this meeting, must be accounted triumphs of palaeoclimatic research.

All our data are, nevertheless, only estimates to which some margin of error applies. And that is actually still the case, despite all refinements of instrumentation, with observations of the present régime today – most particularly as regards rainfall over the oceans, snowfall anywhere, sea surface temperatures and temperatures in the upper air. And in all cases, ancient and modern, we must question the representativeness of the observations available on account of the site and time to which they refer.

Causes and mechanisms, amplitude and phasing of the climatic changes

Spectral analysis of such records as we have seen of Pleistocene climatic variations, reported by J. D. Hays, J. Imbrie and N. J. Shackleton at the Norwich symposium on Long-term climatic fluctuations, published by the World Meteorological Organization (*W.M.O. No. 421*, Geneva 1975), seems now to verify the Milankovitch thesis in so far as the timing (if not the amplitude) of the major climatic fluctuations is dominated by the periods of variation of the Earth's orbital characteristics (ellipticity, obliquity and precession), of about 100000, 40000 and 20000 years.

The phasing of the climatic changes in the northern and southern hemispheres needs further study and explanation. The large-scale long-term variations seem to run nearly in phase in both hemispheres; though we have seen evidence that over the last 5000 years – and this is also true of some of the much shorter-term fluctuations – the histories may be out of phase.

It may seem surprising that the British Isles, which today enjoy such an equable climate, cushioned by the surrounding seas from the severer seasonal shocks and year-to-year variability of the continental interior, emerge as one of the regions of greatest change in the ice age–interglacial cycles. To describe this as increased 'continentality' of the climate in glacial times, though literally true, does not tell the whole story: in particular, it overlooks the substantial contribution from the ocean to the variability of our climate.

The amplitude of year-to-year climatic variability as well as the seasonal range within the year varies with time and differs in different régimes. This is registered by the variability of the yearly ring widths in oak-trees in central Germany since A.D. 1280 in figure 1. Four distinct periods of greatly enhanced variations stand out: about 1300, 1550, 1700 and 1860, coinciding respectively with the end of the medieval warmest régime, the beginning and end of the coldest stage of the so-called Little Ice Age climate and the beginning of the century-long retreat of the glaciers in most parts of the world. There can be little doubt that the enhanced variability in those periods was associated with great prevalence of 'blocking', i.e. of stationary wind circulation systems in middle latitudes, occupying somewhat different longitudes in different years (so that any one place experiences sometimes prevalence of northerly, sometimes southerly winds, sometimes persistent anticyclonic weather and droughts, and sometimes a focus of cyclonic activity and repeated rains). It may be that such marked enhancements of the frequency of blocking patterns, replacing the smooth flow of the middle latitudes westerlies, constitute a regular signal of the end of a régime – whether the ensuing change be great or small.

It also seems likely that the year-to-year variability of temperature in the British Isles in glacial times, with the boundary between the persistent snow and ice surface and much warmer land and ocean surfaces near, would be two to three times as great as now, resembling that in Spitsbergen or Archangel today.

The explanation of the magnitude of the longer-term persistent changes of climate in the British Isles may well lie in the variations of the supply of polar water in the East Greenland

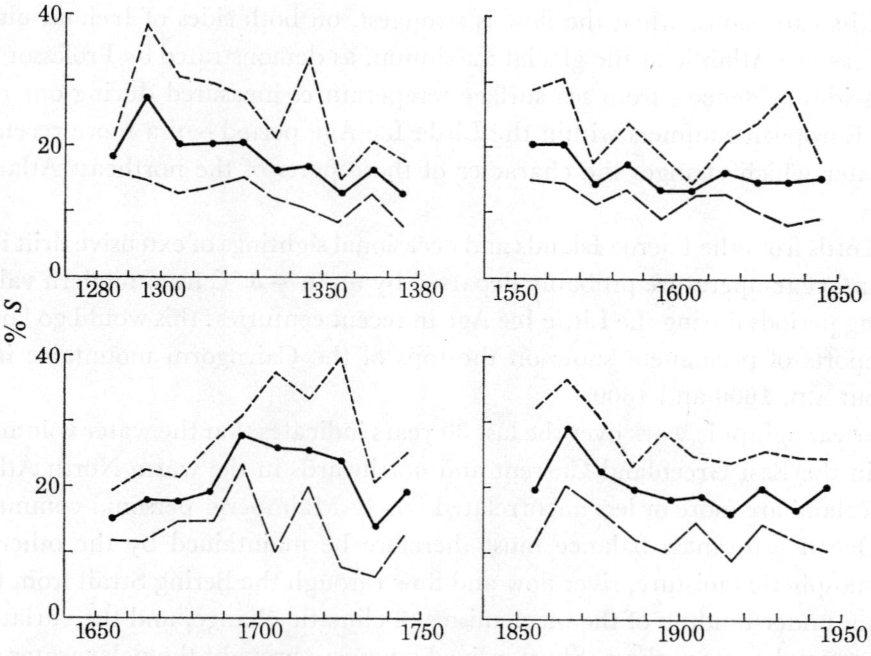

Figure 1. Mean variation of ring width from year to year in oak trees in the Spessart forest area, central Germany in various periods since A.D. 1280. ——, average of many trees; ----, extreme values of the period-averages obtained from individual trees. (From Fürst 1963.)

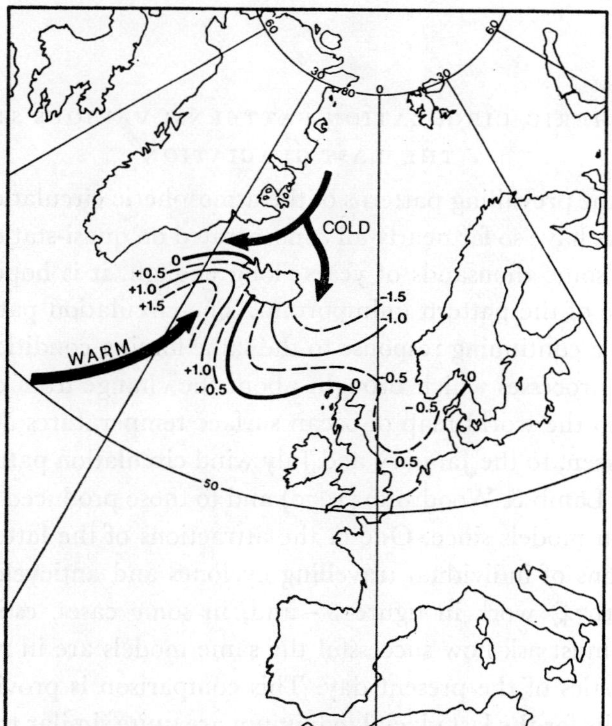

Figure 2. Sea surface temperatures measured in the summer of A.D. 1789, during the voyage of the Stanley expedition to the Faeroes, Iceland and Denmark, as departures (°C) from twentieth-century (1919–38) averages. (Data in West 1970.)

Current and its extensions, when the flow is strongest, on both sides of Iceland ultimately to 42° N in the eastern Atlantic at the glacial maximum, as demonstrated by Professor McIntyre. Figure 2 provides evidence – from sea surface temperatures measured during one of the relasively warm European summers within the Little Ice Age period – of a more recent thrust of this polar water which changes the character of those parts of the northeast Atlantic that it reaches.

Fishery records from the Faeroe Islands and occasional sightings of extensive drift ice indicate that water surface temperature probably departed by up to −5 °C from modern values in that region for long periods during the Little Ice Age in recent centuries: this would go far to explain the many reports of permanent snow on the tops of the Cairngorm mountains in Scotland between about A.D. 1600 and 1800.

Icelandic oceanographic work over the last 30 years indicates that the water volume transport southwards in the East Greenland Current and northwards in the warm North Atlantic Drift water near Iceland are more or less uncorrelated (S.-A. Malmberg, personal communication): the Arctic Ocean watermass balance must therefore be maintained by the other variables, including atmospheric moisture, river flow and flow through the Bering Strait from the Pacific.

We need an understanding of the mechanisms of climatic change, and the variations of this ocean current may be one of them. Shorter-lived advances brought the polar water near to the Faeroe Islands again in 1888 and 1968.

We also need to study the rapidity, and the course, of climatic changes, recognizing that most available fossil evidence – with margins of uncertainty in the dating and disturbances of the surface of deposits – blurs the changes, extending their apparent period of development and reducing their range.

Atmospheric circulation patterns: various stages of the last glaciation

Reconstructions of the prevailing patterns of the atmospheric circulation and of other aspects of past climatic régimes have so far nearly all concentrated on quasi-stationary situations, which apparently lasted for some thousands of years: long enough, it is hoped, to render errors in dating of the evidence of the pattern unimportant. The circulation patterns so developed are those which represent a continuing response to the same forcing conditions. They therefore tell us nothing about the processes which brought about the change from one régime to another. This applies equally to the world map of ocean surface temperatures at the last glacial maximum which we have seen, to the January and July wind circulation patterns first reconstructed for the same epoch by Lamb & Woodroffe (1970) and to those produced by the most sophisticated general circulation models since. One of the attractions of the latter is that they produce proposed daily positions of individual travelling cyclones and anticyclones – as illustrated by the Williams *et al.* (1974) work in figure 3 – and, in some cases, estimates of precipitation amount etc. But one must ask how successful the same models are in reproducing the known circulation characteristics of the present day. This comparison is provided by Williams *et al.* (1974), and their results for the last glacial maximum are quite similar to Lamb & Woodroffe's. But some other circulation models which indicate more easterly surface winds in middle latitudes produce a great deal too much easterly wind (and too strong monsoonal development) in the present-day régime.

Lamb & Woodroffe (1970) also reconstructed mean circulation maps for the Late Glacial warm substage known as the Allerød Interstadial and for the cold substage (Zone III in the European pollen diagrams) which followed. On the interpretation so derived the surface wind régime was generally easterly over Britain in both substages, but with southerly components in both summer and winter in the warm substage and northerly components in the cold substage: this change resulted from a westward shift ('retrogression') of the controlling features of the atmospheric circulation over Europe, as the thermal gradients and circulation strength gradually weakened in middle latitudes over the Atlantic and Europe, implying shortening wavelength in the upper westerlies. This result, however, seems – at least at first sight – to conflict with the

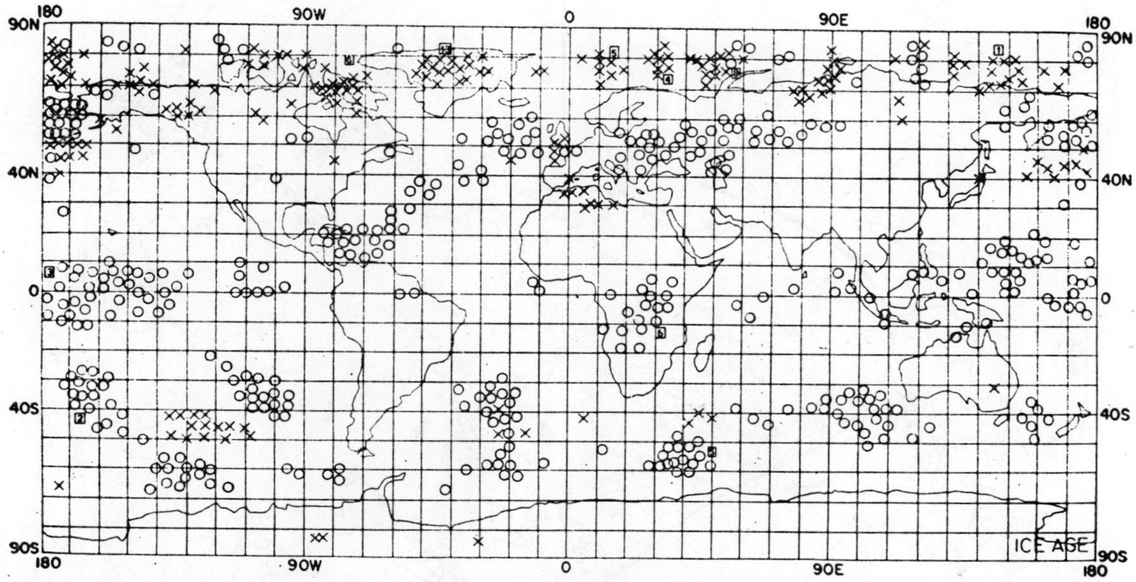

FIGURE 3. Positions of cyclone and anticyclone centres on days 51–80 of integrations with the NCAR globa circulation model: July ice-age situation. (From Williams *et al.* 1974.)

evidence reported by Manley (1959) that the winds which produced the reappearance of corrie glaciers in the Lake District of northwest England were southwesterly. Can it be that the climatic change between the warm and cold substages was brought about quite quickly by an interval in which southwesterly winds prevailed with slow-moving cyclonic centres nearly over the British Isles (probably in the Rockall–Hebrides region) and repeated outbreaks of arctic air and a cold ocean current from the north over the Atlantic just west of these islands?

I believe Professor G. F. Mitchell has put an important proposition before this meeting in his suggested 'frost cycle', distinguishing an onset phase in which abundant snowfall occurs with temperatures not nearly so low as subsequently develop. The winds prevailing in the ensuing cold stage may be quite different from those under which the glaciers and ice sheets formed.

Similarly the variations of strength of the upper westerly winds over Japan, derived by Yoshino & Tabuchi (1975) from survey of the volcanic ash deposits from the ever-active volcanoes, has indicated a phasing in which the westerlies were strongest at times of increasing glaciation in the northern hemisphere and much weaker at the maxima of glaciation.

The proposed circulation pattern of the glacial onset in the early Devensian/Weichselian/ Wisconsin (figure 4), put forward by Lamb & Woodroffe (1970), illustrates the results of similar thinking and would fit the circumstances of the change from Late Glacial Zone II to Zone III proposed above if the features of the circulation over Europe were displaced about 20° of longitude to the west. Such a displacement could very well have occurred at some stages in the Early Glacial also. Both positions would be consistent with massive enhancement of the East Greenland Current and its extension in branches east of Iceland and south of Denmark Strait – the

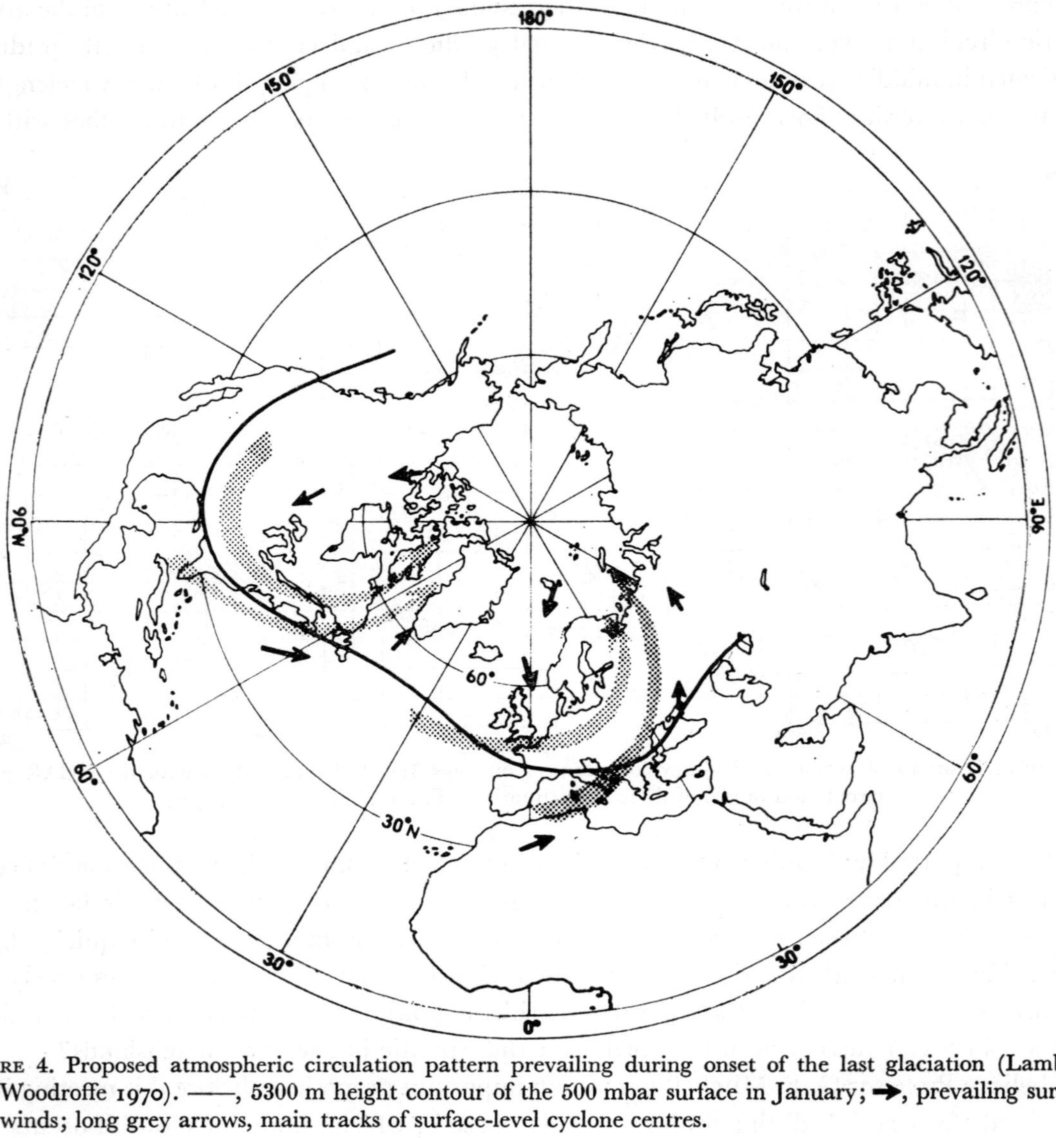

FIGURE 4. Proposed atmospheric circulation pattern prevailing during onset of the last glaciation (Lamb & Woodroffe 1970). ——, 5300 m height contour of the 500 mbar surface in January; ➔, prevailing surface winds; long grey arrows, main tracks of surface-level cyclone centres.

McIntyre–Ruddiman glacial current – carrying the polar water farther south. Even the more modest climatic fluctuations of recent times – the Little Ice Age development and the cold years in the 1880s and 1960s – seem to show a similar pattern, only less extremely developed and of briefer duration. The great postglacial warming and the notable warming of northern climates in the early twentieth century both entailed an equally great predominance of southerly surface wind components and northward thrust of the warm North Atlantic Drift water over the eastern Atlantic, through Denmark Strait and the entire Norwegian Sea.

MID-DEVENSIAN/MID-WEICHSELIAN CLIMATES

This meeting has also produced, in the paper by Professor R. G. West, a notable mustering of data from a network of points over Europe on the nature of the mid-Devensian/mid-Weichselian climates, which for tens of thousands of years were neither fully glacial nor fully interglacial. Dr Coope and others at Birmingham have over the years contributed greatly to our

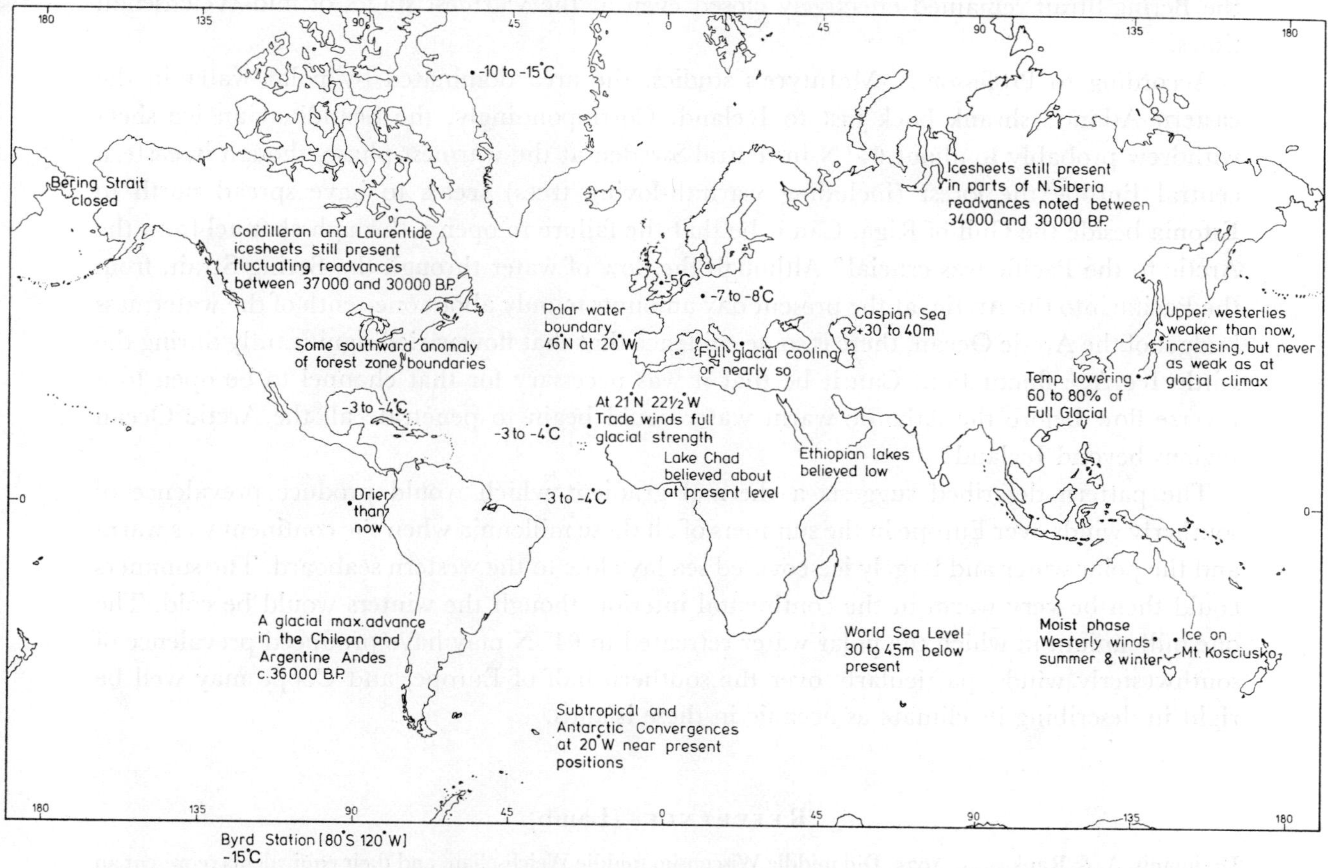

Figure 5. Summary of present knowledge about mid-Weichselian/mid-Wisconsin climate from 40000 to about 25000 years ago. Unlabelled temperatures are departures (°C) from modern averages. Dots mark positions to which the observations refer.

knowledge of the detailed succession over the British Isles, and Dreimanis & Raukas (1975) have lately attempted a hemispheric survey. As a result we are now approaching the possibility of a meteorological interpretation of this very strange climatic problem.

The kernel of the problem seems to be: why did the solar radiation budget, which according to the Milankovitch orbital considerations was particularly around 45000–40000 years ago very similar to what it is today, fail to produce a climate and environment similar to today's?

Figure 5 gives a global survey of conditions during the long period of what may perhaps be described as cold-temperate climates between about 40000 and 25000 a B.P. They may be summarized as intermediate between present and glacial conditions, though very much nearer to full glacial than to the present régime. They had been briefly preceded, apparently about 43000 B.P. in the British Isles, by a régime indicated by the beetle evidence as having warmer summers than now, though with no development of the temperate forest landscape. This was related to the times of maxima of northern-summer radiation receipt in low and high latitudes 5000–15000 a earlier similarly to our position today, except that the summer radiation maxima that time were less strong and the winters produced a deficit by comparison with the millennia leading up to our own times. Thus the integrated total radiation available over the preceding millennia was significantly less than now. To that can surely be attributed the failure of the northern continental ice sheets to melt completely away. Sea level remained low, and the Bering Strait remained effectively closed even at the warmest stages of mid-Weichselian times.

According to Professor A. McIntyre's studies, the area dominated by polar water in the eastern Atlantic shrank back just to Iceland. Correspondingly, the Scandinavian ice sheet withdrew probably to about 64° N in central Sweden at the warmest phase, though in eastern central Europe the forest (including warmth-loving trees) seems to have spread north to Estonia beside the Gulf of Riga. Can it be that the failure to open a through channel from the Arctic to the Pacific was crucial? Although the flow of water through the Bering Strait, from the Pacific into the Arctic, at the present day amounts to only about one-tenth of the watermass budget of the Arctic Ocean, there is some evidence that that flow varied significantly during the Little Ice Age fluctuation. Can it be that it was necessary for that channel to be open to a reverse flow before the Atlantic warm water could begin to penetrate all the Arctic Ocean regions beyond Iceland?

The pattern described suggests a thermal gradient which would produce prevalence of southerly winds over Europe in the summers of all those millennia when the continent was warm and the polar water and largely ice-covered sea lay close to the western seaboard. The summers could then be very warm in the continental interior, though the winters would be cold. The one millennium in which the polar water retreated to 64° N may have produced prevalence of southwesterly winds, particularly over the southern half of Europe, and Coope may well be right in describing its climate as oceanic in these islands.

References (Lamb)

Dreimanis, A. & Raukas, A. 1975 Did middle Wisconsin, middle Weichselian, and their equivalents represent an interglacial or an interstadial complex in the northern hemisphere? pp. 109–120 in *Quaternary Studies* (*Select Papers IX INQUA Congress*) (ed. R. P. Suggate & M. M. Cresswell) Wellington (Royal Society of New Zealand, *Bulletin No. 13*).

Fürst, O. 1963 Vergleichende Untersuchungen über räumliche und zeitliche Unterschiede interannueller Jahrringbreitenschwankungen und ihre klimatologischer Auswertung. *Flora* **153**, 469–508.

Lamb, H. H. & Woodroffe, A. 1970 Atmospheric circulation during the last ice age. *Quat. Res.* **1**, 29–58.

Manley, G. 1959 The Late-Glacial climate of Northwest England. *Liverpool and Manchester geol. J.* **2**, 188–215.

West, J. F. (ed.) 1970 *The journals of the Stanley expedition to the Faroe Islands and Iceland in 1789, vol. 1: Introduction and diary of James Wright.* Torshavn (Føroya Fróðskaparfelag).

Williams, J., Barry, R. G. & Washington, W. M. 1974 Simulation of the atmospheric circulation using the NCAR global circulation model with ice-age boundary conditions. *J. appl. Meteorol.* **13**, 305–317.

Yoshino, M. M. & Tabuchi, H. 1975 Wind variation in the lower troposphere over Japan in the Late Quaternary period, estimated by the distribution of volcanic ashes. Pp. 31–8 in *Proceedings of the WMO/IAMAP Symposium on Long-term climatic fluctuations, Norwich 18–23 August 1975*. Geneva (World Meteorological Organization – *WMO No. 421*).

Discussion

H. T. Mörth (*Climatic Research Unit, University of East Anglia, Norwich NR4 7TJ*)

As a climatologist, I should like to put the following issue to the biologists, especially the botanists. Future explanations of the state of the atmosphere during periods of glaciation may well include major changes in the temperatures and circulations of the stratosphere. Major variations in the amounts and distribution of ozone could be inferred. Since ozone filters out the ultraviolet portion of the solar radiation spectrum, this would imply variations in the amount of ultraviolet radiation received at the Earth's surface. Temporal changes in the plant and animal distribution during the Devensian that were put forward by a number of speakers during this meeting were attributed to temperature variations. Is there any evidence for differentiation according to the need for, or tolerance of, certain levels of u.v. radiation?

A. Dreimanis (*Geology Department, The University of Western Ontario, London, Canada*)

Several of the papers presented during this two-day meeting demonstrated that the Devensian stage experienced a wide range of climatic and environmental changes both in time, and laterally from one area to another, during any particular time interval. This was clearly demonstrated by the CLIMAP palaeoclimatic maps for the Late Devensian maximum and its correlatives 18000 a B.P.

Similar variability, though from a more spotty record, was suggested for the Middle Devensian and its correlatives in Europe: grasslands dominated in Ireland and Britain, possibly due to the cooling effect of the Atlantic Ocean, in spite of the brief very warm episode 43000 a ago. About the same time, forest tundra and forests spread into some parts of the continental Europe and northern Asia, even to higher latitudes than Britain, probably because of sunny warm summers and the dominance of high barometric pressure over the continental interior. The environmental differences between the near-Atlantic area and the interior of Eurasia have been so contrasting during the Middle Devensian and its equivalents, that these intervals have been called interstadials, interstadial complexes, or pleniglacial in the west, but considered to be interglacials, e.g. Karuküla and Karginsk, by some authors in the east.

If the above model of general west–east differences in Eurasia is correct, then even more contrasting mirror-like east–west climatic and environmental differences should be expected on the other side of the Atlantic – in the middle latitudes of North America. Why more contrasting? Because, according to the CLIMAP maps of the Atlantic Ocean, the polar front remained farther south along the eastern Atlantic coast of North America, than its eastern counterpart along the western coast of Europe. About ten days ago, on 19 and 20 March, regional east–west correlations for the last or Wisconsin Glaciation were discussed in Ottawa by the Canadian group of the IGCP Project 73/1/24 on Quaternary glaciations in the northern hemisphere. These correlations suggested the following east–west environmental and climatic change in southern Canada during the Middle Wisconsin substage: (*a*) the Appalachian glacial complex and the Laurentide ice sheet still covered most of the Atlantic Provinces of Canada (except for the coastal areas), and the St Lawrence Lowlands, at latitudes of southern Europe; (*b*) cool boreal climate and open woodlands were present in the Great Lakes Region at the same latitudes; (*c*) a climate little cooler than at present existed in south-central Manitoba, considerably northwest of Great Lakes, at the latitude of southern Ireland and southern Britain; (*d*) the summers were dry and as warm as at present or even warmer in the Peace River area of northwestern Alberta – at the latitude of north-central Labrador and Scotland. Another opposite climatic mirror image developed along an east–west traverse towards the Pacific: dry warm summers and complete deglaciation in south-central British Columbia, but a cooler interstadial climate, as concluded from pollen records, near the Canadian–United States border along the Pacific. As the duration of Middle Weichselian or its correlatives was several tens of thousands of years, the climate changed during this long interval in each of the areas mentioned; therefore the warmest episodes, which occurred before 40000 a B.P., were chosen for the above comparison.

H. H. Lamb. I hope that these suggestions and comments may indicate some profitable lines for further research. And I should like to congratulate those who organized this meeting, and the contributors, on the outcome. The research reviewed in these two days and the ideas and interpretations spoken of surely mark a great advance in our understanding of the development of the environment and of the parts played by the atmosphere and oceans in that development.

Phil. Trans. R. Soc. Lond. B. **280**, 351–372 (1977) [351]
Printed in Great Britain

The palaeogeography of mid- and east Europe during the last cold stage, with west European comparisons

By L. Starkel
Department of Physical Geography, Polish Academy of Sciences, Kraków, Poland

In this review of work on mid- and east European palaeogeography during the last cold stage the author has used a stratigraphic division based on deep-sea sediments in the Atlantic. Difficulties apparent in palaeogeographical reconstructions are discussed. In such reconstructions the sequences of permafrost structures, loesses, fossil soils, slope sheets, fluviatile sediments, vegetation, glacial deposits and land forms are analysed. The consideration of events in both their stratigraphic order (time scale) and their areal distribution in Europe has enabled the author to give a far more accurate indication of temperature and moisture changes in the different parts of Europe. Contrasts revealed by a section through east and west Europe reflect differences in the evolution of permafrost, in the succession of vegetation and soils, and in the history of glaciation. Their source was the different degree of climatic continentality which increased throughout the last cold stage. Continentality was characterized by the contemporaneous decrease in temperature, increase in temperature amplitudes and decrease in precipitation values. The presence of permafrost in Europe was responsible for the different development of biota. This makes for difficulty in reconstructing palaeogeographic conditions on the basis of actualism.

The duration of the last cold stage and the bases of stratigraphy

The stratigraphic divisions by Woldstedt (1958), Gross (1964) and others based on loess sequences with fossil soils, and the curves of van der Hammen, Maarleveld, Vogel & Zagwijn (1967) based on sequences of both alluvia and lake deposits which have been studied palynologically did not draw the lower boundary of the last cold stage (Vistulian, Weichselian) beyond 70000 a B.P. The evidence used was based on radiocarbon dates made mainly in Gröningen. These gave an age of 50000–65000 a B.P. to the Early Glacial stadials. However, dates of sediments older than 50000 years are open to doubt.

At the same time, further evidence was presented by the dating of deep-sea sediments made by Emiliani and others. These revealed the occurrence of a warm phase some 90000–100000 years ago. Moreover, the $\delta\,^{18}O/^{16}O$ determination of age of an ice core from Greenland indicates that the cool Pleniglacial period started as early as 73000 a B.P. (Dansgaard, Johnsen, Clausen & Langway 1970).

Further studies of deep-sea sediments undertaken by Emiliani (1968), McIntyre & Ruddiman (1972), Sancetta, Imbrie & Kipp (1973) and others, based on measurements of the $^{18}O/^{16}O$ ratios in skeletons of planktonic foraminifera, on palaeotemperature reconstructions of water bodies on the basis of foraminifera species, and on terraces on the isle of Barbados (Matthews 1972) indicate that the warmest period occurred some 125000–115000 years ago. This was followed by two separate warm waves thought to be interstadial (Understanding climatic changes 1975). The simplified curves of Kukla & Kukla (1972) and others show a succession of solar radiation cycles lasting about 20000 years each. The lower unit of the last cold stage includes at least two great waves of climatic cooling and of warming with relatively smaller amplitudes. The clearly colder second unit may be subdivided into two very cold phases separated by a longer (30000 a) cool temperate period which shows numerous smaller oscillations. According to the Dutch nomenclature, the above units correspond to the 'Early Weichselian' and to the Pleniglacial with the intervening Interpleniglacial. Wijmstra's (1975) attempt to correlate vegetational changes gives clear indication of their similar rhythm in the different zones, especially in Macedonia. Additional evidence is given by the distinct change in the magnetic declination at the top of the Eemian soil at Modřice in Moravia (Bucha, Horaček, Koči & Kukla 1969) dated at 114000 a ago. The palaeotemperature curve from Greenland remains to be interpreted. The small fluctuations of $\delta\,^{18}O$ values beyond 78000 a B.P. and the short term and abrupt depressions at *ca.* 89500 and 109000 a B.P. may have their origin in the retarded response of the North Atlantic water-body to temperature changes.

On the base of the palaeotemperature curve derived from undisturbed deep-sea sediments (for comparison, see figure 8) the views on the age and velocity of palaeogeographical changes of the continents require thorough revision. These changes were very rapid. Both glaciation (J. Lundquist 1974; Andrews *et al.* 1975) and deglaciation (Flint 1971) took place over a period of less than 10000 years each, whereas the duration of the Early Glacial has to be extended in length. The present author subdivides the last cold stage into two main units, which include the Early Glacial and the Pleniglacial.

The reliability of information on the palaeogeography of mid- and east Europe during the last cold stage

As we go further back into the Pleistocene the available information decreases. During the last cold stage, the most essential part in obscuring older features and destroying sediments was played by the latest cold period, including the glaciation 20000–15000 years ago. The use of radiocarbon dating is limited where deposits are older than 50000 a B.P., so that stratigraphic identification and correlation of the incomplete series of deposits is frequently made impossible. More often, the correlation of events is based on the comparison of sedimentary sequences with type horizons. The following difficulties are apparent in the reconstructions of climatically controlled phenomena: underestimation of the extent of breaks in the depositional sequence, inadequate knowledge of the varied local habitat conditions and topoclimates, and the correlation of data over wide areas in the reconstructions of local events. Non-sequences are most frequently observed in sedimentary sequences that have been exposed either to erosion or to denudation. In the valleys the shift of stream channels that has been recorded from Holocene valley spreads (Klimek & Starkel 1974) is responsible for the lateral contact of rock series of different ages. Because of slope degradation more or less complete series of deposits tend to be preserved only at the foot of the slope. They frequently contain double horizons due to the redeposition of material derived from the upper slope sectors. Woodland periods are reflected in the nature of polygenetic soils whose subsequent alteration was related to the succession of plant communities.

Because other information (excluding pollen spectra) provided by deposits refers only to the site examined and to its neighbourhood, generalizations from results should be made with great caution. As in Alaska and Siberia at present, the former ecosystems and geosystems showed much local variation related to thermal and moisture conditions, and to the thickness of the active layer above permafrost.

The radiocarbon-dated interstadial type horizons found along the North Sea coast are used in correlating horizons in other regions including Poland. However, such correlations are rather uncertain. In palaeoclimatic reconstructions, for instance, in Poland and in Hungary, the fossil floras are related either to the distant type profiles or to the ecological requirements of vegetation as observed at present, although the thermal parameters of air were different in a permafrost environment, with warmer summers and severe winters prevailing in the east.

In Poland and neighbouring countries, the palaeogeographic changes of the Vistulian stage are documented by sequences of the following phenomena: permafrost structures, loesses, fossil soils, slope sheets, fluviatile deposits, vegetation, glacial deposits and land forms.

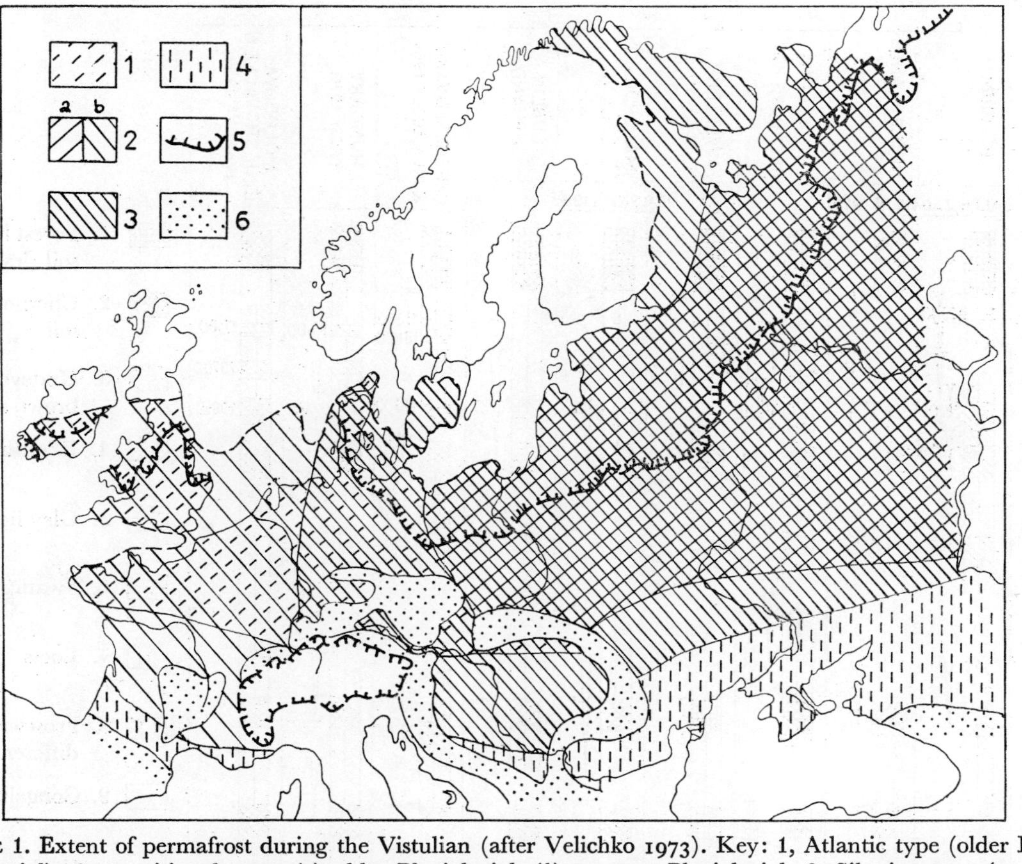

FIGURE 1. Extent of permafrost during the Vistulian (after Velichko 1973). Key: 1, Atlantic type (older Pleniglacial); 2, transitional type: (*a*) older Pleniglacial, (*b*) younger Pleniglacial; 3, Siberian type (younger Pleniglacial); 4, areas of seasonal permafrost (younger Pleniglacial); 5, extent of inland ice; 6, mountains with frost weathering features.

The sequence of permafrost structures

Permafrost surviving over many years was the primary feature of the cold period in Europe (figures 1 and 2). Thus the structures associated with permafrost are the best indicators of former changes of climate (Dylik 1969*a*). In mid- and east Europe permafrost built up slowly. Its earliest traces are materials enveloping the ice cores of fossil pingos (Dylik 1967). Below the horizon of the first Early Glacial interstadial ice-wedge casts also occur, even in Belgium (Paepe & Zagwijn 1972). Two Early Glacial ice-wedge horizons are reported from the Małopolska Upland (Jersak 1973) and from Hessen (Rohdenburg 1967). In west Europe, predominant plastic deformations indicate seasonal freezing (Velichko & Berdnikov 1973).

The very beginning of the older Pleniglacial is marked by a phase of severe climate accompanied by permafrost of Atlantic type, indicated by ice-wedge casts, 2–4 m deep. Temperatures of the ground surface are believed to have dropped to −3 or −4 °C. The thickness of permafrost must have exceeded 100 m, following Velichko's calculations (1973). The mean annual temperature ranged from −6 to −8 °C (this is the annual mean required for the growth of ice wedges (Péwé 1966)). According to Rohdenburg, several generations of ice wedges indicate a discontinuous permafrost with recurrent phases of formation. However, typical 'Siberian' forests of the Interpleniglacial and shallow gley soils give evidence rather of seasonal changes in thickness of the active layer. During the Interpleniglacial permafrost disappeared in the Hungarian Plain.

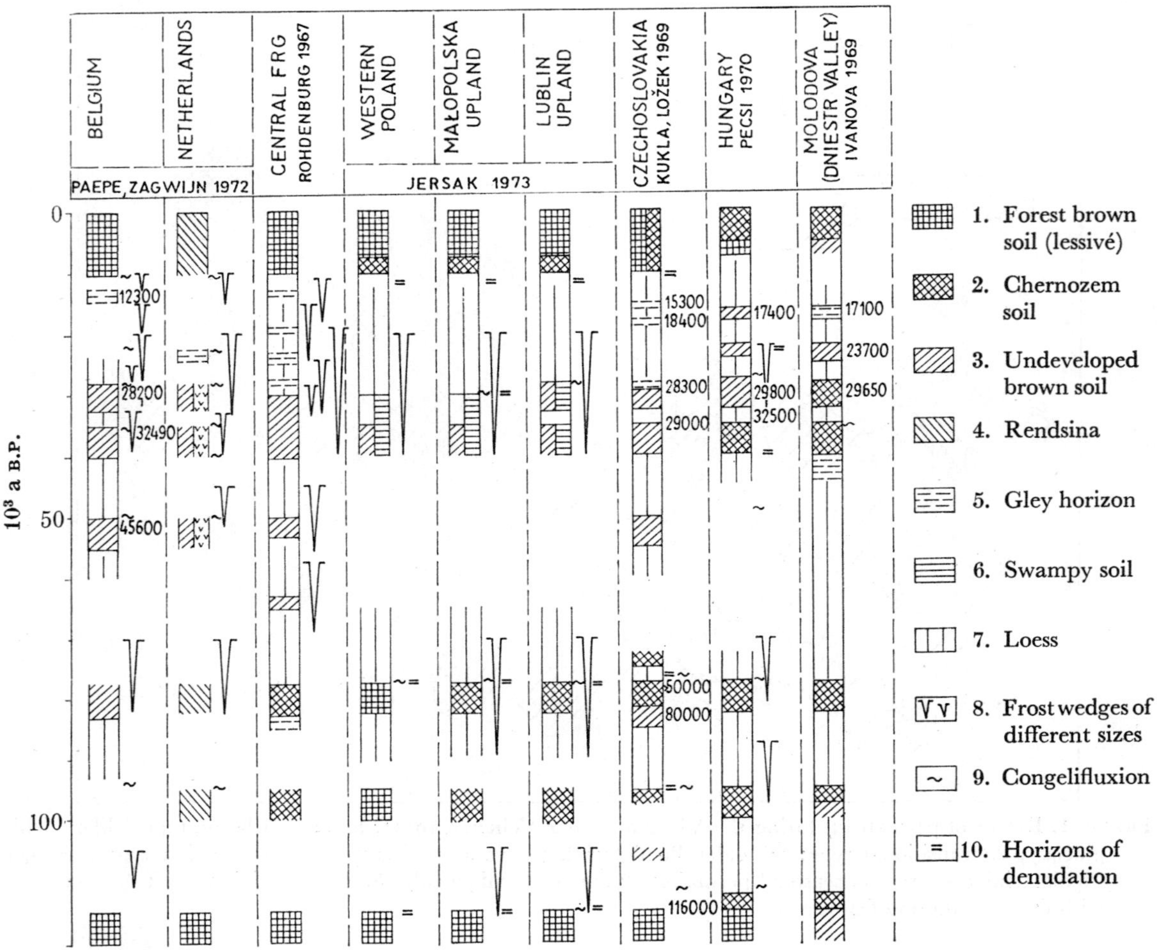

FIGURE 2. Age of fossil soils and permafrost phenomena in different European loess areas during the last cold stage.

The younger Pleniglacial is characterized by a repeated cooling of climate. Permafrost penetrated to greater depths then since the middle horizon of the youngest loess contains ice-wedge casts, 4–6 m deep, and polygons to 20 m in diameter (Rohdenburg 1967; Jersak 1973). Pseudomorphoses which developed from sand wedges devoid of ice give evidence of themarked desiccation of climate (for comparison, see Goździk 1973). The average ground temperature dropped to −6 °C. Permafrost was 200 m thick and extended as far south as latitude 45° N (figure 1). The severity of climate is also indicated by permafrost expanding in areas left freshly exposed by ice retreat (Maruszczak 1960). For example, ice-wedge casts may be found in deposits that have been laid down between two glacial oscillations in Wales (Saunders 1973).

Sand wedges filled with aeolian sand occur even within the boundary as the Pomeranian phase (Kozarski 1974) and point to the marked continentality of climate in east Europe. Kozarski suggests that during the deglaciation the mean annual temperature must have been by 14–17 °C lower than now. The degradation of permafrost was relatively rapid towards the end of the Last Glacial because in central Poland the last phase during which permafrost became reactivated corresponds to pre-Bølling times (Chmielewski 1970). During the Allerød the blocks of ice deeply buried in the subglacial channels were already melting. Degradation of the deeper and drier permafrost proceeded faster under continental climatic conditions than in west Europe. It might have been much slower there since pingo-like forms were still active during the Younger Dryas, for example, in Belgium and the Netherlands (Pissart 1963; Ploeger & Groenman van Waateringe 1964).

The accumulation of loess

Loess is a soil of the cold continental climate (Gierasimov 1962). In mid-Europe its source was the seasonally flooded plains and areas exposed to frost disintegration (Malicki 1950; Maruszczak 1968; Jersak 1973). The distance of transport of the loess dust with increased size-grades to the west was short. In mid- and east Europe loess is found in the hypsometric belt of 100–500 m above sea level (Velichko & Khalcheva 1973). Its lower limit decreases to the west and south. The distribution of loess coincides with the boundary of areas occupied by steppe communities in which the capillary upward movement of water and a dense grass cover favoured deposition and growth of the dust layer (Jersak 1973). Both the distribution and thickness of loesses varied throughout the last cold stage, reflecting the continentality of climate (figure 2). According to Velichko & Khalcheva (1973), the estimated annual rates of loess deposition were 0.07 mm for the older Pleniglacial, and 0.4 mm for the younger Pleniglacial. In southern Poland the Early Glacial loesses are thin or not apparent. The older Pleniglacial loess attains 1 m in the west and 5–7 m to the east. It is clearly gleyed, suggesting wetter conditions. The younger Pleniglacial loess is a typical subaerial deposit, locally 10–15 m thick. Associated with it may be three or four horizons of frost cracks, congelifluxion and gley (Kukla 1969; Jahn 1970; Jersak 1973) which indicate oscillations of climate and its increasing continentality. At that time, the loess belt expanded markedly from the periglacial zone to the border of basins in Bulgaria, Serbia and Italy. The general indication is of a vegetation rather similar in character over wide areas because there were no continuous forest barriers (Velichko & Khalcheva 1973). In Podolia loess deposition continued at least into times postdating a phase of climatic warming some 17000 a ago (Ivanova 1969).

Fossil soils

The sequences of fossil soils are best known from loess profiles in which the particular soil horizons point to pauses in sedimentation (figure 2). Usually there are two different soil complexes recognized, namely the interglacial–Early Glacial complex and the periglacial complex. The older one is called the Mezinski soil complex (Velichko & Morozova 1973) or the Nietulisko type (Jersak 1965). In Czechoslovakia it includes the fossil horizons PK III and PK II (Kukla 1969). The truly interglacial series comprises three horizons which overlie one another: chernozem, brown-earths-lessivé and again chernozem clearly reflecting the succession of vegetation (Jersak 1973). The PK III (uppermost part dated at 114000 a B.P.) is separated either by a hiatus, thin diluvia or loess with frost structures from the overlying soils (PK II in Czechoslovakia) of chernozem type (Mojski 1969 and others). Their two or three horizons indicate three or four oscillations of a dry temperate climate. The brown horizon with mollusc faunas *Brachybaeba triticum* and *Chondrula tridens* (Ložek 1969) is related to the development of forest communities during the Brørup Interstadial. Dating them is beyond the range of the ^{14}C method (at Vestonice > 55000 a B.P., Kukla 1969).

In the lower part of the Pleniglacial loess horizons of initial chernozems or rendsinas can be seen (Kukla 1969). On the River Dniestr (Ivanova 1969) and on the Rhine (Paas 1969) datings of gley horizons suggest an age > 45000–50000 a B.P. The Interpleniglacial period is clearly differentiated. It includes soils dated at 35000–25000 a B.P. which belong to the so-called Paudorf complex (Fink 1969) or Briansk (Velichko & Morozova 1973) or Komorniki (Jersak 1965) and PK I (Kukla 1969). This complex consists of one to three horizons which formed in a permafrost environment and are expressed as pseudogley or arctic brown soils showing poorly developed humus accumulation processes. In the Pannonian Basin under waning permafrost conditions this complex was represented by initial chernozems, or even brown earths (Kukla 1969). Their main horizon is dated at 30000–25000 a B.P. (at Molodova on Dniestr 29650 ± 1320 B.P. (Ivanova 1969); at Mende 29800 ± 600 B.P. (Pécsi & Hahn 1969), at Vestonice 28300 ± 300 B.P. (Kukla 1969)). These correspond in the west to the Denekamp interstadial and to the Zelzate soil 28200 ± 270 a B.P. (Paepe & Zagwijn 1972). The older clearly brown horizon may correspond to the Hengelo Interstadial. The frequently cited dates of the order of 25000 a B.P. (Velichko & Morozova 1973; Mojski 1969) suggesting a younger age of the former Paudorf interstadial are questionable and mark rather the beginning of intense loess deposition which stopped the accumulation of humus in the soil.

Lying on top of this soil complex are initial pseudogley soils, with 3 or even 4 horizons reported, for example, from the loesses at Zwierzyniec by Kraków (Sawicki 1952) and at Molodova on Dniestr. Radiocarbon determinations on the latter indicate an age of 23000–23700 a B.P. and 17100 ± 180 a B.P. (Ivanova 1969). The above horizons are related to periods of climatic warming and increased humidity. In Hungary the chernozem-like humic horizons indicate a forest–steppe community (dates of 18600 ± 150 and 17400 ± 440 (Pécsi 1970)). The sandy Allerød soils (Manikowska 1966) are also of poorly developed brown earth type showing a distinct humic horizon.

The frequently observed distinct boundaries between the buried soils and loess horizons may suggest rather sudden changes of climate in mid- and east Europe (Kukla 1972).

Slope deposits

The slope deposit sequence traced in the loess profiles is complex. At the base of each subaerial loess a solifluxion or deluvial horizon is commonly developed. In the loess-free areas the scheme is simpler. There is evidence of two distinct complexes of slope sheets comprising weathering debris and solifluxion material (for example, in the Carpathians, figure 3) associated with permafrost conditions. This fact is recorded even by devotees to the periglacial denudation cycle that comprises the ascending phase, the climax, and the descending phase (Dylik 1969*a*). Apart from a few sites (e.g. Józefów) distinct evidence for important phases of slope degradation during the Early Glacial is lacking.

The lower Pleniglacial series rests on Early Glacial peats (Sobolewska, Starkel & Środoń

1964) and attains in places a greater thickness than the upper and younger one. This may suggest that a relatively long period of cool climate (Klatka 1962; Starkel 1964) has been available for its formation. Thick layers of scree found in caves, for instance, in the environment of Kraków (Nietoperzowa Cave – Różycki 1967) and in Dobrogea (Adam Cave – Samson 1969) support this interpretation. On the border of the Warta (= Warthe) end moraines, a cryopediment, together with the correlative deposits in depressions predates the Interpleniglacial (Rotnicki 1966, 1974).

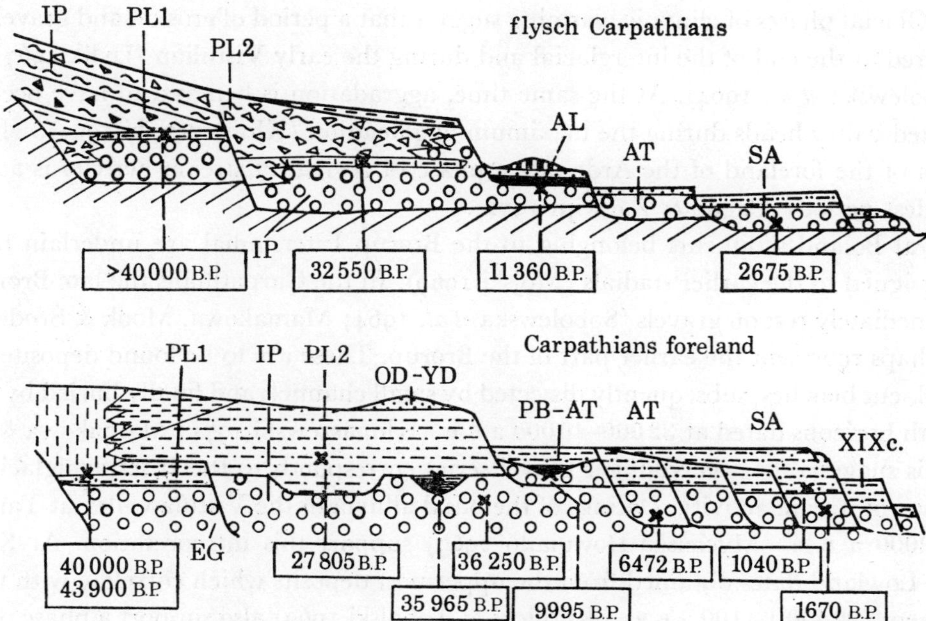

FIGURE 3. Terrace sequence of Vistulian and Holocene age in the Polish Carpathians and their foreland.

The Interpleniglacial phase is well characterized by a weakening of denudational processes. For example, the long section through a hill slope at Zabrodzie in the Polish flysch Carpathians shows changes from solifluxion processes to diluvial processes, and restriction of solifluxion to the steep slope sectors (Dziewański & Starkel 1967).

The younger Pleniglacial appears to have been a time of intensification of slope processes. Colluvial deposits lie on the Interpleniglacial alluvia or soils (figure 3). In the flysch Carpathians solifluxion sheets composed of clay and scree attain locally 10–12 m. At Dobra and Krościenko several scree and clay units alternating with clay units are thought to be associated to warmer oscillations, whereas the intervening clay units may record cooler conditions and shallower summer thaw (Klimaszewski 1971). Evidence may also come from flow structures in the Interpleniglacial horizon (Starkel 1969). It is likely that the scree corresponds to wetter conditions and intense frost weathering, and the creeping clayey shales, to the drier periods. The sedimentary cycle sometimes ends with the latter deposits, indicating a desiccation of climate towards the end of the Pleniglacial (Starkel 1969). During the younger Pleniglacial formation of the younger series of cave scree (Nietoperzowa Cave and others) and of the lower step of cryopediments (Rotnicki 1974) also took place. In the wetter areas to the west solifluxion processes continued into the Younger Dryas (Rohdenburg 1967). Truncated profiles and traces of aeolian processes recorded from the immediate foreland of the last ice sheet suggest that slope processes developed especially in the wetter phase of permafrost aggradation (Dylik 1969*b*).

Fluviatile deposits

In Poland and neighbouring countries the sequences of younger Pleistocene alluvia commonly represent only short spans of time. The most reasonable explanation for this seems to be the shift of stream channels and alternation of phases of either predominant deepening or aggradation (figure 3). Alluvia of the channel facies overlain by sediments which correlate with the Early Glacial phases of climatic warming suggest that a period of erosion and gravel deposition occurred at the end of the interglacial and during the early Vistulian (Dylik 1967; Mojski 1969; Sobolewska *et al.* 1964). At the same time, aggradation is believed to have occurred in the glaciated valley heads during the maximum of glaciation (Klimaszewski 1961), whereas in the valleys of the foreland of the Ardennes a phase of revival of stream activity is attributed to the coldest periods (Paepe & Zagwijn 1972).

In central Poland sediments belonging to the Brørup Interstadial are underlain by sandy deposits assigned to the earlier stadials (Mojski 1969). In the Carpathians the late Brørup sediments immediately rest on gravels (Sobolewska *et al.* 1964; Mamakowa, Mook & Środoń 1975) which perhaps represent the earlier part of the Brørup. These are to be found deposited on the higher rock-cut benches, subsequently dissected by small channels and finally buried by younger alluvia with horizons dated at 32000–40000 a B.P. (Klimaszewski 1971; Mamakowa & Starkel 1974). This suggests that the younger gravel series corresponds to the Interpleniglacial (Paudorf) revival of stream activity. Datings of the basal alluvia in the Vistula valley at Tarnobrzeg (40000 ± 2000 a B.P. – Mycielska-Dowgiałło 1967) support this interpretation. At Kępno in the Polish Lowland dates obtained from the uppermost deposits which correlate with the older cryopediment (29450 ± 1100 a B.P. – Rotnicki & Tobolski 1969) also support a phase of erosion in the Interpleniglacial and indicate the age of valley infilling as prior to that time. The present author is of the opinion that the pre-Interpleniglacial aggradation period is also represented by the upper silty-sandy units of the 'Brørup' profiles at Podgłębokie and Konin in central Poland (Mojski & Rzechowski 1969).

During the younger Pleniglacial, outside the zone of glacially disturbed sediments, aggradation was caused by melt-water floods, as indicated by silty deposits at Brzeźnica dated at 27805 ± 330 a B.P., (Mamakowa & Starkel 1974) and by the thick series of sediments in the drainage basins of the rivers Dniepr, Upper Volga and western Dvina. These predate 25000–20000 a B.P. the maximum of glaciation (Chebotarieva & Makaricheva 1974). As the deglaciation continued the braided rivers of the extraglacial zone cut wider and deeper their channels and finally changed to meandering rivers. On the dissected aggradational plains dunes began to develop. Lying on the surface of the contemporaneous erosional plains in the Carpathian valleys are biogenic deposits considered to belong to the Bølling (Koperowa 1958; Klimaszewski 1969), and gravel and sand also overlain by biogenic deposits dated at 10300–9900 a B.P. (Ralska-Jasiewiczowa & Starkel 1975).

The history of vegetation

Since the floras of the cool stadial periods are very similar to one another the stratigraphy should be based on diagrams for the varied interstadial deposits where age determination is easier. Pollen spectra may be of two types, those typical of the Early Glacial and those representing the Interpleniglacial. The distinct sequence of Eemian interglacial plant communities with the late *Carpinus* and *Picea* phase – marked by the presence of *Buxus sempervivens* and *Ilex aquifolium* and by the absence of *Fagus* – permits the separation of the Eemian from the Early Glacial periods of climatic warming. It also shows that the climate of the Eemian was warmer and more oceanic than that of the Holocene (Środoń 1972). The Early Glacial periods of climatic warming whose age given by ^{14}C dating is more than 40000–52000 a B.P. are thought

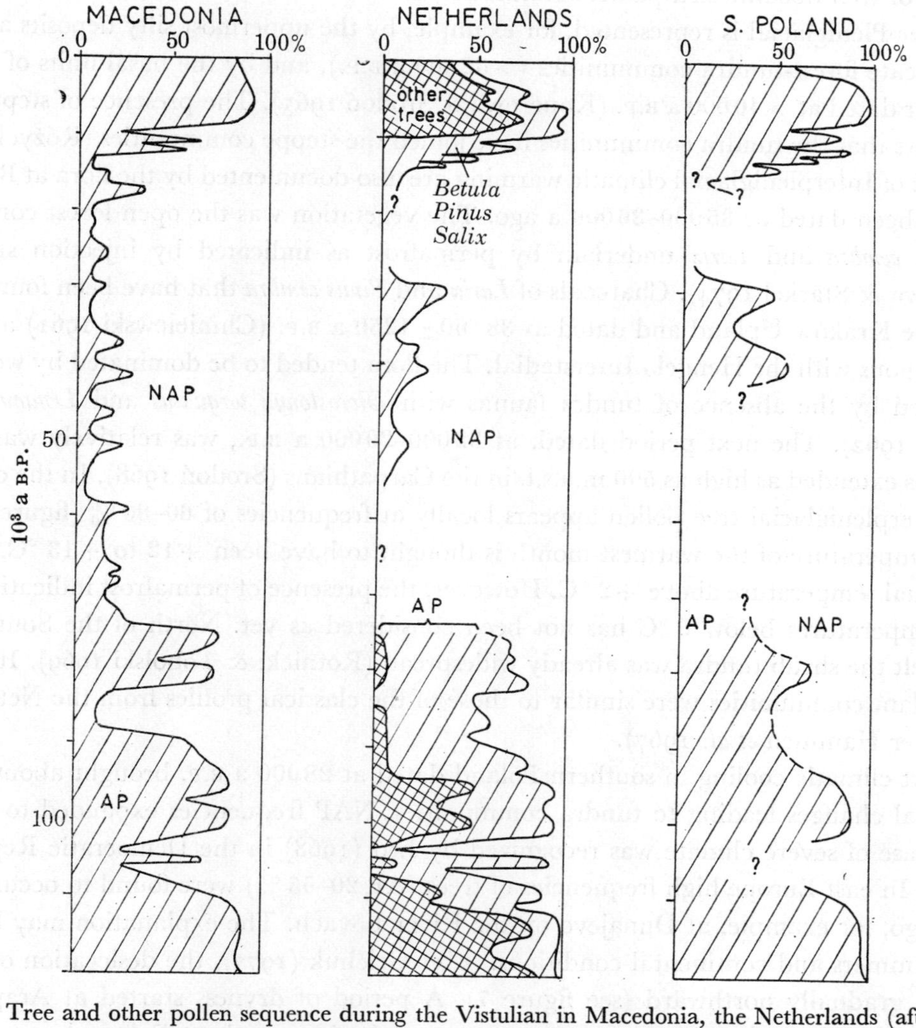

FIGURE 4. Tree and other pollen sequence during the Vistulian in Macedonia, the Netherlands (after Wijmstra 1975) and Poland (after diagrams published by Środoń 1972 and others).

to be synchronous with the Amersfoort (site Józefów, after Dylik 1968) and with the Brørup (for example, at Wadowice – Sobolewska *et al.* 1964; at Konin – Borówko-Dłużakowa 1967). These are characterized by boreal forests showing high frequencies of AP of 60–90 % (figure 4). The Brørup had an admixture of thermophilous trees (*Quercus, Tilia*) and of the characteristic species *Picea omoricoides* (for comparison, see Andersen 1961). The climate of the optimum was slightly cooler than that of today. In the pollen spectra for Poland the younger Odderade Interstadial (Averdieck 1967) has not been distinguished. Menke (1970) synchronized the Odderade with the Dutch Brørup and the classical Danish Brørup with the Amersfoot in the Netherlands. Pollen diagrams from Macedonia show indeed only two great interstadial phases of a warm and wet climate (Wijmstra 1969). Profiles from the Carpathians indicate that a long-term phase of climatic cooling and of solifluxion processes followed immediately a forest phase. The high frequencies of lime (> 20 %) and hornbeam (> 5 %) in the diagram from Konin may suggest the older interstadial, whereas at Wadowice and Podgłębokie (Janczyk-Kopikowa 1969) only the second interstadial of a cooler climate with spruce has been identified. In the U.S.S.R., the division into the Eemian and Early Glacial climatic oscillations remains to be clarified because of absence of well documented palaeobotanical sites.

The older Pleniglacial is represented, for example, by the uppermost silty deposits at Konin which indicate forest-tundra communities (> 52000 a B.P.), and by the basal units of the profile at Zator dated at > 40000 a B.P. (Koperowa & Środoń 1965). The presence of steppe plant pollen shows that the tundra communities have joined the steppe communities (Różycki 1967). The phases of Interpleniglacial climatic warming are also documented by the flora at Brzeźnica which has been dated at 35000–36000 a ago. The vegetation was the open forest community with *Pinus cembra* and *Larix* underlain by permafrost as indicated by injection structures (Mamakowa & Starkel 1974). Charcoals of *Larix* and *Pinus cembra* that have been found in the caves of the Kraków Upland and dated at 38160 ± 1250 a B.P. (Chmielewski 1961) appear to be synchronous with the Hengelo Interstadial. The flora tended to be dominated by woodland, as indicated by the absence of tundra faunas with *Dicrostonyx torquatus* and *Lemmus lemmus* (Kowalski 1962). The next period dated, at 32000–29000 a B.P., was relatively warm since tree clumps extended as high as 500 m.a.s.l. in the Carpathians (Środoń 1968). In the diagrams for the Interpleniglacial tree pollen appears locally at frequencies of 60–80 % (figure 4). The average temperature of the warmest month is thought to have been +12 to +13 °C, and the mean annual temperature above +2 °C. However, the presence of permafrost indicating mean annual temperatures below 0 °C has not been considered as yet. North of the South Polish Upland belt the shrub tundra was already widespread (Rotnicki & Tobolski 1969). It appears that the plant communities were similar to those of the classical profiles from the Netherlands (see van der Hammen *et al.* 1967).

The next climatic cooling in southern Poland dated at 28000 a B.P. brought about distinct vegetational changes leading to tundra communities. NAP frequencies expanded to 90 %. A similar phase of severe climate was recognized by Erd (1968) in the Democratic Republic of Germany. In east Europe high frequencies of trees (AP 20–65 %) were found to occur 30000–21000 a ago, for example, at Dunajevo on the River Lovach. The explanation may be in the still hot summers and continental conditions. After Grichuk (1973), the desiccation of climate proceeded gradually northward (see figure 7). A period of dryness started at Arapovichi – latitude 52° 30′ N – before 24200 ± 1680 a B.P., and farther north at Pokrovskaya – latitude 60° N – after 21410 ± 150 a B.P. but immediately before the advance of ice sheet. Corresponding with this dryness was the arrival of steppe elements *Ephedra distachya* and *Eurotia ceratoides*. An original map by Grichuk, based on several dated sites which document the survival of forest patches in the Russian Lowland (figure 5), shows the reconstruction of plant communities during the maximum of glaciation (20000–18000 a B.P.). The map clearly differs from the earlier maps and profiles presented by Büdel (1951), Frenzel (1960), Vigdortshik, Auslender, Znamienskaya & Dołuchanov (1972). Grichuk recognized three types of plant communities in mid- and east Europe. These were a periglacial-tundra, boreal forests (confined to enclaves in southern Europe, in the East Carpathians and the Ukrainian Upland), and periglacial-steppe. The first and last of these communities existed in a permafrost environment. A very narrow forest belt and the mixing of tundra and steppe are their two characteristics. The existence of refuges and the merging of tundra and steppe communities also was mentioned by Różycki (1967) and Środoń (1968). However, comparison with the present distribution of various plant communities in Yakutia and Mongolia suggests that the tundra-steppe ecosystems occurred at neighbouring sites and were closely adapted to the changing soil, thermal and moisture conditions.

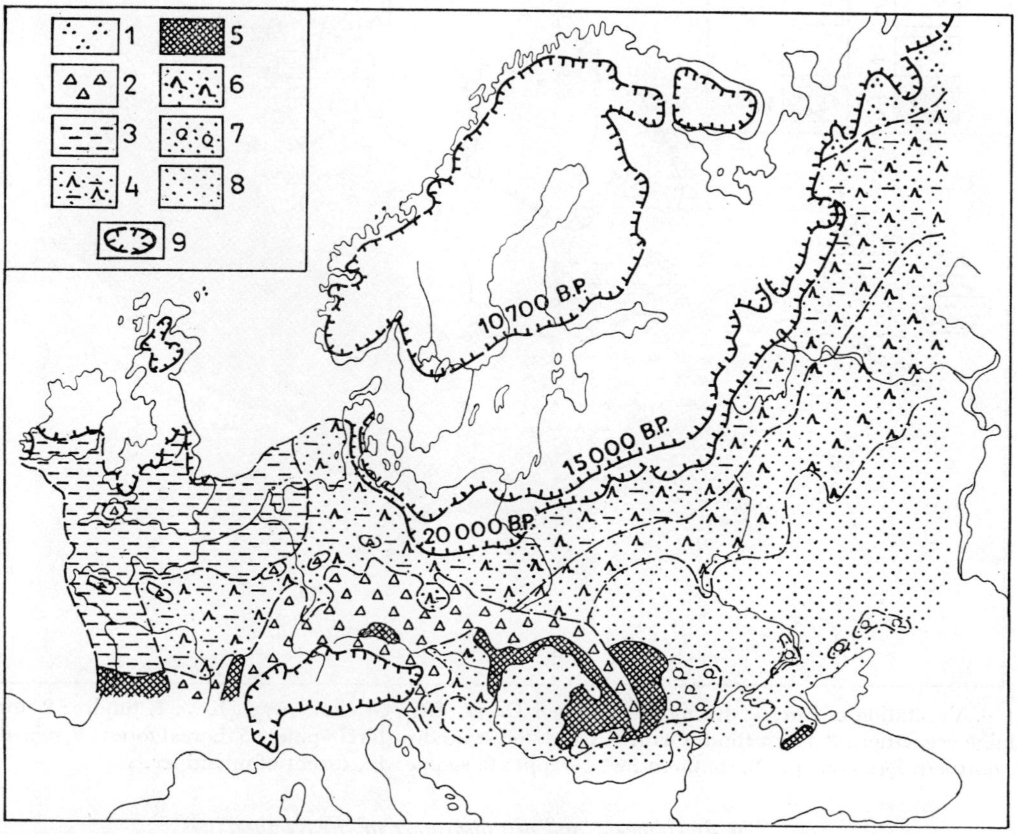

FIGURE 5. Vegetation of Europe about 20000 a B.P. (after Grichuk 1973). 1, polar desert; 2, mountain alpine vegetation; 3, tundra; 4, forest tundra; 5, birch–pine–larch open forest; 6, northern forest-steppe; 7, southern forest-steppe; 8, steppe; 9, margins of ice sheets.

Floras dating from the times of inland-ice retreat (18000–12500 a B.P.) are poorly known. South of the Carpathians there is evidence for patchy distribution of a forest-steppe (Pécsi 1970). The climate of the Mazurian (Halicki 1960) or Somino Interstadial (Chebotarieva & Makaricheva 1974) was perhaps too cool to favour growth of pine–birch forests in northern Poland. In the Soviet Union, fossil floras from the Baltic drainage basins have been dated at some 13500 a ago and indicate a severe climate. More complete profiles are available for the Oldest Dryas and the succeeding periods. For example, at Witów by Łódź (Wasylikowa 1964) tundra communities with steppe patches were replaced by birch forests in the Bølling (some 12300 a B.P.) and by pine forests in the Allerød. The initial forests thrived above permafrost (figure 6). The rapidity of both climatic warming and forest expansion during the glacial retreat is best indicated by pieces of *Pinus* and *Picea* wood contained in Allerød deposits in the environment of Helsingfors (Mölder, Valovirta & Virkkala 1957). The climatic continentality persisted in the Younger Dryas as indicated by the optimum growth of steppe elements in Poland at that time. It is not until then that the Holocene forest communities begin to develop.

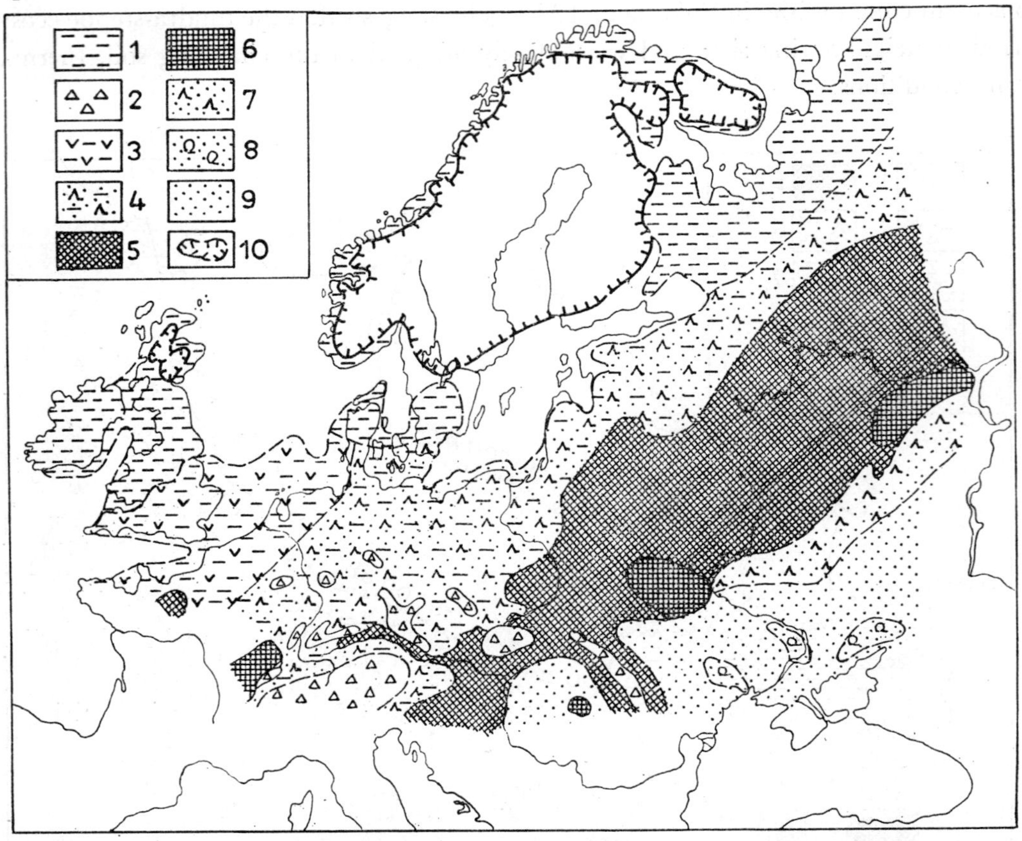

FIGURE 6. Vegetation of Europe during the Younger Dryas (after Grichuk 1973). Key: 1, tundra; 2, mountain alpine vegetation; 3, forest tundra with birch; 4, forest tundra (birch–pine); 5, boreal forest; 6, mixed forest; 7, northern forest-steppe; 8, southern forest-steppe; 9, steppe; 10, extent of inland ice.

The development and disappearance of the ice sheet

The size of the Scandinavian ice sheet before 20000 B.P. is partly of disputed status. The small sizes of the Early Glacial glaciers are indicated by the extent of boreal forests as far north as latitude 62° N during the Brørup (figure 7). Little is known of the ice extent during the older Pleniglacial. J. Lundquist (1971), relating the older boulder clay to this period, suggests that at that time advances brought the ice into southern Sweden. During the Interpleniglacial lowland Sweden and Finland was ice-free, since biogenic deposits have been found in the Lulea area (Fromm 1960), and marine deposits dated at 29000–26700 a B.P. occur near Göteburg (Brotzen 1961). Similar dates were obtained on the tundra deposits on the Kola Peninsula.

The time of maximum ice advance is documented by age determination on organic remains which appear under the moraines marking the maximum phase in the D.R.G. (21160 ± 800 a B.P. – Cepek 1965), and in the environments of Vologda in the U.S.S.R. (21410 ± 159 a B.P. – Vigdortshik *et al.* 1972). Without doubt, the ice advance itself is by 1000–2000 years younger than the above dates. Following Arstanov *et al.* (1971), Chebotarieva & Makaricheva (1974) record the finding of fossil floras under the moraine at Drechaluki by Vitebsk. The youngest date is some 17700 a ago. Because hazel and other pollen are present there it is difficult to accept a maximum of glaciation later than the date cited. The moraines ascribed to the oldest phases: Leszno (= Brandenburg, Bologovo) and Poznań (= Frankfurt, Jedrovo) are ice-pushed features. The deglaciation was of areal type (Roszko 1968 and others) and proceeded under extremely continental and dry conditions. The moraines of the Pomeranian (= Vepsovo) phase include frequently push moraines, indicating a different arrangement of lobes which extended from 0 to 230 km in relation to the position of the ice margin at the maximum phase (cf.

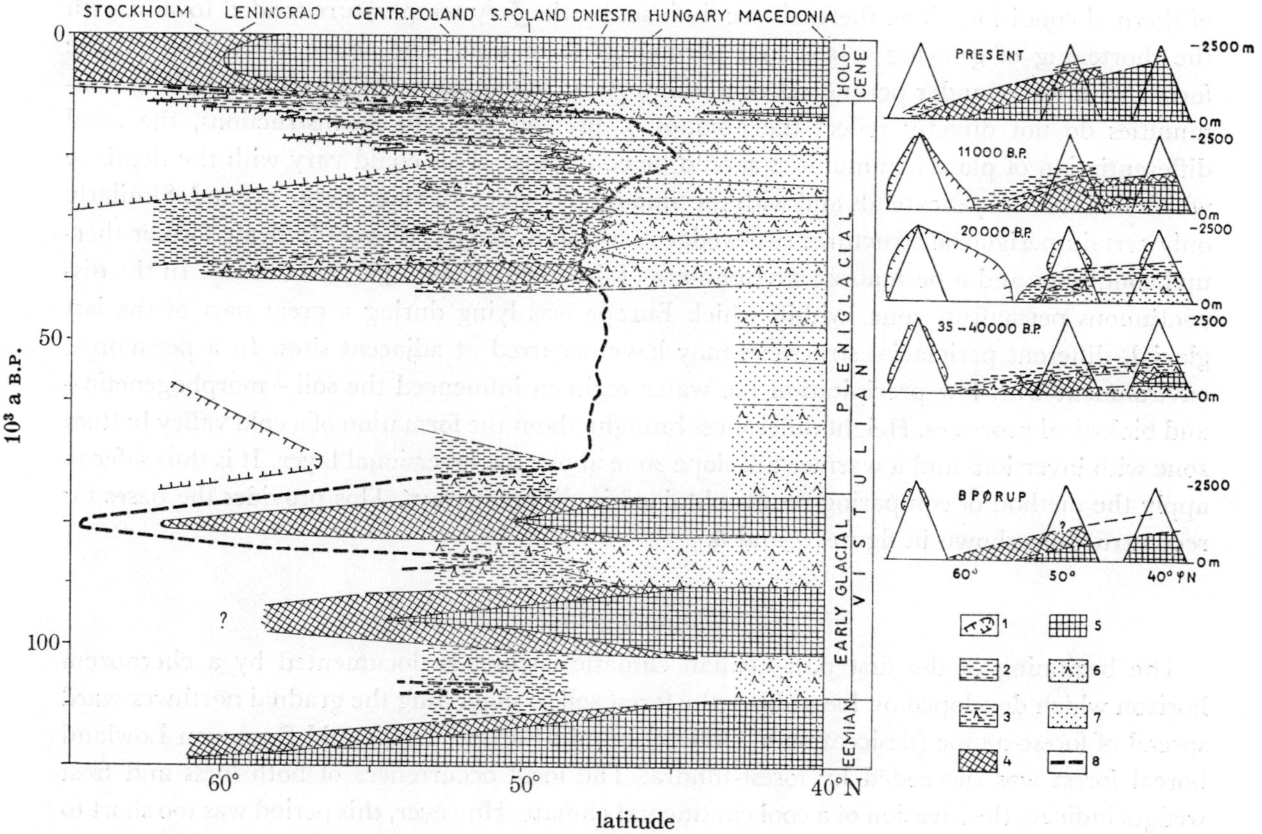

FIGURE 7. Changes of vegetation, permafrost and glaciation during the Vistulian in the meridional transect (Macedonia–Scandinavia) and in vertical belts. Key: 1, inland ice; 2, tundra and arctic desert; 3, forest-tundra; 4, boreal forest; 5, deciduous forest; 6, forest-steppe; 7, steppe; 8, permafrost limit.

Chebotarieva, Faustova *et al* 1973). Thus the preceding interphase was sufficiently long to allow ice readvance from a distant area of at least a few dozens of kilometres. The Pomeranian phase has been dated at some 15000 a ago. The late deglaciation was very rapid. Roszko (1968) drew attention to the fast disintegration of ice leading to the formation of dead ice blocks, and to the dissection of the ice sheet by sub- and supraglacial channels which pass into wide drainage ways (= pradoliny, Urstromtäler). To the east of the River Vistula there were formed systems of extensive ice-dammed lakes. Until the recessional Gardno (= Łuzhska) phase dated at some 13000 a B.P. (Serebryanny 1969), the amount of ice retreat from the Pomeranian end moraines was sometimes 300 km in the east, so suggesting the scarcity of supply of ice and the withdrawal of the ice centre into the mountains (J. Lundquist 1974). The melting of ice blocks was retarded by the expanding permafrost. South of the Baltic Sea coast the thawing of dead ice blocks continued in the early Holocene (cf. Starkel 1966).

The climatic and palaeogeographic changes during the last cold stage

A discussion of the characteristics of both thermal and moisture conditions can be based on the subdivision of the Vistulian stage into two main units. These are the Early Glacial (115000–75000 a B.P.) and the Pleniglacial (75000–10000 a B.P.). In reconstructing past events it is useful to remember that conclusions drawn from the nature of both characteristic plants and type deposits are far from satisfactory. The succession of plants is retarded in relation to changes of thermal conditions. The thermal tree limit set by the July temperature tended to shift with the shortening of growing season and permafrost formation. The active layer supplied the forests with water under permafrost conditions. As a consequence, the Pleniglacial plant communities do not directly reflect the amount of precipitation. In reconstructions, the areal differentiation of plant communities should be considered. This could vary with the depth to which the active layer extends and with the amount of water contained in the ground. Similarly only certain periglacial structures (ice-wedge casts, sand wedges, pingos) indicate former thermal conditions and a permafrost environment (Jahn 1970; Péwé 1966 and others). In the discontinuous permafrost zone, within which Europe was lying during a great part of the last glacial, different periglacial structures may have occurred at adjacent sites. In a permafrost environment with low precipitation the water regimen influenced the soil – morphogenetic – and biological processes. Height differences brought about the formation of a cold valley bottom zone with inversions and a warmer hill slope zone above the inversional layer. It is thus safer to apply the method of comparing results obtained in different ways. This provides the bases for reconstructions shown in figures 7 and 8.

The early glacial

The beginning of the first post-Eemian climatic cooling is documented by a chernozem horizon which developed on loess above the forest soil so suggesting the gradual northwestward spread of forest-steppe (desiccation of climate) in mid-Europe. In the mid-European Lowland boreal forest was succeeded by forest-tundra. The local occurrences of both loess and frost wedges indicate the invasion of a cool continental climate. However, this period was too short to destroy completely *Tilia*, *Carpinus* and *Corylus* which survived in refuges. Consequently the mean temperature of warmest month could not have been much below 10 °C (in west Europe it was below 9 °C, after Paepe & Zagwijn). The main depressions were of the winter temperatures and precipitation values. The increased supplies of slope-derived material and accelerated floods account for the formation of gravel sheets in the Carpathian valleys (Jahn 1968; Starkel 1964; Środoń 1968). In the Brørup the re-establishment of the optimum mixed forests in southern Poland suggests that the temperature of warmest month was 2 °C lower than that of today (July +17 °C) (see figures 7 and 8). The temperature of the warmest month of the Amersfoort is believed to have been lowered by a further 2 °C (Środoń 1972). The Brørup (Odderade?) which was followed by the Pleniglacial period of solifluxion was the second and last great warm oscillation. This may be correlated with the Elevthroupolis interstadial in Macedonia (Wijmstra 1969), and with the warming up of ocean waters some 85000–75000 a B.P. (Sancetta *et al.* 1973).

The Pleniglacial

The rapid world-wide climatic deterioration some 73000 years ago brought about the continentality of climate in mid- and east Europe. The causes suggested include the shift equatorwards of the polar and arctic fronts (McIntyre & Ruddiman 1972; Wijmstra 1975) accompanied by a displacement of the longitudinal depression 'furrow' to the west so involving an increase in strength of the southerly circulation, and the disturbance of jet-streams with the resultant rare penetration of humid air masses to the east (Lamb 1971).

The continental climatic regime was widespread in mid-east Europe embracing all latitudes as indicated by the existence of steppe and forest-steppe in Macedonia and Greece (Wijmstra 1975; Bottema 1974), and by a phase of both aeolian processes (Dylik 1969*b*) and deeply frozen ground. The mean annual temperature must have been at or below −6 °C (figures 7 and 8). The persistence of forest-tundra suggests sufficiently high temperatures of the warmest month and a different seasonal distribution of precipitation than at present. A 'moist' permafrost and gleyed loess indicates rather oceanic conditions in east Europe. Because of the presence of trees it may be suggested that the amount of precipitation was at least half the present one, and winter snowfall was especially low (in Yakutia with an average annual precipitation of 250–300 mm, snowfall is only 20 %). This favoured the deep penetration of freezing into the ground.

The interstadials which constitute the complex Interpleniglacial period of climatic warming (some 50000–29000 a B.P.) were of a short duration. In the Netherlands and Belgium these were nearly treeless periods. However, a conclusion cannot be drawn from the lack of trees that the mean temperature of warmest month was below 10 °C (Paepe & Zagwijn 1972). The waterlogging of the active layer needed higher summer temperatures which however restricted the growth of trees. Forests which existed in southern Poland during the Hengelo Interstadial indicate a deeper downward thawing and summer temperatures perhaps similar to the present ones in Yakutia (mean July temperature 16–18 °C, mean annual temperature −9 °C) (Gavrilova 1973). Velichko & Morozova (1973), reconstructing the climate of the Hengelo Interstadial on the base of both soils and permafrost structures, suggest that in mid-Europe the annual precipitation amount was 200 mm, the mean temperature of coldest month was −30 °C and there were 8 months of severe cold (figure 8). The mean annual temperature must have been −1 or −2 °C. Evidence for clearly wetter conditions is missing. The greater depth of the active layer which provided sufficient moisture controlled the supply of coarse materials into the stream channels. A perennial predominance of westerly circulation might have reduced summer temperatures and restricted tree growth. On the other hand, both the pause in loess deposition and the very low frequencies of steppe elements exclude greater aridity. However, it appears that in the west the waves of climatic warming might have increased winter snowfall nourishing valley glaciers from which ice sheets developed. It has been estimated by Manley (1975) that in the last glacial the mean annual temperature in Britain must have been lowered by 1.3 °C without a decrease in precipitation to produce the ice sheets.

The younger Pleniglacial (29000–10000 a B.P.) corresponds to the time of maximum of glaciation during the last cold stage. The initial expansion of the tundra communities in mid-Europe points to a relatively wet climate. According to Geyh & Rohde (1972), a wetter phase occurred about 23500–20400 a B.P. The masses of humid air must have periodically penetrated to the east and favoured the development of ice sheets in northwest Siberia. These reached their maximum development 24000–21000 a B.P. (Zubakov 1969). The Scandinavian ice sheet reached its maximum some 20000 years ago. In east Europe this event coincides with the spread of steppe plants to the north, and with the progressive desiccation of climate (Grichuk 1973). The presence of sand wedges in Poland is clear evidence of the formation of a 'dry' permafrost now known to occur, for example, in Mongolia. It is probable that the annual precipitation amount dropped below 200 mm. In the immediate foreland of the ice sheet there extended the arctic desert zone with descending winds (Dylik 1969*b*; Maruszczak 1968). Evidence for a wetter climate to the south is lacking. It cannot therefore be concluded that the westerly depression tracks (Liljenquist 1974 and others) moved southwards. The presence of deep ice-wedge casts in west Europe (Rohdenburg 1967; Paepe & Zagwijn 1972 and others) also indicates continentality in the west.

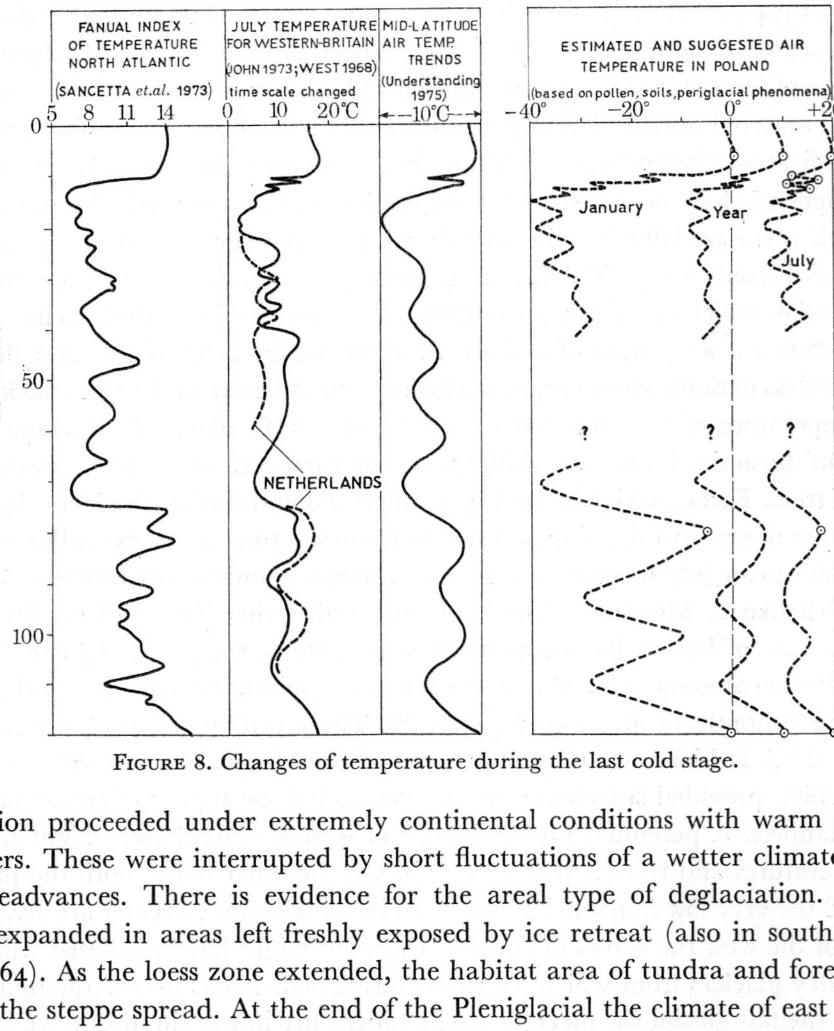

FIGURE 8. Changes of temperature during the last cold stage.

Deglaciation proceeded under extremely continental conditions with warm summers and severe winters. These were interrupted by short fluctuations of a wetter climate, with consequent ice readvances. There is evidence for the areal type of deglaciation. Concurrently permafrost expanded in areas left freshly exposed by ice retreat (also in southern Sweden – Svensson 1964). As the loess zone extended, the habitat area of tundra and forest-tundra was reduced by the steppe spread. At the end of the Pleniglacial the climate of east Siberian type has taken a turn toward the 'Mongolian' type. It seems that the ingress of humid air masses of Atlantic origin must have been short and extremely rare (it is probable that precipitation values did not exceed 200 mm). The progressive warming up of climate may have favoured the development of forest communities in the wetter phases. On the basis of herbaceous plant assemblages the mean temperature of warmest month has been established for the Bølling at +15 °C and that of the Allerød at +16 °C (Wasylikowa 1964). It appears that the mean summer temperature required for the spread of woodland in a permafrost environment must have been higher than that suggested for the present tree line. Moreover, the cause could have been the low ground temperatures and the increased instability of climate, the recurrent ingresses of humid air masses of polar-maritime origin, and of severe arctic air masses from the north which restricted plant growth, especially in spring time. The present deciduous forest zone was occupied by forest-tundra communities in the Younger Dryas and by pine forests in the Pre-Boreal (Środoń 1972). This indicates that in mid-Europe the easterly circulation system may have weakened at the beginning of a wetter climate about 8400 a B.P., with the consequent eastward spread of deciduous forest communities (Starkel 1976).

Final remarks: northwest European comparison

Analysis of the palaeogeographical phenomena recorded in deposits of the last cold stage allows the following conclusions to be made:

(1) The cold Vistulian stage includes two distinct units showing a different rhythm of climatic changes. These are the Early Glacial unit (115000–75000 a B.P.), with greater fluctuations of temperature, generally warmer and wetter, less effective in the refashioning of relief,

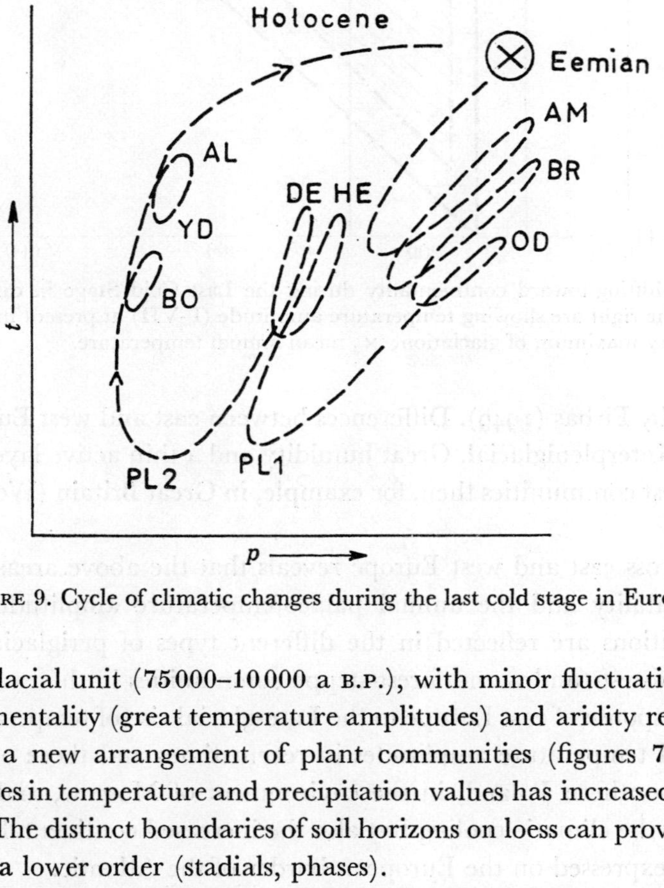

FIGURE 9. Cycle of climatic changes during the last cold stage in Europe.

and the cold Pleniglacial unit (75000–10000 a B.P.), with minor fluctuations of temperature, with a distinct continentality (great temperature amplitudes) and aridity resulting in the relief transformation and a new arrangement of plant communities (figures 7–9). Continentality manifested by changes in temperature and precipitation values has increased almost till the end of the Late-glacial. The distinct boundaries of soil horizons on loess can provide proof of sudden climatic changes of a lower order (stadials, phases).

(2) During the greater part of the last cold period plant communities differing from the present ones developed in a permafrost environment. Thus their occurrence limits the success of direct reconstructions of both palaeotemperatures and precipitation amount based on comparisons with the distribution of plant species and assemblages of today which are not underlain by permafrost. The interglacial species survived in refuge areas whose existence has already

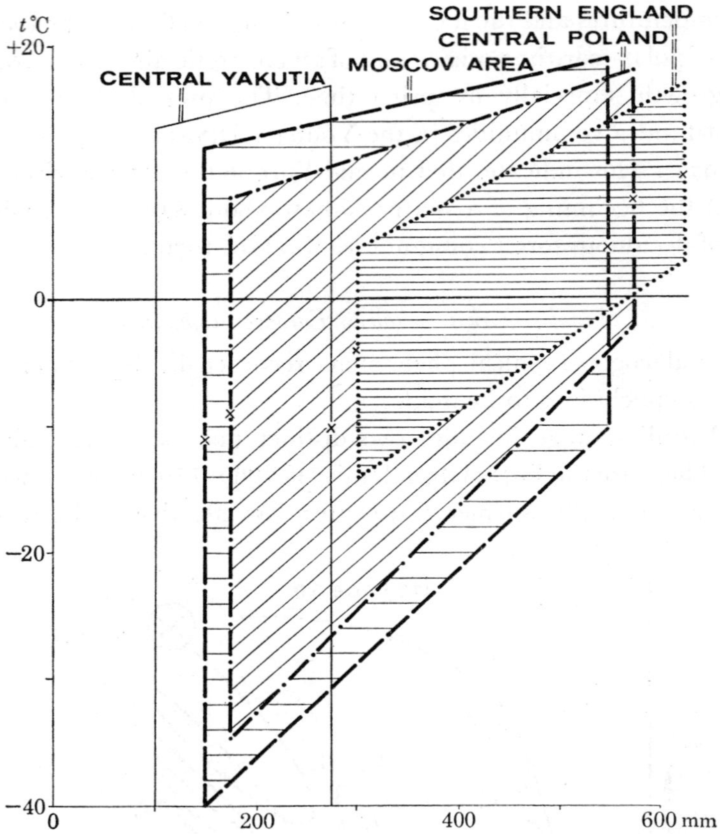

FIGURE 10. Estimated shifting toward continentality during the Last Cold Stage in different parts of Europe. Vertical lines on the right are showing temperature amplitude (I–VII) at present in the 'humid' period, on the left – during dry maximum of glaciation; ×, mean annual temperature.

been stressed early by Firbas (1949). Differences between east and west Europe were especially distinct during the Interpleniglacial. Great humidity and a thin active layer did not favour the development of forest communities then, for example, in Great Britain (West 1969; Pennington 1974).

(3) A section across east and west Europe reveals that the above areas varied by both the degrees of continentality and the annual palaeotemperature amplitudes and precipitation values. These variations are reflected in the different types of periglacial structures, in the different history of forest-tundra and forest-steppe communities in the eastern part of Europe, and in the earlier response of east Europe to the Late-glacial rise of temperature. On a diagram showing the annual temperature amplitudes in recent times and those of the Pleniglacial in Great Britain and southern Poland, in the environment of Moscow, and in central Yakutia (figure 10) not only the climatic cooling but also the increased continentality can be seen. The latter was less well expressed on the European border of the Atlantic.

(4) The formation of the ice sheet was an extreme phenomenon in the scale of the Vistulian as a whole. Ice advanced as continentality increased. This in turn was responsible for the rapid deglaciation of Europe in a slowly warming up climate. The parallel development of ice sheet and continentality may suggest both existence of a barrier, not allowing the westerly circulation system to penetrate to east Europe, and the great instability of climate in northwest Europe. A

cause may be the great fluctuations of water temperature in the Atlantic during the Pleniglacial stated off the shores of the British Isles (McIntyre 1975).

The author thanks cordially Professor A. Środoń of the Institute of Botany, Polish Academy of Sciences in Kraków, and Professor R. G. West, F.R.S., of the Department of Botany, University of Cambridge, for help in composing the problems and for valued suggestions; Dr K. Mamakowa, Dr M. Ralska-Jasiewiczowa and the colleagues of the Institute of Geography, Polish Academy of Sciences in Kraków, for discussion. He is also indebted to Dr S. Gilewska for translating the text into English, and to Mrs M. Klimkowa for drawing the figures.

References (Starkel)

Andersen, S. T. 1961 Vegetation and its environment in Denmark in the Early Weichselian Glacial (Last Glacial). *Denmarks Geol. Undersög.* **2**, 75.

Andrews, J. T., Barry, R. G., Davis, P. T., Dyke, A. S., Mahaffy, M., Williams, L. D. & Wright, C. 1975 The Laurentide ice sheet: problems of the mode and speed of inception. *Proc. of WMO/IAMAP Symp. on Long-term Climatic Fluctuations*, Norwich, pp. 87–94.

Arsłanov, H. A., Boznyachuk, L. N., Velitshkievich, F. U., Kurierova, L. V. & Pietrov, G. S. 1971 Age of maximum extent of the last ice sheet at the Western Dvina–Dniepr interfluve [in Russian]. *Dokl. An SSSR* **196**, 1, M.

Averdieck, F. R. 1967 Die Vegetationsentwicklung des Eem–Interglazials und der Früh-Würm-Interstadiale von Odderade, Schleswig-Holstein. *Fundamenta* (Monographien zur Urgeschichte), Reihe B **2**, 101–125.

Borówko-Dłużakowa, Z. 1967 Badania paleobotaniczne osadów młodoplejstoceńskich (Brørup) w Koninie-Marantowice [English summary]. *Pr. Inst. Geol.* **48**, 81–136.

Bottema, S. 1974 *Late Quaternary vegetation history of Northwestern Greece*. Groningen.

Brotzen, F. 1961 An interstadial (radiocarbon dated) and the substages of the Last Glaciation in Sweden. *Geol. Fören. Förh.* **83**, 144–150.

Bucha, V., Horaček, J., Koči, A. & Kukla, J. 1969 Paläomagnetische Messungen in Lössen. In *Periglazial zone, Löss und Paläolithikum der Tschechoslowakei, Brno*, pp. 123–132.

Büdel, J. 1951 Die Klimazonen des Eiszeitalters. *Eiszeitalter u. Gegenwart*, Bd. 1.

Cepek, A. G. 1965 Geologische Ergebnisse der ersten Radiokarbon-datierungen von Interstadialen im Lausitzer Urstromtal. *Geologie* **14**, 5–6.

Chebotarieva, N. S., Faustova, M. A. *et al.* 1973 Degradation of the last ice sheet in Europe according to palaeogeographical data [in Russian]. In *Paleogeography of Europe during the Late Pleistocene, Moskva*, pp. 41–78.

Chebotarieva, N. S. & Makaricheva, I. A. 1974 Last Glacial Stage in Europe and their chronology [in Russian]. *Izd. Nauka, Moskva.*

Chmielewski, W. 1961 Archaeological investigations in Nietoperzowa Cave in the period 1956–1959. *Archeol. Polon.* **4**, 54–63.

Chmielewski, W. 1970 Późnoplejstoceńskie struktury zmarzlinowe w Witowie [English summary]. *Acta geogr. Univ. lodz.* **24**, 91–99.

Dansgaard, W., Johnsen, S. J., Clausen, H. B. & Langway, C. C. 1970 Ice cores and palaeoclimatology. *Radiocarbon Var. and Absolute Chr. Proc. 12th Nobel Symp. Stockholm*, pp. 337–347.

Dylik, J. 1967 The main elements of Upper Pleistocene paleogeography in Central Poland. *Biul. peryglac.* **16**, 85–115.

Dylik, J. 1968 Najstarszy 'interstadiał' ostatniego piętra zimnego w Polsce (amersfoort), Podział chronologiczny i stratygraficzny oraz terminologia [English summary]. *Bull. Soc. Sci. Lett. Łódź* **19**, 4.

Dylik, J. 1969*a* Slope development under periglacial condition in the Łódź region. *Biul. peryglac.* **18**, 381–410.

Dylik, J. 1969*b* L'action du vent pendant le dernier age froid sur le territoire de la Pologne Centrale. *Biul. peryglac* **20**, 31–44.

Dziewański, J. & Starkel, L. 1967 Slope covers on the middle terrace at Zabrodzie upon the San. *Studia Geomorph. Carp.-Balc., Kraków* **1**, 21–35.

Emiliani, C. 1968 Paleotemperature analysis of Caribbean cores P6304–8 and P6304-9 and a generalized temperature curve for the past 425000 years. *J. Geol.* **74**, 109.

Erd, K. 1967 Eem-Warmzeit und Frühweichsel-Zeit im Richt-profil Kittlitz bei Lübbenau [Niederlausitz], Kurzref. u. Exk.-führer 7. *Treffen der Sec. Quartärgeologie*, pp. 12–13. Berlin.

Firbas, F. 1949–1952 *Spät- und nacheiszeitliche Waldgeschichte Mitteleuropas nördlich der Alpen*. Jena.

Flint, R. F. 1971 *Glacial and Quaternary geology*. New York: Wiley.

Frenzel, B. 1959–1960 Die Vegetations- und Landschaftszonen Nord-Eurasiens während der letzten Eiszeit und während der postglazialen Wärmezeit T. I, T. II. *Abh. math.-naturw. Kl. Akad. Wiss. Mainz*, **13** (1959), **6** (1960).

Fromm, E. 1960 An interstadial peat at Ale near Lulea, Northern Sweden. *Sver. Geol. Unders.* C 574, 1–14.

Gavrilova, M. K. 1973 *Climate of central Yakutia* [in Russian]. Yakutsk.

Geyh, M. A. & Rohde, P. 1972 Weichselian chronostratigraphy, C^{14} datings and statistics. *24th Int. Geol. Congress, Montreal* **12**, 27–36.

Gierasimov, I. P. 1962 Loess and soil formation [in Russian]. *Izwiestia AN SSSR, s. geograf.* **2**.

Goździk, J. 1973 Geneza i pozycja stratygraficzna struktur peryglacjalnych w środkowej Polsce [English summary]. *Acta Geogr. Univ lodz.* **31**, 1–119.

Grichuk, V. P. 1973 Vegetation [in Russian]. In *The paleogeography of Europe during the Late Pleistocene, reconstruction and models*, pp. 182–219. Moscow.

Gross, H. 1964 Das Mittelwürm in Mitteleuropa und angrenzenden Gebieten. *Eiszeitalter u. Gegenwart.* **15**.

Halicki, B. 1960 Zagadnienie interstadiału mazurskiego [English summary]. *Zbiór prac i komunik. treści geol.* 1, Warszawa, Muzeum Ziemi.

Hammen, van der T., Maarleveld, G. C., Vogel, J. C. & Zagwijn, W. H. 1967 Stratigraphy, climatic succession and radiocarbon dating of the Last Glacial in the Netherlands. *Geologie Mijnb.* **46**, 79–95.

Ivanova, I. K. 1969 Les loess de la partie Sud–Ouest du territoire Europeen de L'URSS et leur stratigraphie. In *La strat. des loess d'Europe*, pp. 151–159. Paris.

Jahn, A. 1975 *Problems of the periglacial zone.* 219 pp. Warszawa: PWN.

Janczyk-Kopikowa, Z. 1969 Plejstoceńska flora w Podgłębokiem na Lubelszczyźnie. *Biul. IG. Warszawa* **220**, 51–72.

John, B. S. 1973 Vistulian periglacial phenomena in South-West Wales. *Biul. peryglac.* **22**, 185–212.

Jersak, J. 1965 Stratygrafia i geneza lessów okolic Kunowa [English summary]. *Acta Geogr. Univ. lodz* **20**.

Jersak, J. 1973 Litologia i stratygrafia lessu Wyżyn Południowej Polski [English summary]. *Acta Geogr. Univ. lodz* **32**.

Klatka, T. 1962 Geneza i wiek gołoborzy Łysogórskich [English summary]. *Acta Geogr. Univ. lodz* **12**.

Klimaszewski, M. 1961 Southern Poland. *Guide book of INQUA Excursion*, vol. III. Warsaw.

Klimaszewski, M. 1971 The effect of solifluction processes on the development of mountain slopes in the Beskidy [Flysch Carpathians]. *Folia quatern.* **38**.

Klimek, K. & Starkel, L. 1974 History and actual tendency of flood plain development at the border of the Polish Carpathians. Nachr. Akad., Göttingen. In *Report of the Commission on Present-day Geomorphological Processes IGU*, pp. 185–196.

Koperowa, W. 1962 Późnoglacjalna i holoceńska historia roślinności Kotliny Nowotarskiej [English summary]. *Acta palaeobot, Cracov.* 2/3, 1–57.

Koperowa, W. & Środoń, A. 1965 Pleniglacial deposits of the Last Glaciation at Zator (west of Kraków). *Acta palaeobot, Cracov.* **6**, 3–31.

Kowalski, K. 1962 Paleozoologiczne datowanie osadów jaskiniowych [English summary]. *Folia quatern.* **8**, 1–16.

Kozarski, S. 1974 Evidence of Late-Würm permafrost occurrence in North-West Poland. *Quest. Geographicae, Poznań* **1**, 65–86.

Kukla, J. 1969 Die zyklische Entwicklung und die absolute Datierung der Löss-Serien. In *Periglazialzone, Löss und Paläolithikum der Tschechoslowakei, Brno*, pp. 75–96.

Kukla, J. G. & Kukla, H. J. 1972 Insolation regime of Interglacials. *Quat. Res.* 412–424.

Lamb, H. H. 1971 Climates and circulation regimes developed over the Northern Hemisphere during and since the last ice age. *Palaeogeog., Palaeoclimat., Palaeoecol.* spec. issue, **10**, 125–162.

Liljequist, G. H. 1974 Notes on meteorological conditions in connection with continental land-ice in the Pleistocene. *Geol. Fören. i Stockholm Förhandl.* **92**, 293–298.

Ložek, V. 1969 Le loess et les formations assimilees: correlation entre L'Europe Centrale et la France par la faune de Mollusques. *Etudes sur le Quat. dans le Monde* **2**, 597–605.

Lundquist, J. 1974 Outline of the Weichsel Glacial in Sweden. *Geolog. Fören. i Stockholm Förhandl.* **92**, 327–339.

Malicki, A. 1950 Geneza i rozmieszczenie lessów w środkowej i wschodniej Polsce. *Ann. Uniw. MCS, s.B, Lublin* **4**, 195–228.

Mamakowa, K., Mook, W. G. & Środoń, A. 1975 Late Pleistocene flora at Kąty (Pieniny Mts., West Carpathians). *Acta palaeobot, Cracov.* **16**, 147–172.

Mamakowa, K. & Starkel, L. 1974 New data about the profile of Young-Quaternary deposits at Brzeźnica in Wisłoka Valley, Sandomierz Basin. *Studia Geomorph. Carp. Balc.* **8**, 47–59.

Manikowska, B. 1966 Gleby młodszego plejstocenu w okolicach Łodzi [English summary]. *Acta Geogr. Univ. lodz* **22**, 1–166.

Manley, G. 1975 Fluctuations of snowfall and persistence of snow cover in marginal-oceanic climates. *Proc. of WMO/IAMAP Symp. on Long-term Climatic Fluctuations*, Norwich, pp. 183–188.

Maruszczak, H. 1960 Peryglacjalne utwory pokrywowe na obszarze Wzgórz Szeskich [English summary]. *Biul. peryglac.* **7**, 19–33.

Maruszczak, H. 1968 Przebieg zjawisk w strefie peryglacjalnej w okresie ostatniego zlodowacenia w Polsce [English summary]. *Pr. geogr. IGPAN* **74**, 157–200.

Matthews, R. K. 1972 Dynamics of the Ocean-Cryosphere System: Barbados Data. *Quat. Res.* **2**, 368–373.

McIntyre, A. & Ruddiman, W. F. 1972 Northeast Atlantic Post-Eemian Paleooceanography: a predictive analog of the future. *Quat. Res.* **2**, 350–354.

Mojski, J. E. 1969 The stratigraphy of the Last Glaciation in the territory of Poland. *Geographia Polonica* **17**, 73–91.

Mojski, J. E. & Rzechowski, J. 1969 Plejstocen okolic Podgłębokiego na Polesiu Lubelskim [English summary] *Biul. Inst. geol.* **220**.

Mölder, K., Valovirta, V. & Virkkala, K. 1957 Spätglazialzeit und frühe Postglazialzeit in Südfinnland. *Bull. Comm. géol. Finl. Helsinki* **178**.

Mycielska-Dowgiałło, E. 1967 Formy szczelinowe i inwolucyjne w piaskach i żwirach doliny Wisły koło Tarnobrzega [English summary]. *Biul. peryglac.* **16**, 203–212.

Paas, W. 1969 Les loess de Basse Rhenanie. In *La Strat. des loess d'Europe*, pp. 29–31. Paris.

Paepe, R. & Zagwijn, W. 1972 Possibilites de correlation des depots Weichseliens de la Belgique et des Pays-Bas. *Bull. Ass. franc. pour l'etude du Quat. Paris*, **1**, 59–69.

Pécsi, M. 1970 La division lithostratigraphique des loess du Pleistocene superieur en Hongrie. *Acta Geogr. Univ. lodz* **24**, 343–356.

Pécsi, M. & Hahn, G. 1969 Historique des recherches sur le loess en Hongrie. In *La Strat. des loess d'Europe, Paris*, pp. 85–91.

Pennington, W. (Mrs T. G. Tutin) 1974 *The history of British vegetation*, pp. 1–152. English Universities Press.

Péwé, T. L. 1966 Paleoclimatic significance of fossil ice wedges. *Biul. peryglac.* **15**, 65–73.

Pissart, A. 1963 Les traces de 'pingos' du Pays de Galles (Grande Bretagne) et du Plateau des Hautes Fagnes (Belgique). *Z. Geomorph.* **7**.

Ploeger, P. L. & Groenman van Waateringe, W. 1964 Late glacial pingo and valley development in the Boorne Region near Wijnjerterg, province of Friesland, Netherlands. *Biul. peryglac.* **13**, 199–233.

Pop, E. 1971 Primele determinari de virsta en radiocarbon in turba noastra cauternara. In *Progrese in palinologia romaneasca, Bucureşti.*

Ralska-Jasiewiczowa, M. & Starkel, L. 1975 The basic problems of the Holocene in the Polish Carpathians. *Biul. Geol. Un. Wars. Warszawa*, **19**, 27–44.

Rohdenburg, H. 1967 Eiskeilhorizonte in Südniedersächsischen und Nordhessischen Lössprofilen. *Biul. peryglac.* **16**, 225–245.

Roszko, L. 1968 Recesja ostatniego lądolodu z terenu Polski [English summary]. *Pr. geogr. IGPAN* **74**, 65–100.

Rotnicki, K. 1966 Rzeźba wzgórz Ostrzeszowskich jako rezultat rozwoju stoku podczas Würmu [English summary]. *Pr. Kom. geogr.-geol., Poznań* **5**, 2.

Rotnicki, K. 1974 Development of Riss glaciation end moraines during the Würm, its morphological and geological consequences. *Quest. Geographicae Poznań*, **1**, 109–139.

Rotnicki, K. & Tobolski, K. 1969 Stanowisko interstadiału Paudorf w Kępnie (południowo-wschodnia Wielkopolska) [English summary]. *Bad. fizj. nad Polską Zachodnią*, ser. A, **23**, Poznań.

Różycki, S. Z. 1967 *Plejstocen Polski Srodkowej*, 236 pp. Warszawa: PWN.

Samson, P. M. 1969 La stratigraphie du Quaternaire superieur en Roumanie. *Etudes sur le Quat. dans le Monde* **2**, 629–636.

Sancetta, C., Imbrie, I. & Kipp, N. G. 1973 Climatic record of the past 130000 years in North Atlantic deep sea core V 23–82; correlation with the terrestrial record. *Quat. Res.* **3**, 110–116.

Saunders, G. E. 1973 Vistulian periglacial environments in the Lleyn Peninsula. *Biul. peryglac.* **22**, 257–269.

Sawicki, L. 1952 Warunki klimatyczne akumulacji lessu młodszego w świetle wyników badań stratygraficznych stanowiska paleolitycznego lessowego na Zwierzyńcu w Krakowie [English summary]. *Biul. Inst. geol.* **66**.

Serebryanny, L. R. 1969 L'apport de la radio-chronometrie a l'etude de l'historie Tardi-Quaternaire des regions de glaciation ancienne de la Plaine Russe. *Rev. Géogr. phys. Géol. dyn.* [2], **11** (3), 293–302.

Sobolewska, M., Starkel, L. & Środoń, A. 1964 Młodoplejstoceńskie osady z florą kopalną w Wadowicach [English summary]. *Folia Quat. Kraków*, **16**, 1–60.

Środoń, A. 1968 O roślinności interstadiału Paudorf w Karpatach Zachodnich [English summary]. *Acta palaeobot., Cracov.* **9**, 1–27.

Środoń, A. 1972 Roślinność Polski w czwartorzędzie. *Szata Rośl. Polski t.I, Warszawa*, 527–569.

Starkel, L. 1966 Post-glacial climate and the moulding of European relief. *Proc. of Symp. World Climate from 8000–0* B.C., *London*, pp. 15–33.

Starkel, L. 1969 L'evolution des versants des Carpates a'flysch au Quaternaire. *Biul. peryglac.* **18**, 349–379.

Starkel, L. 1977 *Paleogeografia holocenu.* Warszawa: PWN. (In the press.)

Svensson, H. 1964 Traces of pingo-lake frost mounds. *Lund Stud. Geog.*, A. *Phys. Geogr.* **30**.

Understanding climatic change; A program for action 1975 National Academy of Sciences, U.S.A.

Velichko, A. A. & Berdnikov, V. V. 1973 Forms and phases of the old cryogenesis [in Russian]. In *The Paleogeography of Europe during the Late Pleistocene, reconstruction and models, Moscow*, 146–161.

Velichko, A. A. & Khalcheva, T. A. 1973 Location and thickness of Upper Pleistocene Loesses [in Russian]. In *The Paleogeography of Europe during the Late Pleistocene, Reconstruction and models, Moscow*, 133–146.

Velichko, A. A. & Morozowa, T. D. 1973 Old soils [in Russian]. In *The Paleogeography of Europe during the Late Pleistocene, Reconstruction and models, Moscow*, 161–182.

Vigdortshik, M. E., Auslender, W. G., Znamienskaya, O. M. & Dołuchanov, P. M. 1972 New data on evolution of the Waldai ice sheet and its marginal zones [in Russian]. In *Krajewyje Obraz. Materikowych Oledienieni, Moskva*, pp. 166–176.
Wasylikowa, K. 1964 Roślinność i klimat późnego glaciału w środkowej Polsce na podstawie badań w Witowie koło Łęczycy [English summary]. *Biul. peryglac.* **13**, 261–417.
West, R. G. 1969 *Pleistocene geology and biology*. London: Longmans.
Wijmstra, T. A. 1969 Palynology of the first 30 metres of a 120 metre deep section in northern Greece. *Acta bot. neerl.* **18**, 511–528.
Wijmstra, T. A. 1975 Palynology and palaeoclimatology of the last 100000 years, *Proc. of WMO/IAMAP Symp. on Long-term Climatic Fluctuations*, Norwich, pp. 5–20.
Woldstedt, 1958 *Das Eiszeitalter*, 2nd edition, Stuttgart.
Zagwijn, W. H. 1974 Vegetation, climate and radiocarbon datings in the Late Pleistocene of the Netherlands. *Meded. Rijks geol. Dienst.* (*N. serie*) **25**, 101–111.
Zubakov, V. A. 1969 La chronologie des variations climatiques au cours du Pleistocéne en Sibérie Occidentale. *Rev. Géogr. phys. Géol. dyn.* [2], **11** (3), 315–324.

Phil. Trans. R. Soc. Lond. B. **280**, 373–374 (1977) [373]
Printed in Great Britain

Concluding remarks

By Sir Harry Godwin, F.R.S.
The Botany School, Cambridge, CB2 3EA

The conclusion of this two day meeting finds us with a very great deal on which we may congratulate ourselves. In the first place there is the extremely large attendance, embracing scientists of all ages, and graced and illuminated by the attendance of many overseas colleagues of experience and distinction.

In the second place we have the great range of scientific disciplines that are now applied to our field of study, many now extremely sophisticated, and the corresponding extension of Quaternary Studies into fields of evidence not hitherto exploited. In the early days of palynology of laminated lake sediments one could write of deciphering the 'annals of the lakes', but beginning by reading the record of lakes, peat bogs, coastal, fluviatile, glacial and periglacial geology, we have progressed to translating the long and detailed records of the deep oceans, and now the encapsulated history of the Arctic and Antarctic ice sheets. We have been introduced to the marvellous potential of the great CLIMAP Project, and all biologists in the British Isles at least will now have to consider whether their hypotheses of past biotic history satisfy the new principle that we can all see emerging as 'McIntyre's Gate'.

Thirdly, and importantly, we have been cautioned by *many* speakers against proceeding from inadequate or outdated assumptions, and to these cautions I would myself add the suggestion that much time, and some emotion, may be wasted in recognizing and correlating so-called 'interstadia': when everyone defines an interstadial for himself, little progress in communication between us is possible. Likewise it will help from time to time to consider the end to which our research is being directed: is it in effect the ancient and entirely proper goal of geological study, correlation, or is it the definition, equally proper, of the nature of former environmental conditions and the magnitudes of factors concerned therein? Or are we unwittingly and prematurely confusing these objectives? And if the evidence from different groups of organisms appears to be in conflict is not the conclusion that we need to know much more of the intimate biology and physiological ecology of the species themselves in natural conditions, and if it is objected that Devensian climates no longer obtain anywhere, surely the answer is to make use of phytotron techniques and reimpose artificially such conditions upon them?

All the convergent evidence on which we rightly congratulate ourselves has revealed how, by comparison with the rather clear and detailed familiarity we now have with the second half of the Devensian, over the first half there is still a deplorable cloak of ignorance broken only by shafts of illumination such as for the Chelford interstadial and the continuity of record of isotopic composition and fauna in deep oceanic sediments. What went on in this long period in the British Isles, a glacial period without glaciations, a sequence of quasi-interstadial episodes without woodlands, a time of insularity or of unity with the European land-mass?

It is not surprising that we remain so ignorant when subsequent glacial readvances and periglacial activity have destroyed so much, but we know from the Dutch and other mainland European records that in certain circumstances the field evidence remains: there is ample indication in this meeting of the enthusiasm to seek out such evidence here, and of the scientific ability fully to exploit it. When our next symposium is held we may hope to hear a fuller story...it could hardly be a more eventful or stimulating one than that to which we have just listened and for which we should express our gratitude to the organizers, speakers, participants and hosts who have made it so successful and have so well inaugurated the scientific collaboration of our two learned Societies.